Multifunctional Ultrawideband Antennas

Trends, Techniques, and Applications

Multifunctional Ultrawideband Antennas

Trends, Techniques, and Applications

By
Chinmoy Saha, Jawad Y. Siddiqui, and
Yahia M.M. Antar

CRC Press
Taylor & Francis Group
Boca Raton London New York

CRC Press is an imprint of the
Taylor & Francis Group, an **informa** business

CRC Press
Taylor & Francis Group
6000 Broken Sound Parkway NW, Suite 300
Boca Raton, FL 33487-2742

Printed on acid-free paper

International Standard Book Number-13: 978-1-138-55354-5 (Hardback)

Visit the Taylor & Francis Web site at
http://www.taylorandfrancis.com

and the CRC Press Web site at
http://www.crcpress.com

Contents

About the Authors

Chinmoy Saha (M'06–SM'15) received his B.Tech, M.Tech., and PhD degrees in Radio Physics and Electronics, University of Calcutta, Kolkata, India in 2002, 2005, and 2012, respectively. He is currently working as an Associate Professor at the Department of Avionics, Indian Institute of Space Science and Technology, Department of Space, Government of India and visiting the Royal Military College of Canada, Kingston, Ontario, Canada as a post-doctoral fellow. Prior to his present affiliation, he was associated with various reputed engineering colleges in India as a lecturer and assistant professor. He was also associated with Jadavpur University, Kolkata, India as visiting faculty from 2008–2012. He has been a member of the Institute of Electrical and Electronics Engineers (IEEE) since 2006, promoted to a senior member grade in 2014, and served in various positions in the IEEE AP-MTT Kolkata Chapter and IEEE Kolkata Section. He was secretary of the IEEE AP-MTT Kolkata Chapter in 2011–2012. He served as a member of the organizing committee of the IEEE Applied Electromagnetics Conference (AEMC) in 2007, 2009, 2011, and 2013. He was also the Organizing Chair of the first IEEE Indian Antenna Week in Puri, India in 2010. He is also a Life Member of the Institution of Electronics and Telecommunication Engineers (IETE). Currently he is Chairman of the Antennas and Propagation Chapter of the IEEE Kerala Section. He was the recipient of a national scholarship from the Indian government's Ministry of Human Resource Development in 1999 for excellence in his B.Sc. Physics (Hons) degree from the University of Calcutta. He also received an Outstanding Contribution Award from the Antennas and Propagation and Microwave Theory and Techniques Chapter, IEEE Kolkata Section in 2010. He was awarded "Best Contribution Award for Notable Services and Significant Contributions toward the Advancements of IEEE and the Engineering Profession" from the IEEE Kolkata Section in 2013. His current research interests include microwave circuits, engineered materials, metamaterial-inspired antennas, and circuits, reconfigurable and multifunctional antennas for modern wireless applications, dielectric resonator antennas, and THz antennas. He has more than seventy publications in peer-reviewed national and international journals and conference proceedings. He is on the board of reviewers of several international journals of repute, including IEEE Transactions on Antennas and Propagation, IEEE Antennas and Wireless Propagation Letters (IEEE AWPL), IET Microwaves, Antennas and Propagation, Electronic Letters, etc.

Jawad Y. Siddiqui (S'01–M'04–SM'14) is an Associate Professor at the Institute of Radio Physics and Electronics, University of Calcutta, Kolkata, India. He received his Master of Technology and Doctor of Philosophy degrees in Radio Physics and Electronics from the University of Calcutta in 1999 and 2005, respectively. He worked as a post-doctoral fellow and then as a visiting scientist at the Royal Military College of Canada at different periods between 2008–2014. He has more than 100 publications in peer-reviewed journals and conferences. His research areas include ultrawideband antennas, frequency reconfigurable antennas, tapered slot antennas, and multifunctional antennas for cognitive radio application. He is on the board of reviewers of several international journals of repute including IEEE Transactions on Antennas and Propagation, IEEE Antennas and Wireless Propagation Letters (IEEE AWPL), IET Microwaves, Antennas and Propagation, Electronic Letters, etc. He is a Co-Principal Investigator on the Stratosphere Troposphere (ST) Radar Project at the University of Calcutta, Kolkata, India.

He is the Co-General Chair for the IEEE International Microwave and RF Conference 2018. He is a senior member of the IEEE and is currently serving as Chair for the AP-S/MTT-S Jt. Chapter, IEEE Kolkata Section and Chair for the SIGHT Kolkata Chapter.

Yahia M.M. Antar (S'73–M'76–SM'85–LF'00) received his B.Sc. (Hons.) degree in 1966 from Alexandria University, Alexandria, Egypt, and M.Sc. and PhD degrees from the University of Manitoba, MB, Canada, in 1971 and 1975, respectively; all in electrical engineering. In 1977, he held a Government of Canada Visiting Fellowship at the Communications Research Centre in Ottawa, and in May 1979 he joined the Division of Electrical Engineering at the National Research Council of Canada. In November 1987, he joined the Department of Electrical and Computer Engineering at the Royal Military College of Canada, Kingston where he has held the position of Professor since 1990. He has authored or coauthored about 250 journal papers, several books and chapters in books, over 450 refereed conference papers, holds several patents, has chaired several national and international conferences, and has given plenary talks at many conferences. He has supervised and co-supervised over 90 PhD. and M.Sc. theses at the Royal Military College of Canada and at Queen's University, Kingston, several of which have received the Governor General of Canada Gold Medal Award, the Outstanding PhD. Thesis of the Division of Applied Science, as well as many Best Paper Awards in major international symposia. He served as the Chair of the URSI Canadian National Commission (1999–2008), Commission B (1993–1999), and has a cross appointment at Queen's University in Kingston.

Dr Antar is a Life Fellow of the Institute of Electrical and Electronic Engineers (IEEE), a Fellow of the Engineering Institute of Canada (FEIC), a Fellow of the Electromagnetic Academy, and a Fellow at the International Union of Radio Science (URSI). He serves as an Associate Editor of many IEEE and IET Journals and as an IEEE-APS Distinguished Lecturer. In May 2002, he was awarded a Tier 1 Canada Research Chair in Electromagnetic Engineering which was renewed in 2016. In 2003, he was awarded the Royal Military

College of Canada "Excellence in Research" Prize, and the RMCC Class of 1965 Teaching Excellence award in 2012. He was elected by the URSI Board as Vice President in August 2008 and in 2014, and to the IEEE AP AdCom in 2009.

On January 31, 2011, he was appointed Member of the Canadian Defence Advisory Board (DAB) of the Canadian Department of National Defence. In October 2012, he received the Queen's Diamond Jubilee Medal from the Governor General of Canada in recognition for his contribution to Canada. He is the recipient of the 2014 IEEE Canada RA Fessenden Silver Medal for "Ground Breaking Contributions to Electromagnetics and Communications" and the 2015 IEEE Canada J.M. Ham outstanding Engineering Education Award. In May 2015, he received the Royal Military College of Canada Cowan Prize for excellence in research. He is the recipient of the IEEE Antennas and Propagation Society prestigious Chen-To-Tai Distinguished Educator Award for 2017.

Preface

Multifunctional antennas are a comparatively new area of antenna research. Cognitive radio (CR) in software-defined radio (SDR) technology and MIMO technology are some of the new technologies employing these antennas, which include integration of multiple antennas with narrowband and ultrawideband (UWB) characteristics. Designing of UWB antennas for CR and MIMO technology is another area of vibrant research. This book is an attempt to assimilate works done by different researchers and authors over the last few years on antennas that can perform multiple antenna functionalities. Chapter 1 focuses on an overall introduction to the UWB system, including an historical perspective through to the most recent applications of UWB technology. Chapter 2 deals with the discussion on the design and development of UWB antennas and the discussion on various classes of UWB antenna designs proposed by research groups over the last two decades. Chapter 3 deals with techniques of designing UWB antennas with frequency stop or frequency notch characteristics. Several antenna designs are described which provide frequency notch response and the techniques involved. Chapter 4 deals with the design and development of antennas capable of performing multiple antenna functionalities. As CR in SDR technology is the major application of these antennas, a section focuses exclusively on antenna requirements for SDR and CR technologies. Various arts and techniques of designing multifunctional antennas and their advantages are thoroughly discussed. Chapter 5 deals with arts and techniques of design and the realization of reconfigurable UWB antenna design with a clear focus on recent and upcoming technological requirements. A section on a recent technique of designing dual reconfigurable printed antenna, which can work as a tunable notched UWB antenna and reconfigurable narrowband antenna, is discussed. In addition, filter integrated UWB reconfigurable antenna, popularly known as reconfigurable filtenna, is also discussed. Fundamental concepts and physical insight on the design principles and contemporary contributions by various researchers are also highlighted. The last chapter, Chapter 6, focuses on MIMO antennas for UWB applications. As MIMO systems deal with multiple antennas installed in a common platform, their performance metrics in terms of additional parameters such as envelope correlation coefficient (ECC), diversity gain (DG), mean effective gain (MEG), total active reflection coefficient (TARC), etc. are also discussed. After a brief overview on MIMO antenna parameters, design of UWB-based MIMO antennas is thoroughly discussed. In addition, the chapter includes two sections focusing on frequency notched MIMO antenna design and isolation-enhancement techniques in MIMO antennas.

This book is based on dissertations/theses and many papers published by several research graduate students. We are grateful to Prof. Al Freundorfer (Queen's University, Kingston, Ontario, Canada), Prof. Debatosh Guha (Institute of Radio Physics and Electronics, University of Calcutta, Kolkata, India), Dr Sudhakar Rao (Northrop Grumman, Falls Church, Virginia), Dr Goutam Chattopadhyay (NASA-Jet Propulsion Laboratory, California Institute of Technology, Pasadena, California), Nandan Bhattacharyya (RCCIIT, Kolkata, India), Chittajit Sarkar (SVIST, Kolkata, India), and Debarati Ganguly (Royal

Military College of Canada, Kingston, Ontario, Canada) for their valuable input/suggestions and support by various means while preparing the manuscript. Finally, we would like to express our thanks to our families for their encouragement and support.

Chinmoy Saha
Jawad Y. Siddiqui
Yahia M.M. Antar

1

Introduction to UWB Systems and Applications

1.1 Introduction

As the name suggests, the term ultrawideband (UWB) system corresponds to a commercial (military/civilian) communication system which utilizes a very wide frequency spectrum. While the basic definition of bandwidth* in a general electrical engineering context is applicable to an UWB system, it can be quantified in terms of absolute $(f_h\text{-}f_l)$, percentage $(f_h\text{-}f_l)/f \times 100$ or ratio bandwidth (f_h/f_l), where f_h and f_l are upper and lower frequencies of the UWB spectrum. Using the complementary relation of time and frequency domain, governed by Fourier transform, the UWB system can also be treated as a communication system employing very narrow pulses of duration in the range of nanosecond and sub-nano second. Thus, communicating (transmit/receive) over a very wide frequency range employing extremely narrow pulses (time duration) is the key feature of a UWB system in general. This feature enables the UWB technology to support transmission of high data rates of the order of 100 MBPS or even higher. Apart from providing faster communication with high data rates, UWB technology has various interesting features which have made it a unique tool for various applications, ranging from UWB radar to ground-penetrating radar (GPR) and biomedical imaging. Very large bandwidth of the UWB signal offers various advantages such as high data rate, lower power consumption, high time resolution, obstacle penetration capability, and most importantly, low-cost implementation. Since the UWB system occupies a very wide bandwidth, its spectrum naturally overlaps with various traditional narrowband services. To support these narrowband services simultaneously, the UWB system is operated at a very low power level. Simultaneous operation of UWB and various narrowband inside this UWB band is, recently, being used for dynamic spectrum sharing in cognitive radio (CR) of software-defined radio (SDR). This CR-based SDR and multiple input and multiple output (MIMO) systems are very recent applications of UWB technology.

In this chapter, we focus on an overall introduction to the UWB system, including an historical perspective through to the most recent applications of UWB technology. Chronological growth of the UWB technology, starting from the spark-gap realization of electromagnetic waves by Heinrich Hertz [1], is systematically presented with a special emphasis on landmark inventions/achievements in the area. An introductory definition of UWB systems and various schemes to realize UWB technology, followed by a discussion on spectrum regulation and applications, provide readers, in particular those new to this area, with solid background information.

* The range of frequency over which a system provides its desired operation and complies with respect to a certain predefined figure of merit

1.2 An Overview of the UWB Systems

One of the motivations behind and benefits of the UWB systems can be excellently corroborated by Shannon's channel capacity formula [2] for a band-limited channel with Gaussian noise,

$$\text{Channel capacity}, C = B\log_{10}\left(1+\frac{S}{BN_0}\right) \tag{1.1}$$

where:
- B is the bandwidth of the channel (Hz)
- S is the signal power (Watt)
- N_0 is the channel spectral density of Noise (Watt/Hz)

Shannon's formula of Equation 1.1 indicates that for a given N_0, channel capacity C can be maintained to a fixed value by decreasing signal power level and increasing the bandwidth. On the other hand, for a given (S/N_0), channel capacity C increases with the bandwidth. Though Shannon's capacity theorem is more widely used in information theory [3], its implications are extremely encouraging for wideband systems, such as spread spectrum techniques and UWB.

Formal Definition of UWB Systems

- UWB system was first formally defined in 1990 by the Defense Advanced Research Project Agency (DARPA) [4]. According to this definition, any system occupying a fractional bandwidth greater than or equal to 0.25 is categorized as a UWB system. Thus, according to DARPA, a system operating over a bandwidth (BW),

$$BW = \frac{f_h - f_l}{f_c} = \frac{2(f_h - f_l)}{f_h + f_l} \geq 0.25 \tag{1.2}$$

  is called a UWB system. Here, f_h and f_l are the higher and lower end of UWB band respectively and $f_c = (f_h + f_l)/2$ is the center frequency.
- In 2003, the Federal Communications Commission (FCC) [5] modified the fractional bandwidth limit to 0.2 instead of 0.25 defined by DARPA. Thus, according to the FCC, a UWB system should have a fractional bandwidth given by,

$$BW = \frac{f_h - f_l}{f_c} = \frac{2(f_h - f_l)}{f_h + f_l} \geq 0.2 \tag{1.3}$$

  Alternatively, as per FCC regulation, any system having an absolute bandwidth greater than or equal to 500 MHz is referred to as a UWB system. This definition of a UWB system,

$$BW = f_h - f_l \geq 500\ \text{MHz} \tag{1.4}$$

  is applicable when the center frequency is above 6 GHz.

In the definition of a UWB system, f_h and f_l are the upper and lower end of the UWB band respectively, where the radiated power is 10 dB down on the peak level.

In 2002, the FCC designated a 7.5 GHz band ranging from 3.1 to 10.6 GHz in the USA for UWB communications. Even though various other agencies/bodies adopted different bands for UWB communication, 3.1 to 10.6 GHz is very popular and almost universally accepted as the modern UWB band. Spectrum allocation for UWB and associated regulation will be discussed in more detail in Section 1.2.3.

1.2.1 History of UWB

Research on UWB systems including UWB radar, antennas and communication have drawn a great deal of attention over the last two decades, more specifically since 2002 after the FCC's regulation on exclusive spectrum allocation for UWB. However, the origin of UWB signal and UWB technology dates to 1888 when Heinrich Hertz (1857–1894) generated a short pulse using his famous spark-gap transmitter [1]. Though the short pulse generated by Hertz was the first laboratory-generated UWB signal, at that time it was not very useful because the focus of radio engineers was more on the narrowband systems to exploit maximum benefit of frequency division multiplexing. Before we start our journey on modern UWB systems, it would be interesting to briefly look back on some of the pioneers of radio wave. Table 1.1 summarizes the landmark contributions by Heinrich Hertz, Oliver Lodge, Jagadish Chandra Bose and Guglielmo Marconi, who are considered the four pioneers of electromagnetic engineering. Though Hertz's spark-gap excited short pulses are inherently UWB signals in nature and some of the antennas proposed by Lodge and Bose are popular modern wideband antennas, UWB technology was left almost dormant for several decades. The obstacles or key challenges for UWB systems were:

- *Popularity of narrowband systems*: After Marconi's successful trans-Atlantic radio wave propagation, narrowband communication emerged as a reality. Also, considering the less spectral demand and existing concept of frequency division multiplexing, narrowband transmission was more than sufficient to serve the demand. Thus, the general interest of radio engineers was completely focused on narrowband communication, eclipsing the initial spark of the UWB technology.
- *Low spectral efficiency of the UWB signal*: Radio signals generated by spark-gap transmitters were extremely short pulse (wideband in frequency domain), the UWB signal lacked in spectral efficiency: in other words, it had a low bit rate, occupying a large bandwidth. Also, huge dispersion associated with spark-gap radiated UWB signal was considered another important factor.
- *Commercial implications*: Even after most of the early challenges and constraints were mitigated by UWB researchers, frequency regulators and industries were not willing to adapt to the new technology. Due to spectrum overlapping of UWB systems, with already existing narrowband communication services distributed over a wideband, service providers and spectrum regulatory bodies were not keen to accept proposals for a dedicated spectrum for UWB bands.

In 2002, the FCC dedicated a frequency range from 3.1 GHz to 10.6 GHz for UWB with certain restrictions and limitations on the emitted power spectrum. This FCC regulation was mainly a result of the following two aspects:

1. The UWB community established the fact that UWB systems can coexist with preexisting narrowband services (low power level of the UWB signal will not mask the narrowband signal).

TABLE 1.1

Pioneers of Modern UWB Systems and Radio Engineering

Pioneers of Radio Engineering	Landmark Contribution	A Brief Discussion
 Heinrich Hertz (1857–1894) German Physicist	• Discovery of radio waves • Estimation of velocity of generated radio waves and demonstration of phenomenon like reflection, refraction and diffraction	• Hertz's radio wave generation was based on an experiment, famously known as the spark-gap resonator • Stored energy across capacitors was relieved and discharged through resonant circuits and antennas • Introduced half wave dipole for transmitting the energy and loop antenna for detecting the radiated energy
 Oliver Lodge (1851–1940) British Physicist	• Invented bow tie and biconical antenna • First patented the syntonic radio • Introduced the concept of monopole antenna with earth as ground	• Oliver Lodge designed and developed the first practical radio system in which the transmitter and receiver was tuned to the same frequency in 1898 • Radiating structures preferred by him were "cones or triangles or other such diverging surfaces with the vertices adjoining and their larger areas spreading out in space" • He used triangular/bowtie elements in the transmitter and receiver section of his syntonic radio
 Jagadish Chandra Bose (1858–1937) Indian Scientist	• Experimentally demonstrated the first millimeter wave communication at 60 GHz in 1895, over 23 meters • Invented horn antennas	• Bose designed, developed and experimentally demonstrated the first millimeter wave communication at 60 GHz. He named the transmitting antenna as "funnels" which is nothing but modern horn antenna • Using millimeter wave, he demonstrated ringing of a remote bell
 Guglielmo Marconi (1874–1937) Italian Engineer	• First to experimentally demonstrate long-range radio communication (trans-Atlantic)	• Marconi realized the commercial implications of long-range radio communications • First demonstrated real long-range radio link in a trans-Atlantic experiment

2. Requirements for high-speed data rate communications offered by UWB systems along with numerous applications.

Before we start the detailed technical discussion on UWB signals and systems, we will consider the chronological development of the UWB system. Here we include only those works/contributions which impacted most on the growth and development of the UWB system:

1950s: In 1957, Astanin developed X-band 0.5 ns duration transmitters for waveguide study at the A. Mozhaisky Military Space Engineering Academy in the USSR [6]. Almost at the same time, Kobzarev and his group conducted indoor tests of UWB radar at the Radio Electronics Institute of the USSR Academy of Science [7].

1960s: Research and development of UWB technology experienced unprecedented growth as a result of the pioneering contribution of Harmuth (The Catholic University of America), Ross and Rabins (Sperry Rand Corporation), and Van Etlen (United States Air Force (USAF)) during this decade [6–8]. Their work on time domain electromagnetics, pulse transmitters, receiving system, and antennas contributed immensely in the design and development of military radar with high spatial resolution. Growth of research on UWB systems and technology was further accelerated by the development of the sampling oscilloscope by Tektronix and Hewlett Packard [9] in the 1960s. This sampling oscilloscope allowed the UWB researchers to visualize and analyze short duration pulses in time domain. In 1967, Cook and Bernfeld published the first book [10] on UWB technology that summarized the contemporary contributions on pulse compression techniques, match filtering, and correlation techniques.

1970s: In 1972, the invention of sensitive pulse receivers by Rabins led to the first patented design of UWB communication systems by Ross and the Sperry Rand Corp. [11]. In 1974, the first ground-penetrating radar using UWB communication was commercialized by Morey at Geophysical Survey Systems Corp.

1990s: The biggest limitation of UWB technology, multiple access interference (MAI), was solved by Win and Schultz [12–14]. They introduced time-hopping impulse radio (TH-IR) and showed that it can cater to many users by assigning pseudorandom transmission time to the pulses from the different users. In 1993, Robert Schultz developed the multiple access technique in UWB communication. Allocating each user a unique spreading code that allows a user to transmit at a specified time instance, Schultz demonstrated the potential of UWB as future viable technology for wireless communication. In 1994, a compact and inexpensive UWB system was developed by McEwan [15] using micropower impulse radar (MIR).

2002: Based on three key factors: proven superiority of UWB in impulse radar, enormous potential for future wireless communications, and convincing demonstration that UWB emission under a constrained power level doesn't mask/interfere with narrowband services, in 2002 the FCC allocated a wide range of band from 3.1 to 10.6 GHz for UWB communications. This particular spectrum allocation for UWB technology is considered the greatest milestone for UWB systems as this emerged as the center of attraction for EM and communication researchers. Various modern concepts of wireless radios, such as orthogonal frequency division multiplexing (OFDM), direct sequence code division multiple access (DS-CDMA), and recently introduced cognitive radio (CR) in software-defined radio (SDR) environments, are direct/indirect descendants of UWB technology.

1.2.2 UWB Signals and Systems

Although including detailed discussion on UWB signals, coding schemes, and associated analysis is beyond the scope of this book, having a general idea of various UWB schemes and signaling techniques is important for UWB antenna researchers. Design specification of UWB antennas is inherently related to UWB signaling and protocols associated with a particular scheme. UWB technology can be broadly classified into two categories:

1. *I-UWB*: I-UWB or impulse UWB is based on sending very short duration pulse train to convey information. Thus, like base band communication, I-UWB scheme doesn't use any modulated carrier to send information.

2. *MC-UWB*: MC-UWB or multi carrier UWB uses multiple carriers simultaneously and can be further classified into various categories. Orthogonal frequency division multiplexing (OFDM) is a popular UWB scheme under MC-UWB. Before we discuss various practical UWB schemes, let us have a quick look at various popular UWB pulses.

Gaussian pulse, its various derivatives, and pulse derived from Gaussian pulse are popularly used in various forms of UWB schemes. A Gaussian pulse is mathematically expressed as:

$$p(\mathrm{t}) = \frac{1}{\sqrt{2\pi\sigma^2}} e^{(t-\mu)^2/2\sigma^2} \quad (1.5)$$

where:

- σ is standard deviation of the pulse
- μ determines the midpoint of the pulse

Figure 1.1a–c shows the time domain plot of the Gaussian pulse and its first and second derivative, respectively. The Gaussian pulse modulated by sinusoidal carrier shown in Figure 1.1d is another very important pulse for UWB technology.

Based on two broad categories of UWB, I-UWB and MC-UWB just introduced, UWB technology can be further classified as:

- UWB impulse radio (UWB-IR)
- Direct sequence spread spectrum (DS-SS)
- Orthogonal frequency division multiplexing (OFDM)
- Frequency hopping (FH)

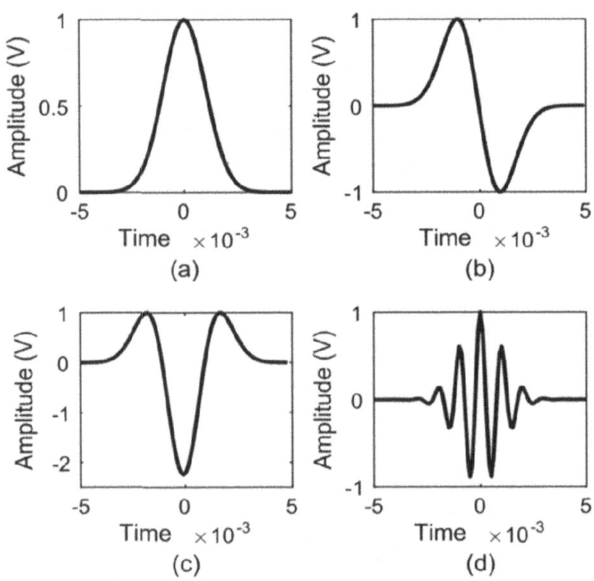

FIGURE 1.1
Various UWB pulses: a) Gaussian pulse, b) first derivative of Gaussian pulse (monocycle pulse), c) second derivative of Gaussian pulse and d) sinusoidal Gaussian pulse.

1.2.2.1 UWB Impulse Radio (UWB-IR)

UWB impulse radio, commonly known as I-UWB or time hopping (TH), is one of the most popular UWB schemes. Waveform in this scheme is characterized by periodic pulses of ultra-short duration, typically in the range of nano- or pico-seconds. The transmission interval between two successive pulses is defined as pulse repetition frequency (PRF) where typical duration of a pulse is 2 ns. To present a particular data sequence from the user, the relative position of the pulse is shifted in time with respect to a reference pulse; because of this, the scheme is known as time hopping. Thus, UWB-IR essentially uses a pulse position modulation. Figure 1.2 shows a simplified block diagram of UWB-IR [16]. Extremely short duration of the UWB pulse along with low power spectral density, makes this scheme quite useful to establish a secure network. Low duty cycle of I-UWB pulses is exploited in precise localization of the object. By choosing a proper pulse, precision of less than 1 m is achievable. Also, with sufficient short duration pulses, I-UWB radar is capable of monitoring movement of various bio-medical organs inside a human without direct contact of the human body.

1.2.2.2 Direct Sequence Spread Spectrum (DS-SS)

Though generating short pulses, as done in the I-UWB scheme, is the most fundamental technique of broadbanding a signal in frequency domain, there are other ways to do so. DS-SS is one such technique in which the original data sequence is multiplied by a second signal having a very large bandwidth. The second signal is basically a high data rate pseudorandom sequence. This technique is popularly known as direct sequence spread

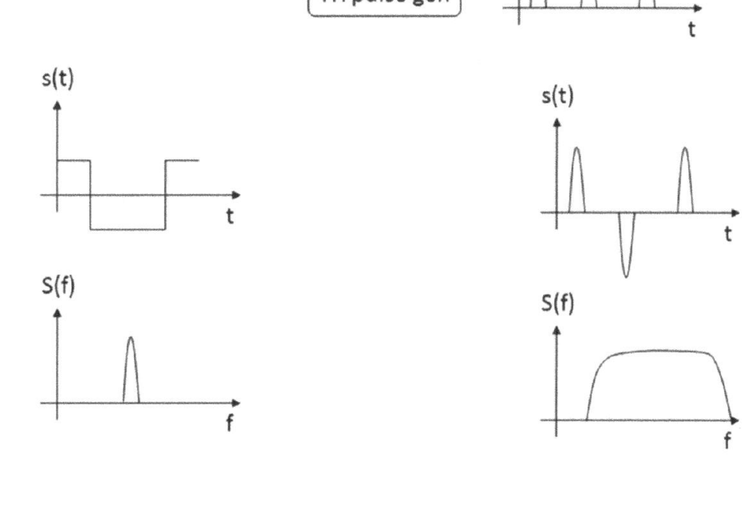

FIGURE 1.2
Simplified block diagram of UWB-IR [16].

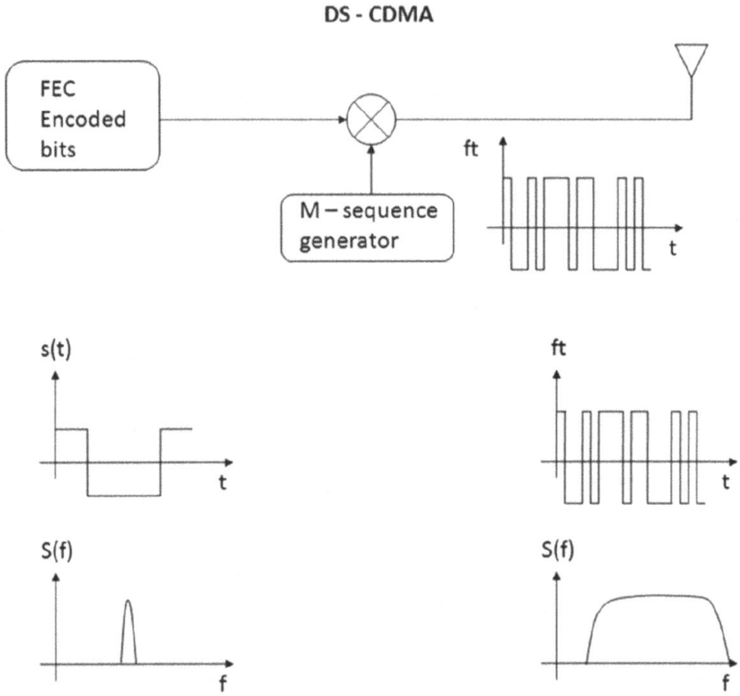

FIGURE 1.3
Simplified block diagram of DS-SS UWB [16].

spectrum technique (DS-SS). Bandwidth of the resultant signal in this scheme is approximately equal to the bandwidth of the wideband spreading signal. Figure 1.3 shows the schematic block diagram of the direct sequence spread spectrum technique. DS-SS signals are extremely attractive for high data rate secured communication due to their attractive features, for example, high immunity against interference, low probability of being intercepted, robustness on multi-path propagation signal processing, and supporting various spreading codes to generate pseudorandom signals. As indicated in Figure 1.3, the sequence generator, which generates the sequence of ± 1s, plays the most important role in this scheme. Designing a proper spreading code/pseudorandom sequence generator is extremely crucial in this scheme. The spreading code should have low auto correlation side lobes to generate a flat spectrum of the signal. For popular digital code, like code division multiple access (CDMA), the spreading codes should have low cross-correlation to reduce the inter-user interference.

1.2.2.3 Orthogonal Frequency Division Multiplexing (OFDM)

In this scheme, UWB waveform is generated by multi-carrier modulation with a large number of sub-carriers. This scheme evolved due to the practical requirement of separating the 3.1–10.6 GHz UWB band and multiple narrowband coexisting within this band. Thirteen sub-bands, each occupying 525 MHz bandwidth, are defined in the UWB spectrum of 3.1–10.6 GHz. In general, the sub-carriers are separated by frequency spacing of W/N, where W is the total transmission bandwidth and N is the number of sub-carriers.

The modulation process is performed either with analog technique or using digital means. In digital technique, the modulation process involves the inverse fast Fourier

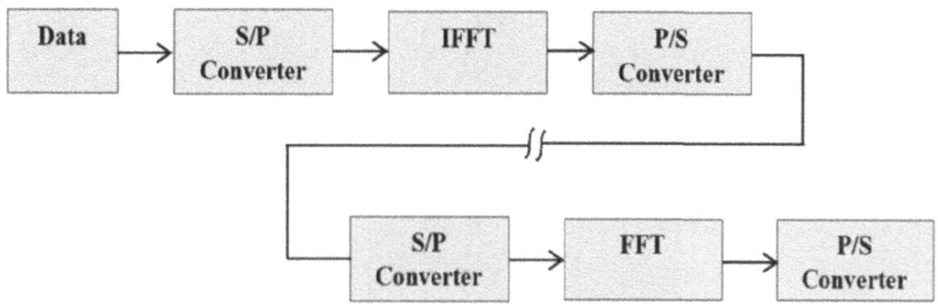

FIGURE 1.4
Simplified block diagram of OFDM.

transform (IFFT) of the data. Figure 1.4 shows the schematic block diagram of the digital implementation of the OFDM. As demonstrated in the diagram, the data is initially split using serial to parallel converter, followed by IIFT on it. It is again converted back to serial data using parallel to serial converter (P/S) before being sent into the channel. At the receiving end, the exact reverse operation (S/P→FFT→P/S) is performed and transmitted data is retrieved.

1.2.2.4 Frequency Hopping

This is another elegant technique to generate spread spectrum signal. In this technique, carrier frequency hops within a certain frequency band. Figure 1.5 shows the basic block diagram of this scheme. As indicated in the figure, a hopping sequence, acting as the input of a voltage-controlled oscillator (VCO), controls the carrier frequency. The frequency hopping (FH) technique can be sub-divided into two categories:

1. Fast frequency hopping: Here, carrier frequency is changed several times even during transmission of one symbol. This results in the spreading of each separate symbol over a large bandwidth.
2. Slow frequency hopping: In this case, one or more number of symbols are transmitted over a single frequency. Here, spreading of the signal is relatively narrower.

Advantages of FH schemes are as follows:

- Low probability of being intercepted
- Inherent immunity against jamming and interference
- Can be sued for multiple access schemes like CDMA

1.2.3 Spectrum Regulation for UWB

Even though an exclusive spectrum was allocated for UWB technology by the FCC in 2002, the FCC ensured that UWB radiation satisfied the strict frequency regulation and emission power level. This regulation was imposed to ensure that existing narrowband services, co-existing within the licensed UWB spectrum of 3.1–10.6 GHz, are not affected. The limitation on the power level, indicated by effective isotropic radiated power (EIRP), is expressed in dBm. This restriction on UWB signal, commonly known as the "frequency mask", varies based on the application of UWB; it also varies from country to country depending on the spectrum regulation. The spectrum regulation on this "frequency mask" is more relaxed

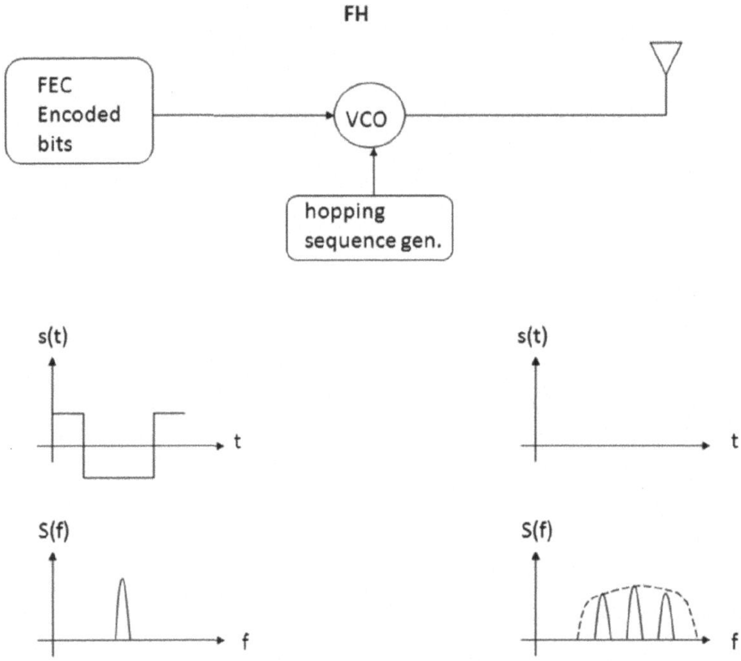

FIGURE 1.5
Simplified block diagram of a FH UWB scheme [16].

TABLE 1.2
UWB EIRP Emission Level Authorized by the FCC in 2002 for Indoor and Outdoor Applications [17]

	Indoor Emission Mask		Outdoor Emission Mask	
Frequency (MHz)	**(dBm/MHz)**	**(W/MHz)**	**(dBm/MHz)**	**(W/MHz)**
960–1610	–75.3	2.95×10^{-11}	–75.3	2.95×10^{-11}
1610–1900	–53.3	4.67×10^{-9}	–63.3	4.67×10^{-10}
1900–3100	–51.3	7.41×10^{-9}	–61.3	7.41×10^{-10}
3100–10600	–41.3	7.41×10^{-8}	–41.3	7.41×10^{-8}
Above 10600	–51.3	7.41×10^{-9}	–61.3	7.41×10^{-10}

on indoor/short-range applications, such as GPR, through wall imaging, medical applications, security applications, etc. For these applications, a power spectral density of −41.3 dBm/MHz is allowed in the frequency band of 3.1–10.6 GHz. On the other hand, for outdoor communication applications, more restriction is imposed on emitted UWB signals. Table 1.2 shows the emission mask defined by the FCC for indoor and outdoor applications.

The following important points from the indoor and outdoor emission mask table are noteworthy:

- The "Part 15" limit corresponds to the limit tolerated by the FCC for non-intentional emissions, that is, radiation from electric household appliances.
- Outside the 3.1–10.6 GHz frequency band, UWB transmission is allowed for certain specific applications and authorized power levels are 10 dB lower than the tolerable level inside the building.

- A special protection is provided to GPS and cellular services in the frequency range of 900 MHz to 1.6 GHz by lowering the maximum allowable power level to –75 dBm/MHz.

It is worth noting that in specific situations, UWB emission needs to allow "detect-and-avoid" protocol. According to this protocol, the UWB system must be intelligent enough to detect coexisting narrowband service in its vicinity with overlapping frequency spectrum. If such channel coexists, UWB devices should refrain from radiation in the band already being used by narrowband service.

1.3 Applications of UWB Technology

Ever since the exclusive spectrum was allocated for UWB technology by the FCC in 2002, a plethora of applications has emerged, exploiting the attractive features of UWB signals. The applications of UWB technology are widespread as various sectors such as communication, civilian, defense, medical, etc. have used UWB technology for many purposes. As a UWB researcher, we must have a broad idea of these applications, as the design of the UWB system, including UWB antennas, vastly depends and varies on the applications. For certain applications, the design may be quite straightforward whereas for some applications, the design may be quite challenging as the system has to work in extremely harsh environments.

A common feature that a UWB antenna, irrespective of what the applications must have, are

- Very compact size
- Ease of integration
- Mechanical robustness
- Potential to work in harsh environments

Here, we will summarize the major applications of UWB technology including all sectors. It is interesting to note that for some cases these applications are mutually dependent.

1.3.1 Ranging and Localization

This is a major application of UWB technology where the UWB system provides the localization information and effectively performs the task of an "indoor" GPS [18]. Extremely large bandwidth of UWB signals contributes to high precision in ranging. With multi-GHz bandwidth, UWB technology can lead to localization accuracy of the order of cm or even lower than this. Applications such as the precise positioning of a robot for a certain automated function to be carried out inside a building, demand such high precision localization. UWB-based localization employs a unidirectional communication from transmitter to receiver with antenna arrays. While it has been proven to be extremely effective for line-of-sight (LOS) localization, where transmitter and receiver can "see" each other, non-LOS (NLOS) localization is very challenging and needs more attention from UWB researchers [19]. UWB-based localization is a very attractive choice over other conventional localization tools, such as ultrasonic or laser-based systems, due to its re-usability for high data rate communication, good penetration properties, and operational safety.

1.3.2 High-Speed Data Link

This is one of the most popular applications of UWB technology due to the recent surge in UWB communication and demands driven by modern requirements. With enormous available bandwidth, high data rate links with speeds of the order of hundreds of MBPS or even higher can be easily established for short-range applications. The range of such high-speed links is constrained to mainly indoor applications covering a few tens of meters due to small spreading factors and limited allowable power spectral density of UWB signals. Such short-range data links, popularly known as personal area network (PAN), are becoming increasingly popular in applications such as consumer electronics and personal computing applications. The most popular examples of such applications are

- High data rate link between HDTV (high-definition television) and set top box or DVD players
- Wireless USB (universal serial bus) in which a link with a high data rate of 480 MBPS can be established between different components of a computer

Wireless local area network (WLAN) is another potential technology to support such high data rate link. However, for short-range applications, UWB has the edge due to cost effectiveness and higher data rates over WLAN. Current research on integrating multiple-input `1 multiple-output (MIMO) technology with UWB, which deploys multiple UWB antennas in a common platform, will further increase the data rate supported by UWB technology.

1.3.3 Wireless Sensor Network

Wireless sensor network, which consists primarily of a large number of spatially distributed sensors across a network, relies on data communication from various sensor nodes to its central server and/or in between the sensor nodes. Such networks are quite useful to monitor an environment and have shown potential applications within surveillance, healthcare, etc. Most of these applications deal with a low volume of data with an average data rate of a few KBPS. UWB technology has the potential for establishing the required communication link to maintain such networks. The main challenge UWB researchers face in sensor network design constitutes the following critical aspects:

- Compact design of the antennas and associated accessories due to the stringent size restrictions as the tran-receiver is to be co-housed within the compact sensor
- Since the sensor is deployed with a predefined battery life, the system has to be designed with the minimum power consumption possible
- Ensure faithful functionalities and data communication in challenging propagations conditions and harsh EM environment

A sudden change in environment can impose an added challenge on the communication link of the wireless sensor network and the UWB system must be robust enough to handle such conditions.

1.3.4 Body Area Network (BAN)

Body area network, popularly known as BAN, is another important area where UWB technology has potential applications. In BAN, numbers of nodes or sensor units are placed on the human body or on clothing. A current wired BAN is quite inconvenient for the user. An alternative to this is to use a special garment to be worn by the user which is made of a smart textile (pre-wired). However, this may conflict with the user's personal preference. Due to this, wireless BAN is gaining in popularity. UWB technology providing a precise geo-location of the sensors compared to the conventional narrowband schemes can be successfully used to establish the short-range communication link in the BAN network.

1.3.5 UWB Radar

UWB-based radar is one of the most potent applications of UWB technology. A variety of applications including ground-penetrating radar (GPR), through wall imaging, electronic surveillance monitoring (ESM), electronic counter measures (ECM), and directed energy weapons (DEW), long-range UWB radar, etc. can serve the civilian and defense industries. ESM is extremely useful to the army and military groups for urban warfare and land-mine detection. Another important commercial application of UWB radar is vehicular collision avoidance systems. Each of these applications is quite unique in nature and requires individual attention. For example, GPR catering to important applications, such as demining, utility location and road inspection, archaeological and forensic studies, built-structure imaging, demands the widest possible bandwidth and high peak gain. Unlike conventional radar (operating in air), GPR operates in complex RF environments (ground, walls, etc.) with unfavorable propagation characteristics and inherent heterogeneity of the medium. Designing a wide bandwidth system and maintaining other specifications under such complicated scenarios is a great challenge to UWB engineers.

1.3.6 Bio-Medical Imaging

This is a very important medical application of UWB technology. In principle, this is somewhat similar to GPR; however, there are some major differences. The basic principle of medical diagnosis using radio waves is to identify the changed refractive index of the structure of interest compared to the surrounding area. However, high attenuation of the radio waves in human tissue, the small size of the object to be diagnosed and the cluttered, inhomogeneous RF environment, all pose challenges for RF engineers. UWB technology is rapidly emerging as a potential solution for bio-medical imaging, for example, in cancer detection. UWB-based bio-medical imaging has shown great potential for breast cancer detection [19] and has the edge over currently used X-ray mammography which suffers from relatively high false negative and false positive detection rates [19].

References

1. Heinrich Hertz, *Electric Waves: Being Researches on the Propagation of Electric Action with Finite Velocity Through Space*, Dover Publications, 1893.
2. C.E. Shannon, "A Mathematical Theory of Communication", *Bell Syst. Techn. J.*, vol. 27, 379–423, 623–656, July, October, 1948.
3. Chen Xiaodong, Ling Cong, Daniel Valderas, *Ultrawideband Antennas: Design and Applications*, Imperial College Press, London, UK, 2010.
4. OSD/DARPA, Ultra-Wideband Radar, Review Panel, "Assessment of Ultra Wide-Band (UWB) Technology", Arlington, VA: Defense Advanced Research Project Agency (DARPA), 1990.

5. US Federal Communications Commission (FCC), October 2003 - Part 15 [Online]. Available: http://www.fcc.gov/oet/info/rules.
6. T.W. Barrett, *History of UltraWideBand (UWB) Radar and Communications: Pioneers and Innovators, Proc. Progress in Electromagnetics Symposium 2000*, Vienna, VA, 2000.
7. T.W. Barrett, "History of ultrawideband (UWB) communications and radar: Pioneers and innovators: Part 2, UWB radars and sensors", *Microwave J.*, 44(2), 22–52, 2001.
8. H.F. Harmuth, *Nonsinusoidal Waves for Radar and Radio Communications*, Academic Press, 1981.
9. R.J. Fontana, "Recent System Applications of Short-Pulse UWB Technology," *IEEE Trans on Microwave Theory and Techniques*, Vol. 52, No. 9, pp. 2087–2104, Sep. 2004.
10. C.E. Cook and M. Bernfeld, *Radar Signals: An Introduction to Theory and Application*, New York, Academic Press, 1967.
11. G.F. Ross, "Transmission and Reception System for Generating and Receiving Base-Band Duration Pulse Signals without Distortion for Short Base-Band Pulse Communication Systems", U.S. Patent 3,728,632, issued April 17, 1973.
12. R.A. Scholtz, "Multiple Access with Time-Hopping Impulse Modulation", *Proc. IEEE MILCOM*, vol. 2, 447–450, 1993.
13. M.Z. Win and R.A. Scholtz, "Impulse Radio: How it Works", *IEEE Comm. Lett.*, vol. 2, 36–38, 1998.
14. M.Z. Win and R.A. Scholtz, "Ultra-Wide Bandwidth Time-Hopping Spread-Spectrum Impulse Radio for Wireless Multiple-Access Communication", *IEEE Trans. Comm.*, vol. 48, 679–691, 2000.
15. T.E. McEwan, "Ultra-Wideband Radar Motion Sensor", U.S. Patent 5,361,070, issued November, 1994.
16. Moe Z. Win, Davide Dardari, Andreas F. Molisch, Werner Wiesbeck, and W. Jinyun Zhang, "History and Applications of UWB", Institute of Electrical and Electronics Engineers, 2009.
17. FCC, Revision of part 15 of the commission's rules regarding Ultra Wide Band transmission systems, First report and order, and Docket 98-153, FCC 02-03, adopted/released, February 14/ April 22 2002.
18. W.C. Liu, F.M. Yeh, and M. Ghavami, "Miniaturized Implantable Broadband Antenna for Biotelemetry Communication", *Microwave Opt. Technol. Lett.*, vol. 50, no. 9, 2407–2409, 2008.
19. B. Allen, M. Dohler, E. Okon, W. Malik, A. Brown, and D. Edwards, eds. Ultra-Wideband Antennas and Propagation: For Communications, Radar and Imaging. John Wiley & Sons, 2006.

2

Design and Developments of UWB Antennas

2.1 Introduction

In Chapter 1, we had a brief overview of UWB technology, various techniques of UWB signaling and associated coding, spectrum regulation on UWB technology, followed by its applications. This introductory overview provides the reader with the key concepts and challenges of UWB technology before we shift our attention to the scope of this book. The performance of the UWB system largely depends on the performance of the antennas installed in the UWB devices. To ensure faithful system operation, UWB antennas must satisfy a set of predefined criteria in terms of a wide impedance matching, phase linearity, spectral efficiency, and the radiation characteristics which are determined by various parameters, such as directivity, gain, efficiency, etc. Unlike the traditional antenna designs which are characterized in terms of reflection coefficient and gain only, UWB antenna design is much more challenging as it must qualify in terms of additional parameters in time domain, such as group delay, fidelity, etc. Moreover, UWB antenna design has to strictly comply with the stringent size specification as per the applications and sometimes has to be functional in very harsh EM environments. Another notable difference between the traditional antenna and UWB antennas is the perspective from which their figures of merit are quantified. Traditional antennas are studied based on their "steady state" or time averaged response and therefore looked upon from a "frequency domain" point of view. On the other hand, since UWB antennas are supposed to deal with narrow pulses (in time domain) due to their wideband nature, studying their transient response, defined as "impulse response", is extremely crucial. Thus, UWB antennas, in addition to the classical "frequency domain" studies, are qualified with respect to "time-domain" studies. Even though time domain and frequency domain views are mathematically related by Fourier transform, there is a significant difference between the two approaches of antenna characterization. Frequency domain studies dealing with average quantities, i.e. in steady state, treat the antenna as radiating "power", while time-domain analysis, dealing with transient response, considers the antenna as radiating "energy". Though "energy" and "power" are again interrelated as "energy" can be obtained by time integrating "power", in general time-domain studies are much more robust and reveal extra information on dispersion characteristics of the radiated pulse by the UWB antennas.

In this chapter, we will deal with an elaborate discussion on the design and development of the UWB antenna, considering various families of the antenna. To begin with, the first part of the chapter will focus on a quick review of antenna parameters such as bandwidth, directivity, efficiency, and gain, followed by additional parameters such as group delay and fidelity, which are exclusive requirements for characterizing a UWB antenna. This is followed by a discussion on different classes of UWB antenna designs proposed by various research groups over the last two decades. While discussing these UWB antennas, special emphasis is placed on the physical understanding and extra effort is devoted to explaining their operation principle using the fundamental concepts of antenna engineering. Another section is dedicated to comparing various UWB antennas. This gives a clear idea to UWB antenna engineers when selecting the antenna type for a desired application. Lastly, a section on UWB antennas for special applications is included in this chapter.

2.2 Review of Fundamental Antenna Parameters

Before going into the details of UWB antennas and their characterization techniques, let us quickly review the fundamentals of the antenna. Here, we will confine our discussion to the general concepts of radiation pattern, polarization, and antenna matching. We are already familiar from fundamental courses on antenna engineering that an antenna is a transducer which converts guided electrical energy from a signal source into radiated EM energy in the medium around it or converts incoming electromagnetic energy into electrical energy. Since an antenna does not distribute energy uniformly in space, but rather does so with a particular spatial distribution, it is extremely important to understand and even visualize this spatial distribution of energy. We try to describe this spatial distribution of radiated field/energy or power by a 3-D plot or set of 2-D plots on various special 2-D planes, called the radiation patterns. Thus, the radiation pattern of an antenna corresponds to the spatial distribution of the radiated fields or power as a function of polar coordinates (θ,ϕ).

To get exact information on the radiation distribution, 3-D patterns are the most ideal. While modern EM software provides excellent color versions of such 3-D plots, for most of the antennas, suitable 2-D cuts are sufficient to describe their radiation patterns. Based on whether the field (normally E-field) or power radiated by the antenna is plotted, the patterns are classified as field pattern or power pattern. Since in the normalized scale, the power is the square of the E- field, these two patterns are related and the power pattern can be easily derived by squaring the field pattern in the absolute scale. For 2-D patterns, selecting proper planes while plotting the pattern is important. Normally two special planes, namely the E-plane and the H-plane, are preferred for this. Since the E- and H-plane of the antenna can change depending on the orientation and nature of the antenna, let us have a quick look on the E- and H-plane along with the associated radiation pattern for a few popular antennas. We recall that the E-plane of the antenna is defined as the plane containing the E-field vector and the direction of maximum radiation. Similarly, the plane constituted by the H-field and the direction of maximum radiation is known as the H-plane.

Dipole Antenna

A dipole antenna has maximum radiation along its equatorial plane. In the far field, it has θ-component of E-field and ϕ-component of H-field. Thus, any vertical plane, containing the dipole is the E-plane of a dipole antenna. On the other hand, the equatorial plane (x-y plane, dipole is z oriented) perpendicular to the dipole is the H-plane of a dipole antenna. Figure 2.1 shows a z-oriented dipole antenna along with the E- and H-planes of the antenna.

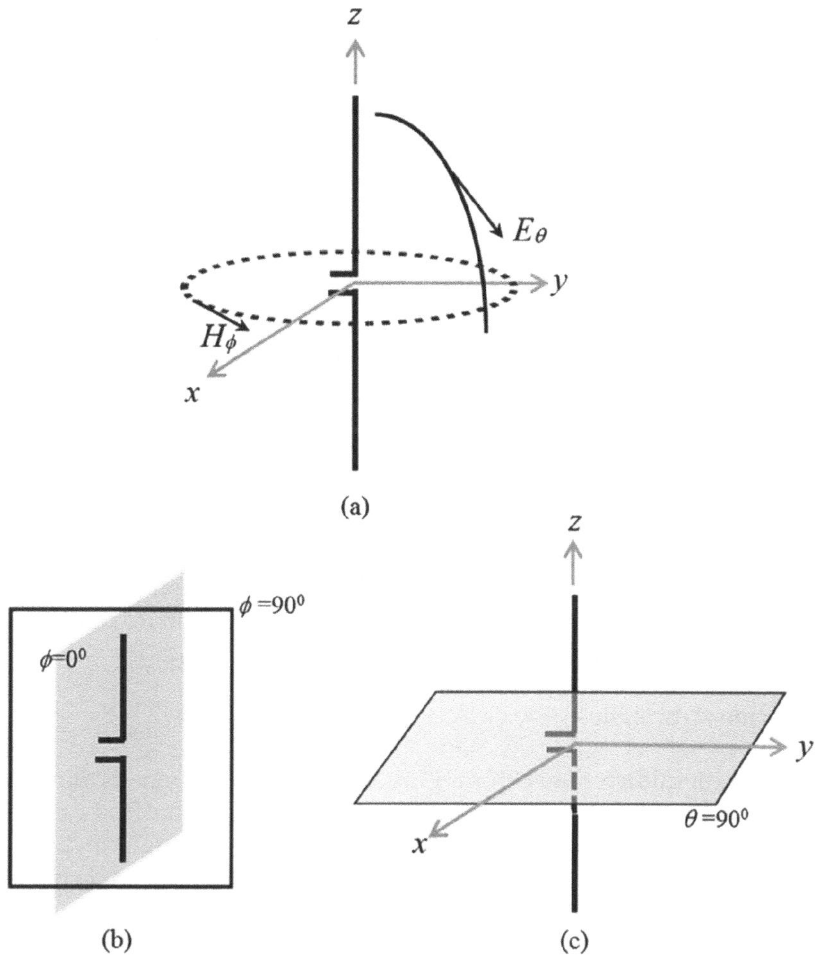

FIGURE 2.1
(a) Dipole antenna with its far-field components, (b) E-plane and (c) H-plane of the antenna.

Horn Antenna

A horn antenna is one of the most fundamental and widely used aperture antennas. Figure 2.2 shows a pyramidal horn antenna along with its aperture field. Here, the *y-z* plane is the E-plane and *x-z* plane is the H-plane.

Rectangular Microstrip Patch Antenna

In the rectangular microstrip patch antenna (Figure 2.3a), the net component of the fringing E-field is along the *x*-axis. Maximum radiation is toward boresight, i.e. along the *z*-axis. Thus, the *x-z* plane is the E-plane of the antenna. The magnetic field forms a closed loop on the *y-z* plane due to the *x*-directed currents. Hence, the *y-z* plane acts as the H-plane of the antenna. Figure 2.3d clearly shows the E- and H-plane of the antenna along with the radiating patch printed on the grounded substrate.

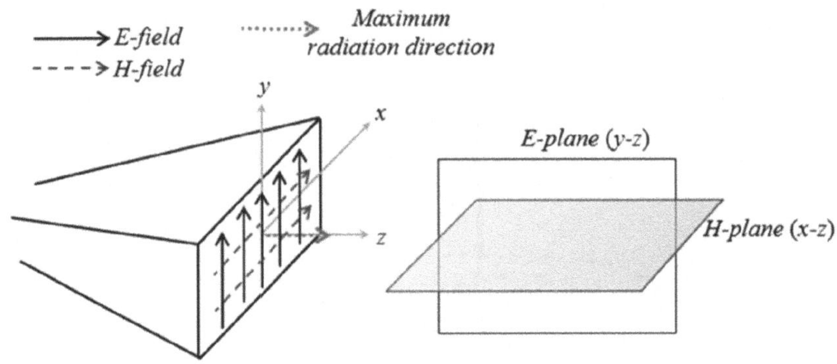

FIGURE 2.2
(a) Pyramidal horn antenna and its (b) E- and H-planes.

An ideal antenna pattern is an isotropic pattern. An isotropic antenna is an ideal antenna or a point source which radiates EM energy uniformly in all directions (θ,ϕ). That is, the radiated power from a point source for a fixed radius is the same, i.e. the radiation is in the form of a sphere. However, such an antenna does not exist and this is just a theoretical concept used for characterizing directivity or gain of any standard antenna. In a more practical sense, antennas can be classified into the following categories:

Omnidirectional Antennas

Radiation from omnidirectional ("omni" means "one") antennas is directive in one plane while it radiates uniformly in its transverse plane. As the far field of an antenna is described by (θ,ϕ)-dependence, for omnidirectional antenna having directionality in the vertical plane, the radiated fields or power is a function of elevation angle (θ) and is independent of the azimuthal angle (ϕ).

The most common example of an omnidirectional antenna is a dipole antenna. For a vertically erected dipole, as shown in Figure 2.4, the electric field has component only in the elevation plane, i.e. $\vec{E} = E_\theta\hat{\theta}$, whereas the H-field is oriented along the azimuthal plane, i.e. $\vec{H} = H_\phi\hat{\phi}$, which ensures $\vec{P} = \vec{E}\times\vec{H} = \vec{P} = \vec{E}\times\vec{H} = E_\theta H_\phi\hat{r}$ oriented radially outward. For a λ/2 dipole, the power pattern of the antenna is given by a factor:

$$F \rightarrow \frac{\cos^2\left(\dfrac{\pi}{2}\cos\theta\right)}{\sin^2\theta} \tag{2.1}$$

This factor, as indicated in Figure 2.4, implies that maximum radiation is toward θ = 90° for any value of ϕ whereas nulls of the antenna are toward θ = 0° or 180°. In 3-D, the pattern looks like a "doughnut" with dips/nulls oriented along the dipole axis. Omnidirectional antennas are very useful in covering a particularly wide area uniformly. For example, antennas in cell phones or radiation from cell phone towers ideally should be omnidirectional.

Directional Antennas

For a directional antenna, radiated energy is confined within a very narrow solid angle. Thus, the radiated field or power of such antennas is a function of both θ and ϕ. Directional antennas are used for establishing a point-to-point link. For example, antennas used in

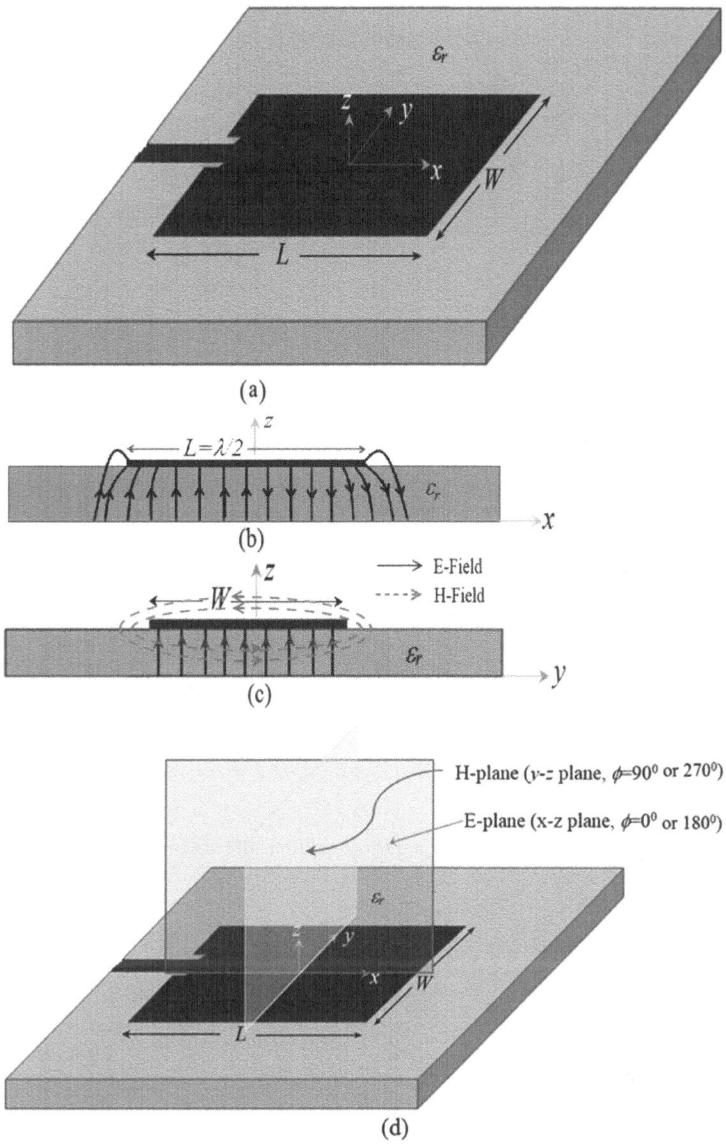

FIGURE 2.3
A rectangular microstrip patch antenna (a) structure, (b) *x*-*z* cut, (c) *y*-*z* cut, and (d) E-plane and H-plane of the antenna.

a satellite link between satellite transponder and ground stations are highly directional antennas. Horn antennas, tapered slot antennas are a few examples of such directional antennas. Figure 2.5 shows typical examples of directional antennas along with their 3-D, E-, and H-plane radiation patterns.

2.2.1 Radiation Power Density

Radiation power density is defined as the average value of the Poynting vector contributed in space due to the radiation of an antenna. Normally, it is measured in the far field of an antenna and has a unit of W/m^2.

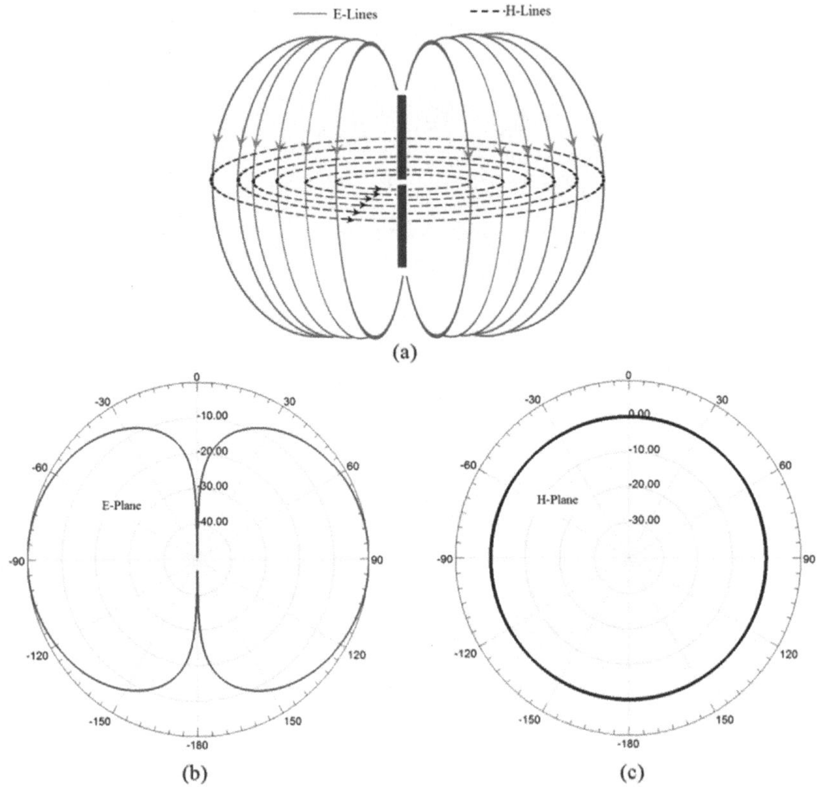

FIGURE 2.4
Typical example of an omnidirectional antenna: (a) Dipole with its 3-D fields, (b) E-plane, and (c) H-plane radiation pattern.

We know Poynting vector: $\vec{P} = \vec{E} \times \vec{H}$
Thus, radiation power density,

$$\vec{W}_{\text{rad}} = \frac{1}{2}\text{Re}\left\{\vec{E} \times \vec{H}^*\right\} \tag{2.2}$$

Since an antenna radiates a directive beam, radiation intensity of an antenna is a function of space coordinates (θ and ϕ) and also varies $1/r^2$ against the radial distance r.

The total power radiated by an antenna can be obtained by integrating $\vec{W}_{\text{rad}}$ over a closed surface surrounding the antenna.

Thus,

$$\begin{aligned} P_{\text{rad}} &= \oint_s \vec{W}_{\text{rad}} . \vec{d}s \\ &= \frac{1}{2}\oint_s \text{Re}\left\{\vec{E} \times \vec{H}^*\right\} . \hat{n} ds \end{aligned} \tag{2.3}$$

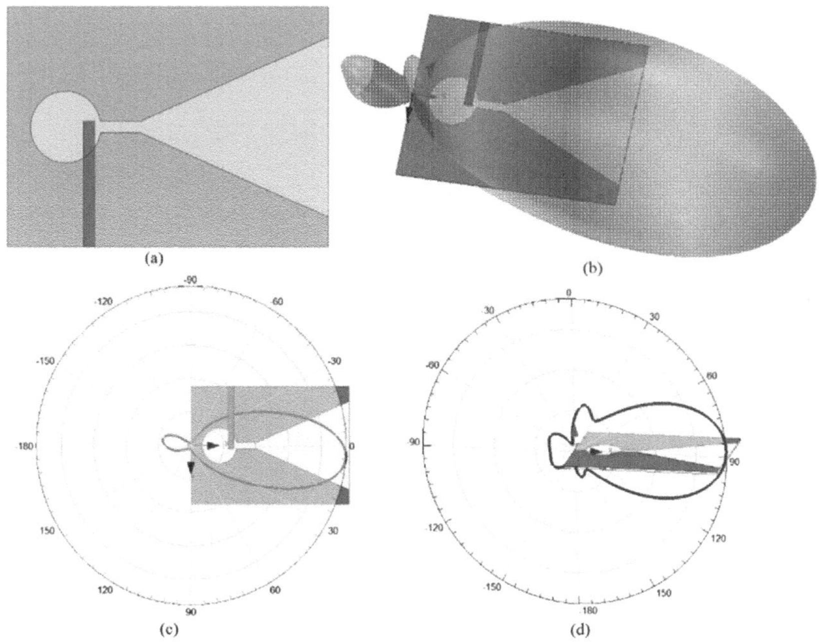

FIGURE 2.5
Typical example of an omnidirectional antenna: (a) Vivaldi with its 3-D pattern, (b) E-plane, and (c) H-plane radiation pattern.

For an isotropic radiator $\vec{W}_{rad}$ is not a function of a spherical coordinate (θ and ϕ). Thus, considering a spherical surface surrounding the antenna at its center,

$$P_{rad} = \oint_S \vec{W}_{rad}.\vec{d}s$$

$$= W_{rad} \oint_S ds = W_{rad}\left(4\pi r^2\right)$$

2.2.2 Radiation Intensity

This is another directional quantity associated with the radiation from an antenna which is defined as the amount of power radiated per unit solid angle. We know that an elemental surface ds and solid angle associated $d\Omega$ are related by,

$$d\Omega = dsr^2$$

Thus, surface area of unit solid angle ($d\Omega = 1$), $ds = r^2$

Thus radiation intensity,

$$U \cdot (\theta, \phi) = r^2 W_{rad} \tag{2.4}$$

Using (2.2) and (2.4),

$$\begin{aligned} U(\theta,\phi) &= r^2 \frac{1}{2} \mathrm{Re}\left|\vec{E}\times\vec{H}\right| \\ &= r^2 \frac{1}{2\eta}\left|\vec{E}(r,\theta,\phi\right|^2 \\ &= r^2 \frac{1}{2\eta}\left[\left|E_\theta(r,\theta,\phi\right|^2 + \left|E_\phi(r,\theta,\phi\right|^2\right] \\ &= r^2 \frac{1}{r^2}\frac{1}{2\eta}\left[\left|E_\theta(\theta,\phi)\right|^2 + \left|E_\phi(\theta,\phi)\right|^2\right] \\ &= \frac{1}{2\eta}\left[\left|E_\theta(\theta,\phi)\right|^2 + \left|E_\phi(\theta,\phi)\right|^2\right] \end{aligned} \tag{2.5}$$

Radiation intensity $U(\theta,\phi)$ integrated over the full space, i.e. entire solid angle contributes to the total radiated power of an antenna.

Thus,

$$\begin{aligned} P_{\mathrm{rad}} &= \oint_S U(\theta,\phi)d\Omega \\ &= \int_{\phi=0}^{2\pi}\int_{\theta=0}^{\pi} U(\theta,\phi)\sin\theta d\theta d\phi \end{aligned} \tag{2.6}$$

For an isotropic radiator $U(\theta,\phi) = U_0$, is independent of θ and ϕ.

$$\begin{aligned} P_{\mathrm{rad}} &= \oint_S U(\theta,\phi)d\Omega \\ &= U_0 \oiint d\Omega = 4\pi U_0 \\ \therefore U_0 &= \frac{P_{\mathrm{rad}}}{4\pi} \end{aligned} \tag{2.7}$$

2.2.3 Directivity

This is an important parameter describing the directional nature of the radiation beam of an antenna. It is defined as the ratio of the radiation intensity $U(\theta,\phi)$ in a given direction (θ,ϕ) to that of an isotropic source, U_0. Thus, directivity $D(\theta,\phi)$ is expressed as,

$$D(\theta,\phi) = \frac{U}{U_0} = \frac{U}{\frac{P_{\mathrm{rad}}}{4\pi}} = \frac{4\pi U(\theta,\phi)}{P_{\mathrm{rad}}} \tag{2.8}$$

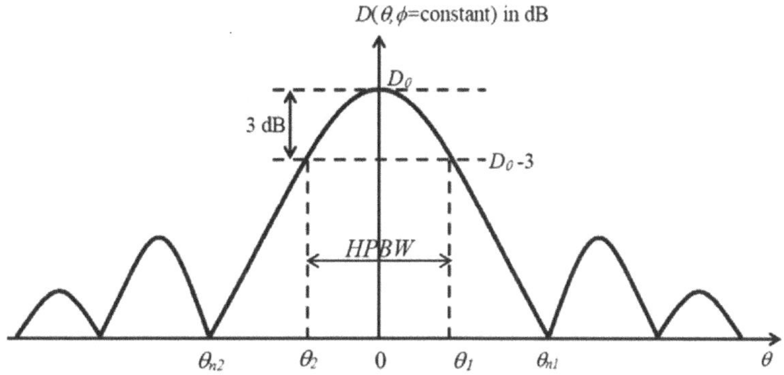

FIGURE 2.6
Typical radiation pattern of an omnidirectional antenna in Cartesian coordinate.

Maximum directivity of an antenna is obtained by considering maximum radiation intensity corresponding to the direction of maximum radiation. Thus,

$$D_{\max}(\theta,\phi)=\frac{U_{\max}}{U_0}=\frac{U_{\max}}{\dfrac{P_{\text{rad}}}{4\pi}}=\frac{4\pi U(\theta,\phi)|_{\max}}{P_{\text{rad}}} \tag{2.9}$$

For an antenna radiating two orthogonal electric fields, or circular polarized antenna, directivity is separately estimated for each component of the electric field. Such directivities are normally termed as partial directivities, normally expressed as D_θ and D_ϕ, corresponding to the θ- and ϕ-component of the fields respectively.

To get a more comprehensive idea on the directional nature of radiation from an antenna, let us plot the directivity of an antenna as a function of elevation angle θ for a constant ϕ. For an omnidirectional antenna, the choice of any value of ϕ would result in the same plot.

Figure 2.6 shows a typical example of such a plot in dB scale. The pattern consists of a main beam, called the major lobe, through which most of the power of the antenna is radiated. In addition, there might be various other lobes, named as side lobes of the antenna. θ_{n1} and θ_{n2} are the angles corresponding to the first null on both sides of the radiation maxima. Angular separation between these two nulls are known as first null beam width (FNBW). θ_1 and θ_2 are the angles at which the directivity is 3 dB down (half in absolute scale) the peak directivity D_0. Thus $\theta_1 \sim \theta_2$ is the half power beam width (HPBW) of the antenna.

2.2.4 Efficiency and Gain of an Antenna

An antenna in transmit mode accepts power from a source through a transmission line and radiates the power in free space. Let us consider a simplified block diagram of a transmitting antenna as shown in Figure 2.7. Here, P_g is the amount of power delivered to the transmission line by the signal generator, P_{in} is the amount of power coupled to the antenna, and P_{rad} is the amount of power radiated by the antenna. The most favorable and desired condition is $P_g=P_{in}=P_{\text{rad}}$. This condition is achieved when there is perfect impedance matching between the feeding transmission line and the antenna impedance which ensures $P_{in}=P_g$, and power accepted by an antenna (P_{in}) is fully radiated, i.e. $P_{\text{rad}}=P_{in}$.

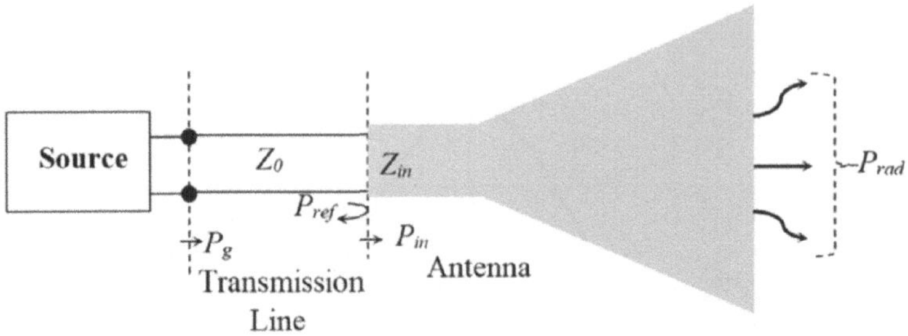

FIGURE 2.7
A generalized block diagram of a transmitting antenna.

However, in reality, both of these conditions are extremely difficult to achieve. An extremely important figure of merit for an antenna, known as antenna efficiency, describes its performance in this regard.

2.2.4.1 Radiation Efficiency

Radiation efficiency of an antenna is defined as the ratio of the total power radiated (P_{rad}) by an antenna to the power coupled to the antenna (P_{in}) through a transmission line. Thus, radiation efficiency,

$$\eta_{\text{rad}} = \frac{P_{\text{rad}}}{P_{\text{in}}} \tag{2.10}$$

Due to various intrinsic losses inside the antenna, such as conduction loss, dielectric loss, surface wave loss, etc., the radiated power by the antenna is always less than the power accepted by it. Thus, radiation efficiency is always less than 1.

Power loss in the antenna,

$$\begin{aligned} P_{\text{loss}} &= p_{\text{in}} - p_{\text{rad}} \\ &= p_{\text{in}} - \eta_{\text{rad}} p_{\text{in}} \\ &= (1 - \eta_{\text{rad}}) p_{\text{in}} \end{aligned} \tag{2.11}$$

2.2.4.2 Reflection Mismatch Efficiency

As indicated in the block diagram in Figure 2.7, impedance mismatch between the feeding transmission line and the input impedance of the antenna causes some power to be reflected back which doesn't allow the entire power in the transmission line to be coupled to the antenna. This loss in an antenna feed is modeled as the reflection mismatch efficiency.

Reflection coefficient,

$$\Gamma = \frac{Z_{\text{in}} - Z_0}{Z_{\text{in}} + Z_0} \tag{2.12}$$

where $Z_{in} = R_A + jX_A$ is the antenna input impedance and Z_0 is the characteristic impedance of the feeding line and normally equates to 50 Ω.

Amount of power reflected due to impedance mismatch,

$$P_{\text{ref}} = |\Gamma|^2 P_g \tag{2.13}$$

Thus, power coupled to the antenna,

$$P_{\text{in}} = P_g - P_{\text{ref}} = \left[1 - |\Gamma|^2\right] P_g \tag{2.14}$$

Reflection mismatch efficiency,

$$\eta_{\text{ref}} = \frac{P_{\text{in}}}{P_g} = 1 - |\Gamma|^2 \tag{2.15}$$

2.2.4.3 Overall Efficiency

Overall efficiency is defined as the ratio of the power radiated to the power coupled to the feedline of the antenna. It is a product of η_{rad} and η_{ref} (in linear scale). Overall efficiency,

$$\eta_0 = \frac{P_{\text{rad}}}{P_g} = \frac{P_{\text{rad}}}{P_{\text{in}}} \frac{P_{\text{in}}}{P_g} = \eta_{\text{rad}} \eta_{\text{ref}} = \left[1 - |\Gamma|^2\right] \eta_{\text{rad}} \tag{2.16}$$

Apart from these, for certain antennas, especially when used in receiving mode, there might be additional losses due to polarization mismatch of the antenna with respect to the incoming electric field. This additional loss in antenna due to polarization mismatch is modeled as polarization efficiency, η_{pol}. By reducing cross-polar radiation, polarization efficiency of an antenna, η_{pol}, can be made close to unity.

According to the IEEE standard, efficiency of an antenna should be mainly determined based on radiation efficiency as the antenna alone is not responsible for losses due to feed-reflection and polarization mismatch and can be mitigated by an efficient feed design and ensuring proper orientation of the antenna.

2.2.4.4 Gain

Similar to directivity, gain of an antenna also describes the directional nature of the radiated beam of an antenna. It is very closely related to the directivity of an antenna with a subtle difference. It is defined as the ratio of the radiation intensity $U(\theta,\phi)$ in a given direction (θ,ϕ) to that of the radiation intensity that would have been obtained if the power accepted by an antenna (P_{in}) was radiated isotropically.

$$\text{Thus, Gain}, G(\theta,\phi) = \frac{U}{\frac{P_{\text{in}}}{4\pi}} = \frac{4\pi U(\theta,\phi)}{P_{\text{in}}} \tag{2.17}$$

$$G(\theta,\phi) = \eta_{\text{rad}} \frac{4\pi U(\theta,\phi)}{P_{\text{rad}}} = \eta_{\text{rad}} D(\theta,\phi) \tag{2.18}$$

2.2.4.5 Realized Gain or Absolute Gain

Realized or absolute gain of an antenna considers the losses due to reflection mismatch between the feedline and the antenna, and therefore is expressed as,

$$G_{\text{realized}}(\theta,\phi) = \frac{U(\theta,\phi)}{\frac{P_g}{4\pi}} = \frac{4\pi U(\theta,\phi)}{\frac{P_{\text{rad}}}{\eta_{\text{rad}}\eta_{\text{ref}}}} = \eta_{\text{rad}}\eta_{\text{ref}}\frac{4\pi U(\theta,\phi)}{P_{\text{rad}}} = \eta_{\text{rad}}\eta_{\text{ref}}\, D(\theta,\phi) \tag{2.19}$$

Even though gain of an antenna is a function of (θ,ϕ), it is almost common practice to refer to "maximum gain" as the gain of an antenna. In practice, gain of an antenna is expressed in dBi, where the lower case "i" refers to an "isotropic" radiator. Thus, gain in dBi,

$$G(\text{dBi}) = 10\log G \tag{2.20}$$

2.2.5 Beam Efficiency

Beam efficiency is another important figure of merit for antennas, especially for high gain antennas with narrow main lobe and negligible side lobes.

Beam efficiency,

$$\eta_{BE} = \frac{\text{Power radiated/received through cone with semi-vertical angle } \theta_1}{\text{Total radited/received power by the antenna}}$$

where θ_1 is the angle through which a high % power is radiated.

If $U(\theta,\phi)$ is the radiation intensity and the antenna's peak radiation is toward the z-axis ($\theta=0$),

$$\eta_{BE} = \frac{\int_{\phi=0}^{2\pi}\int_{\theta=0}^{\theta_1} U(\theta,\phi)\sin\theta d\theta d\phi}{\int_{\phi=0}^{2\pi}\int_{\theta=0}^{\pi} U(\theta,\phi)\sin\theta d\theta d\phi} \tag{2.21}$$

For example, for an antenna with $\theta_1 = 10°$, $\eta_{BE} = 90\%$ means that 90% of the total radiated power of the antenna passes through a cone with semi vertical angle 10°.

2.2.6 Effective Aperture

This is an important concept, especially for receiving antennas. Since a receiving antenna is exposed to a uniform plane wave with a given power density (say, S Watt/m^2), we can express the amount of power delivered to the load by the antennas as,

$$P_L = A_e S$$

$$\text{Or}, A_e = \frac{P_L}{S}$$

where A_e is the effective aperture of the antenna having a unit of m^2. For some antennas, e.g. horn, parabolic reflector, etc., it is possible to make a relationship between effective aperture and the physical area of the radiating aperture. However, in general, effective aperture has no correspondence with its physical area, since there are many antennas with extremely small physical aperture (e.g. rod antennas), yet they have good potential to radiate or receive EM energy, indicating a good effective aperture.

It can be shown that peak directivity D_0 and effective aperture of antenna are related by,

$$D_0 = \frac{4\pi A_e}{\lambda^2} \tag{2.22}$$

2.2.7 Front-to-Back Ratio

The front-to-back ratio (f/b) is a parameter describing the directional radiation patterns for antennas. If an antenna has maximum gain in a particular direction, the front-to-back ratio is the ratio of the maximum gain in that particular direction to the gain in the opposite (180 degrees) direction. The parameter is given in dB.

2.2.8 Input Impedance and Matching

Input impedance of an antenna is the most important figure of merit as it determines the amount of power that is accepted by an antenna from a signal source through its feedline. A good impedance matching between the feedline and input impedance ensures that almost the entire amount of incoming power is coupled to an antenna with much less reflected back toward the source. Input impedance is defined as the impedance presented by an antenna at its terminals and normally is determined by taking the ratio of the appropriate components of electric and magnetic field.

The input impedance of the antenna in general is expressed as $Z_A = R_A + jX_A$, where R_A and X_A are antenna resistance and reactance, respectively. Like any resonant circuit, Z_A, and hence R_A and X_A, are function of frequency. The frequencies for which $X_A = 0$ is normally known as resonance frequency. At these frequencies, antenna impedance is fully resistive and the antenna shows an excellent matching if $R_A \approx 50\,\Omega$, the typical impedance of the feeding line.

Matching between an antenna with its feeding line can be quantified in terms of reflection coefficient, or voltage standing wave ratio or S_{11} in dB. Since these parameters are related and almost synonymous, understanding their mutual relationship is very important. To quantify the antenna matching, let us consider a simplified block diagram as shown in Figure 2.8. A transmission line of characteristic impedance $Z_0 = 50\,\Omega$ is connected to the antenna with input impedance $Z_{in} = R_A + jX_A$.

2.2.8.1 Reflection Coefficient

Reflection coefficient at the antenna input terminal is given by,

$$\Gamma = \frac{Z_{in} - Z_0}{Z_{in} + Z_0} \tag{2.23}$$

At resonance $X_A = 0$; hence reflection coefficient at resonance,

$$\Gamma_{res} = \frac{R_A - Z_0}{R_A + Z_0} \tag{2.24}$$

Thus, an antenna with $R_A = Z_0 = 50\,\Omega$ ensures zero reflection, i.e. perfect matching. Figure 2.9 shows a typical plot of antenna impedance as a function of frequency.

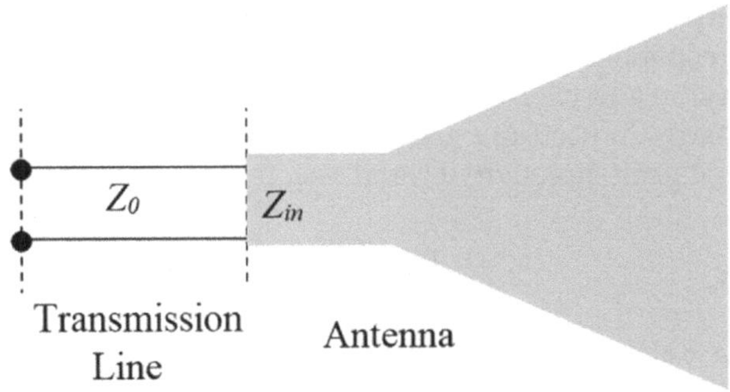

FIGURE 2.8
Matching of antenna and feeding transmission line.

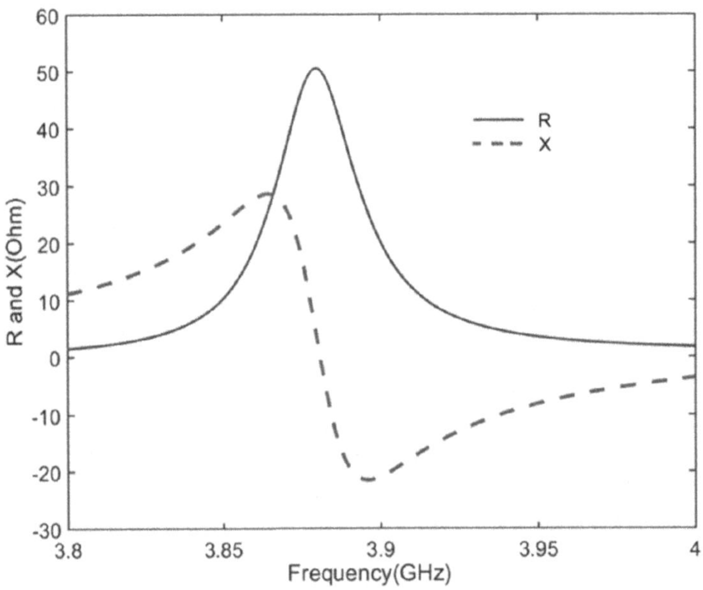

FIGURE 2.9
Typical plot of antenna impedance as a function of frequency.

It is important to understand that $|\Gamma|^2$ indicates the amount of fractional power reflected back toward the generator while $(1-|\Gamma|^2)$ indicates the fraction of power coupled to the antenna.

To understand the implication of Z_A on antenna matching, let us consider a few cases as indicated in Table 2.1.

2.2.8.2 Voltage Standing Wave Ratio (VSWR)

Antenna matching can be described by observing the VSWR at the input side. VSWR is defined as,

$$\text{VSWR} = \frac{1+|\Gamma|}{1-|\Gamma|} \tag{2.25}$$

TABLE 2.1

Reflection Coefficient for Various Cases of Antenna Input Impedance when Fed by a 50 Ω Line

Z_0 (Ω)	Z_A (Ω)	Γ	$\lvert\Gamma\rvert$	% of Power Reflected Back ($\lvert\Gamma\rvert^2 \times 100$)	% of Power Coupled to the Antenna $(1-\lvert\Gamma\rvert^2)\times 100$
50	50	0	0	0	100
	50+10j	0.0099+0.0990i	0.0995	9.95	91.05
	50–10j	0.0099–0.0990i	0.0995	9.95	91.05
	25	1/3	1/3	11.11%	88.89
	100	1/3	1/3	11.11%	88.89

$$\therefore \lvert\Gamma\rvert = \frac{\text{VSWR}-1}{\text{VSWR}+1} \tag{2.26}$$

Thus, perfect matching ($\Gamma = 0$) corresponds to a VSWR = 1, while a high mismatch, i.e. higher $\lvert\Gamma\rvert$ ($\lvert\Gamma\rvert$ approaches 1), indicates a high value of VSWR. Thus, VSWR ranges between 1 and ∞. For most of the applications, VSWR ≤ 2 is considered a good match. At resonant condition $X_A = 0$, which implies that antenna resistance should be confined between 25 Ω and 100 Ω to get VSWR ≤ 2. Table 2.2 presents a comprehensive summary of antenna matching for various cases of VSWR.

2.2.8.3 S_{11} in dB

This is the most widely used and preferred parameter by antenna engineers to describe antenna impedance matching. It is very closely related to the reflection coefficient expressed in log scale.

$$\begin{aligned} S_{11}(\text{dB}) &= 10\log\lvert\Gamma\rvert^2 \\ &= 20\log\lvert\Gamma\rvert \end{aligned} \tag{2.27a,b}$$

S_{11} = –10 dB corresponds to 10% of the applied power reflected back and 90% power is coupled to the antenna. Thus, S_{11} less than or equal to –10 dB is considered as a good matching for an antenna.

From Equation 2.27b,

$$\begin{aligned} \lvert\Gamma\rvert &= 10^{\frac{S_{11}(\text{dB})}{20}} \\ \text{or, VSWR} &= \frac{1+\lvert\Gamma\rvert}{1-\lvert\Gamma\rvert} = \frac{1+10^{\frac{S_{11}(\text{dB})}{20}}}{1-10^{\frac{S_{11}(\text{dB})}{20}}} \end{aligned} \tag{2.28a,b}$$

Table 2.3 presents a brief summary of antenna matching for various level of S_{11} of an antenna.

Note that since VSWR = 2 corresponds to S_{11} = –9.5 dB, i.e. very close to S_{11} = –10 dB, either of these two conditions are considered as good matching. Both of these conditions ensure

TABLE 2.2

Comprehensive Summary of Antenna Matching for Various Cases of VSWR

VSWR	$\|\Gamma\|$	% of Power Reflected Back ($\|\Gamma\|^2 \times 100$)	% of Power Coupled to the Antenna $(1-\|\Gamma\|^2)\times 100$
1	0	0	100
∞	1	100	0
1.5	1/5	4	96
2	1/3	11.11	88.89
3	1/2	25	75

TABLE 2.3

Comprehensive Summary of Antenna Matching for Various Levels of S_{11}

S_{11}	*VSWR*	$\|\Gamma\|$	% of Power Reflected Back ($\|\Gamma\|^2 \times 100$)	% of Power Coupled to the Antenna $(1-\|\Gamma\|^2)\times 100$
0	∞	1	100	0
−9.5	2	1/3	11.1	88.9
−10	1.92	0.31	10	90
−15	1.43	0.178	3.1	96.9
−20	1.222	0.1	1	99

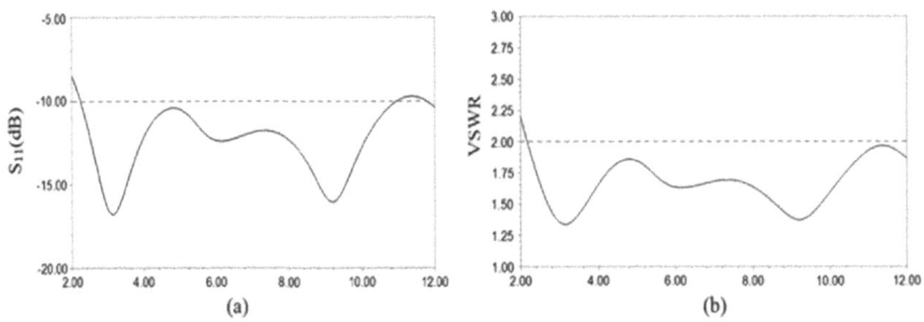

FIGURE 2.10
Typical (a) S_{11} and (b) VSWR plot for a UWB antenna.

nearly 90% of the input power is coupled to the antenna. The range of frequency over which VSWR ≤ 2 or $S_{11} \leq -10$ dB, is known as impedance bandwidth of antenna.

Figure 2.10a,b shows the plot of S_{11} (dB) and VSWR of an antenna. As revealed form the plots, the antenna exhibits a very wide impedance bandwidth ranging from 2.5 GHz to 11 GHz.

2.2.9 Polarization

Polarization of an EM wave indicates the orientation of the electric field of the propagating wave. Since EM waves are transverse in nature, the electric field (and also the magnetic field) of the wave lies in a plane perpendicular to the direction of propagation. For example, for a plane wave propagating in *z*-direction, E- and H-fields must lie in the *x*-*y* plane. Since an antenna radiates or receives EM waves with a particular orientation of the E-field,

TABLE 2.4

Identifying Polarization: Various Cases of Linear Polarizationw. Arrow-Headed Vectors in the Figures Indicate E-Field Component of the EM Wave Propagating in Z-Direction.

Electric Field	Schematic of the E-field	Polarization	Comments
$\vec{E} = E_0 e^{-j\beta z}\hat{x}$	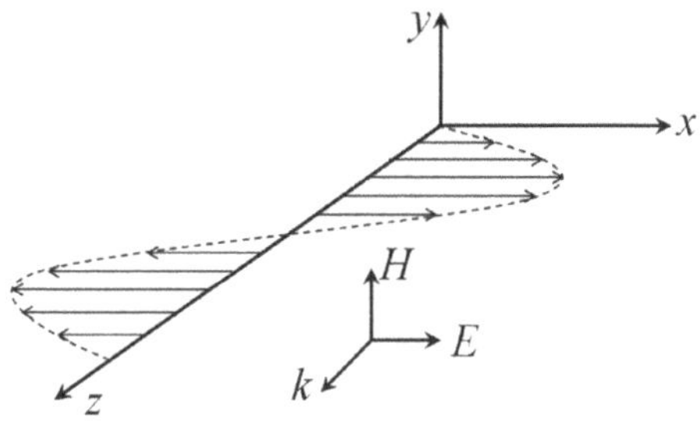	Linear	Electric field is oriented along $\hat{x}$ and $\hat{y}$
$\vec{E} = E_0 e^{-j\beta z}\hat{y}$	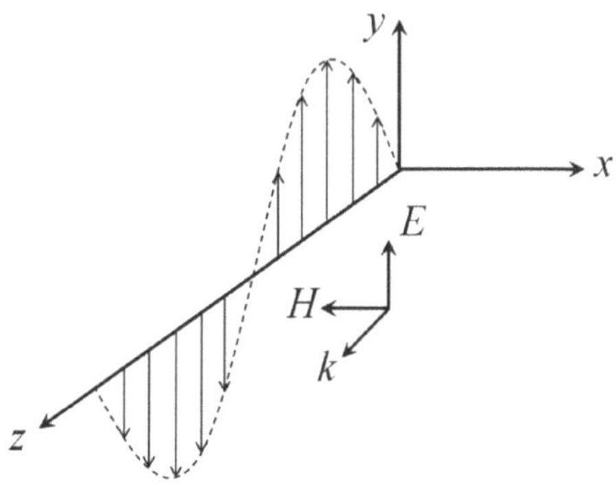	Linear	Electric field is oriented along $\hat{x}$ and $\hat{y}$
$\vec{E} = E_0 e^{-j\beta z}(-\hat{x} + \hat{y})$	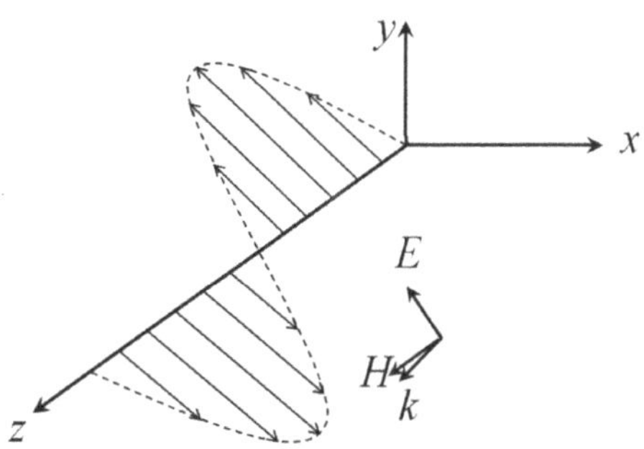	Linear	Electric field is oriented along $\hat{x}$ and $\hat{y}$

understanding its polarization and performance with respect to any undesired orthogonal component of electric field is extremely important.

2.2.9.1 Linear Polarization

In this case, the E-field of the antennas is oriented along a particular direction. Table 2.4 shows three different cases of linear polarization. Here the wave is propagating along the *z*-axis. In the first and second case, the E-field is *x*-directed linearly polarized and *y*-directed linearly polarized, respectively. In the third case too, the wave is linearly polarized but it has two orthogonal E-field components in the same phase. An antenna radiating a linearly polarized signal is known as a linearly polarized antenna. Polarization of an antenna is fundamentally determined by the orientation of the current causing radiations. Most of the wire antennas, dipoles, monopoles, loops, etc. produce linearly polarized EM signals.

2.2.9.2 Circular Polarization

In the case of circular polarization, orientation of the electric field is not fixed along a particular direction. In this case, the tip of the electric field vector rotates along a circle oriented perpendicular to the direction of propagation. The sense of rotation of the E-field when viewed in a way such that the wave goes away from the observer, determines the nature of circular polarization. If the movement is in a clockwise direction, the resultant circularly polarized wave is called a clockwise circularly polarized signal (CWCP) or right-hand circularly polarized signal (RHCP). On the other hand, if the tip of the electric field rotates in a counter-clockwise direction, the resultant wave is called counter-clockwise circularly polarized signal (CCWCP) or left-hand circularly polarized wave (LHCP).

An antenna that produces circularly polarized (CP) EM wave, is known as a CP antenna. Circular polarization in an antenna can be achieved by exciting two orthogonal components of the E-field with phase quadrature. However, producing perfect CP wave is very difficult as it demands two components having equal amplitude. In reality, two components of the E-field might vary slightly in magnitude which converts the wave into elliptic polarization. Table 2.5 shows three different cases of circular polarization. The quality of the CP signal is determined by the axial ratio.

2.2.9.3 Axial Ratio

The nature of polarization, especially the quality of circular polarization, is defined by the axial ratio (AR) of the ellipse formed by two possible components of the rotating electric field. AR is obtained by taking the ratio of two orthogonal components of the electric field which are in phase quadrature. While taking this ratio, if the two components are not equal, we take the larger component in the numerator. In a perfect circularly polarized EM wave, the two orthogonal components in phase quadrature have equal amplitude. Thus, for perfect circularly polarized signal, AR is 1. In a linear polarization, since there is no quadrature component of field, AR is infinity. Thus, AR ranges between 1 and infinity, i.e.,

$$1 \leq AR \leq \infty \tag{2.29}$$

Any intermediate value of AR between 1 and infinity, corresponds to elliptical polarization. For most practical applications, AR is expressed in dB using,

$$AR(\text{in dB}) = 20\log 10(\text{AR}) \tag{2.30}$$

Since perfect CP is difficult and practically impossible to generate, applications requiring CP signal allow a margin on the AR value for which the signal is treated as almost CP. Typically $AR \leq 3$ dB is allowed and termed as a circularly polarized signal in a broad sense. It should be noted from Table 2.5 that $AR = 3$ dB corresponds to $AR = \sqrt{2}$ in a linear scale which means one of the orthogonal components of the electric field is $\sqrt{2}$ times higher than the other component in phase quadrature. In another sense, $AR = 3$ dB corresponds to the fact that power associated with the stronger component of the E-field is double than that of the other component.

TABLE 2.5

Identifying Polarization: Various Cases of Circular Polarization and Associated AR Values

Electric Field	Polarization	Axial Ratio	AR in dB
$\vec{E} = E_0\left(\hat{x} - j\hat{y}\right)e^{-j\beta z}$	RHCP or CW	1	0
$\vec{E} = E_0\left(\hat{x} + j\hat{y}\right)e^{-j\beta z}$	LHCP or CCW	1	0
$\vec{E} = E_0\left(\hat{x} - \frac{j}{\sqrt{2}}\hat{y}\right)e^{-j\beta z}$	Elliptical RH sense (CW)	$\sqrt{2}$	+3 dB
$\vec{E} = E_0\left(\hat{x} - 2j\hat{y}\right)e^{-j\beta z}$	Elliptical RH sense (CW)	2	+6 dB

2.2.9.4 Co- and Cross-Polarized Radiation

Co-polarized radiation from an antenna refers to the desired radiation contributed by a desired polarization of the E-field. On the contrary, the undesired radiation, contributed by an orthogonal component of the E-field (orthogonal with respect to the desired component) is known as cross-polarized radiation of the antenna. For example, an undesired horizontal component of the E-field from an antenna designed to radiate vertical component of the E-field is cross-polar radiation. A vertically (+z axis) erected linear dipole antenna has E_θ-component of E-field. Thus, for this antenna E_ϕ, the azimuthal component of the E-field, contributes to cross-polar radiation. For a CP wave, undesired generation of the opposite sense of rotation of the E-field contributes to a cross-polar radiation. Undesired LHCP wave from an antenna, designed for generating RHCP wave, is cross-polar radiation. Since a receiving antenna receives EM wave with a particular polarization, cross-polar radiation is basically wastage of power and causes interferences to other sources. Polarization purity of an antenna is described by evaluating the radiation intensity due to co-polar radiation over that due to cross-polar radiation in dB scale. Typically, a co- to cross-polar discrimination of 15–20 dB is considered good for almost all the antennas. An undesired orthogonal component of current in the antenna contributes to cross-polar radiation in an antenna.

2.2.9.5 Polarization Loss Factor

Another important aspect in connection with polarization of an antenna, especially for receiving antenna, is to ensure that the antenna is aligned properly so that its polarization is matched with that of the incoming wave. A mismatch in polarization results in weak or even null reception.

Polarization loss factor, commonly known as PLF, is expressed as,

$$\text{PLF} = \left|\hat{\rho}_i . \hat{\rho}_a\right|^2 = \cos^2 \Psi p \tag{2.31}$$

where $\hat{\rho}_i$ and $\hat{\rho}_a$ are unit vectors associated with electric field of the incoming wave and antenna radiation and ψ_p is the angle between them. $\psi_p = 0$ corresponds to perfect matching yielding PLF = 1, i.e. no polarization loss. On the other hand, if $\psi_p = 90°$, i.e. antenna electric field is orthogonal to the electric field of the incoming wave, PLF = 0, corresponding to null reception.

To get a more practical idea of the impact of polarization mismatch and PLF, let us consider a transmitting pyramidal horn antenna radiating z-polarized E-field and a dipole antenna in receiving mode with two different cases: vertically oriented, i.e. E-field having z-polarization, and horizontally oriented, i.e. y-polarized E-field. The arrangement of the antennas for two cases is shown in Figure 2.11. In the first case, the polarization of the incoming wave perfectly matches that of the receiving dipole. So, in this case a strong reception would be observed (PLF = 1). In the second case, the incoming wave has orthogonal polarization with respect to the receiving antenna. So here a very weak reception (PLF $\approx$ 0) would be observed. The cross-polar component of the radiated E-field of the horn (along the y-axis) would contribute to a weak reception.

2.3 Characterization of UWB Antennas

UWB antennas are characterized by conventional impedance and radiation characteristics, used in frequency domain plus time-domain studies to evaluate the spectral response of the antenna. Traditionally, an antenna is characterized by fundamental figure of merits such as input impedance, directivity, efficiency, and gain. In addition, for certain applications, we are interested in additional antenna parameters such as axial ratio (to characterize polarization purity of the antenna), beam efficiency, front-to-back ratio, etc. UWB antennas, in addition to these parameters, are characterized with respect to time-domain parameters such as phase linearity, group delay, and fidelity. In this section, we will briefly define all of the antenna parameters. Before going into that, we need to be familiar with various quantifications of bandwidth of an antenna.

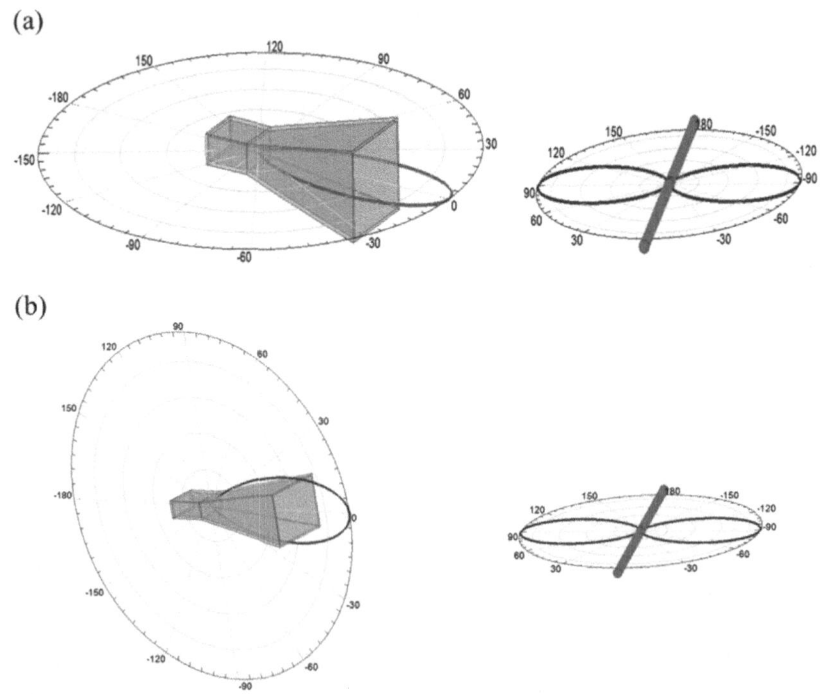

FIGURE 2.11
Horn antenna and a dipole antenna in (a) co-polarized configuration and (b) cross-polarized configuration.

2.3.1 High Bandwidth

As UWB antennas deal with a very large bandwidth, we need to have a thorough idea of the definition of antenna bandwidth. The bandwidth of an antenna is defined as the range of frequency over which it meets certain predefined standards in terms of its impedance and radiation characteristics. Bandwidth is quantified by various means as follows:

2.3.1.1 Absolute Bandwidth

Absolute bandwidth is defined as the difference between the upper (f_h) and lower (f_l) operating frequency of an antenna. Thus, absolute bandwidth of an antenna,

$$(BW)_{abs} = f_h - f_l \tag{2.32}$$

For the FCC-defined UWB band, f_l and f_h are 3.1 GHz and 10.6 GHz respectively, corresponding to (10.6–3.1) GHz = 7.5 GHz absolute bandwidth.

2.3.1.2 Ratio Bandwidth

Radio bandwidth is defined as the ratio of the upper (f_h) and lower (f_l) operating frequency of the antenna. Thus, ratio bandwidth of an antenna is expressed as,

$$(BW)_r = \frac{f_h}{f_l} \tag{2.33}$$

A UWB antenna operating frequency from 3–12 GHz has a ratio bandwidth of 12/3 = 4 or 4:1. An antenna exhibiting a 2:1 bandwidth is called an octave bandwidth antenna, while a 10:1 bandwidth antenna is called a decade bandwidth antenna.

2.3.1.3 Fractional Bandwidth of Antenna

An alternative and very popular definition of wideband antenna quantifies the bandwidth as the ratio of absolute bandwidth to its central frequency (f_c). Thus, fractional bandwidth of an antenna is expressed as,

$$(BW)_f = \frac{(BW)_{abs}}{f_c} = \frac{f_h - f_l}{f_c} \tag{2.34}$$

Now, the central operating frequency f_c can be defined as the arithmetic or geometric mean of antennas upper and lower operating frequency. Thus,

$$f_c = \left.\begin{matrix} \dfrac{f_h + f_l}{2} \\ \sqrt{f_h f_l} \end{matrix}\right\} \begin{matrix} \text{considering arithmatic mean} \\ \text{considering geometric mean} \end{matrix} \tag{2.35}$$

Substituting f_c from (2.34) into (2.33),

$$\begin{aligned} (BW)_f^a &= \frac{2(f_h - f_l)}{f_h + f_l} \\ (BW)_f^g &= \frac{f_h - f_l}{\sqrt{f_h f_l}} \end{aligned} \tag{2.36a,b}$$

For an UWB antenna, having $f_l = 3$ GHz and $f_h = 12$ GHz, using arithmetic mean in calculating (f_c), the fractional bandwidth is:

$$(BW)_f^a = \frac{2(12-3)}{12+3} = \frac{2\times 9}{15} = 1.2$$

while with geometric mean,

$$(BW)_f^g = \frac{2(12-3)}{\sqrt{12\times 3}} = \frac{2\times 9}{6} = 3$$

Since geometric mean is always smaller than the arithmetic mean, fractional bandwidth obtained with geometric mean as center frequency is always larger than that obtained with arithmetic mean.

2.3.1.4 Percentage Bandwidth

Fractional bandwidth of an antenna when expressed as a percentage is known as percentage bandwidth of an antenna. Thus, percentage bandwidth of an antenna is expressed as,

$$(BW)_f^a = \frac{2(f_h - f_l)}{f_h + f_l} \times 100\%$$

$$(BW)_f^g = \frac{f_h - f_l}{\sqrt{f_h f_l}} \times 100\% \qquad (2.37a,b)$$

The two expressions are obtained with two different notations of calculating center frequency. From Equation 2.37a, one important observation is that maximum percentage bandwidth of an antenna is always less than 200%. This is due to the fact that $f_l = 0$ leads to a mathematically maximum possible value of 200%, which is not physically supported as $f_l = 0$ corresponds to a DC signal, which cannot be radiated.

As discussed previously, f_l and f_h are determined based on input impedance matching, compatibility with respect to desired pattern or based on the gain versus frequency plot of the antenna. This gives rise to various bandwidth such as "impedance bandwidth", "pattern bandwidth", and "gain bandwidth". For a circularly polarized (CP) antenna, in addition to this, we can introduce "axial ratio bandwidth" to measure the quality of CP radiation.

2.3.2 Dispersion and Distortion of UWB Pulse

Although enormous bandwidth of UWB antennas sounds very attractive, signals emitted by such antennas are more likely to suffer from significant dispersion or distortion if not properly taken care of in the design stage. To understand the amount of dispersion, which is basically spreading of the signal in time domain, UWB antennas are characterized by their temporal behavior which is popularly known as time-domain analysis. Since UWB antennas radiate a broad frequency electromagnetic wave, the radiated wave of different frequency might suffer a different amount of time delay in reaching the receiving antenna. This is the fundamental reason of pulse spreading for UWB antennas. This effect of varying time delay is more prominent for UWB antennas for which the phase center of the

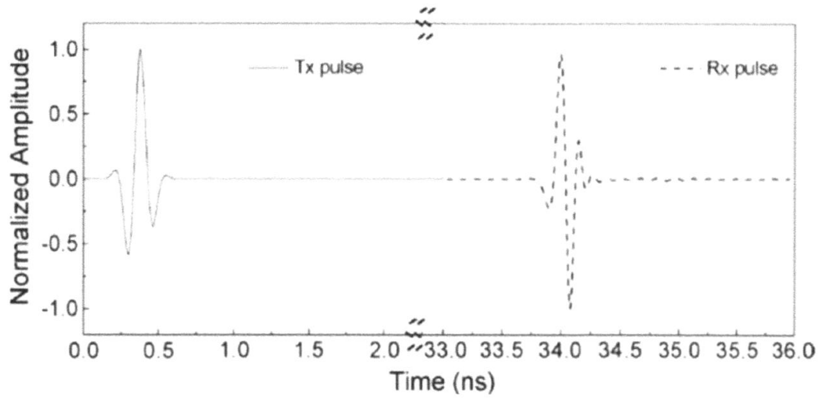

FIGURE 2.12
Measured time domain response of a tapered slot antenna [2]. Reproduced by permission @ 2011 IEEE [2]

antenna moves with frequency. For example, a tapered slot antenna, for which the phase center moves along the axis of the slot, is more dispersive than that of a printed monopole antenna which is more compact and has an almost stable phase center [1]. Figure 2.12 shows the time-domain response of a pair of a tapered slot antenna [2]. Figure reveals that the received pulse is a time derivative of the transmitted pulse. The received pulse could be a slightly spread indicating the presence of dispersion. In reality, presence of a dynamic channel between the transmitter and receiver and varying look angle between them makes the dispersion issue much more complicated.

This temporal dispersion characteristic of pulse emitted from a UWB antenna can be nicely interpreted from the phase response between two pairs of identical antennas oriented with good receiving condition. A linear phase-vs-frequency plot indicates a constant amount of delay for all the frequency components, i.e. almost zero dispersion.

In general, in addition to the conventional frequency domain characterization in terms of impedance matching and radiation pattern, etc., UWB antennas are characterized in time domain. When executing this time-domain characterization, choosing a proper excitation pulse is crucial. Typically, a Gaussian pulse or its first derivative, known as a monocycle pulse of proper pulse width [1], has been proven to be the most suitable for this time-domain characterization of UWB antennas. Time-domain characterization of UWB antennas is performed by studying the following parameters:

2.3.2.1 Peak Value

The peak value $p(\theta,\phi)$ of the radiated or received pulse in a particular direction is observed. It is a direct measure of the power associated with the pulse. Hence a high peak value of the pulse is desired.

2.3.2.2 Envelope Width

Envelope width describes the broadening of the radiated pulse and according to the basic principles of Fourier's relation (narrower in time, wider in frequency), should not exceed a few hundred picoseconds to ensure high data rates and high-resolution UWB applications.

2.3.2.3 Ringing

Ringing refers to the unwanted undulations near the tail of the radiated pulse after the main peak. Ringing is mainly a result of the structural reflections in the antenna. The duration of the ringing is defined as the time elapsed before the pulse falls from its peak value $p(\theta,\phi)$ to a certain lower limit $\infty\ p(\theta,\phi)$. Since energy contained in ringing is of no use, duration of the ringing should be very small.

2.3.2.4 Phase Response and Group Delay

Since a UWB antenna radiates a wide range of frequency, it is important to ensure a minimum amount of differential time delay over the entire band. This characteristic of UWB antennas can be conveniently interpreted from the phase-frequency (ϕ-ω) plot of the antenna. A linear phase response is desired for the dispersion of less radiation. Group delay is obtained from the slope of the phase-frequency plot of the antenna and expressed as,

$$T_g(f) = -\frac{d\phi}{d\omega} = -\frac{1}{2\pi}\frac{d\phi}{df}\bigg|_{f=f_0} \tag{2.38}$$

An ideal UWB antenna should have a constant $T_g(f)$, i.e. perfectly linear phase-frequency plot. Practically, deviation in group delay of the antenna is measured by relative group delay defined as,

$$T_{g.\mathrm{rel}} = T_g(f) - \overline{T_g(f)} \tag{2.39a,b}$$

where $\overline{T_g(f)} = \frac{1}{f_2 - f_1}\int_{f_1}^{f_2} Tg(f)df$ = average group delay

For a good UWB antenna, relative group delay should be as small as possible. Too many undulations in the group-delay curve indicate undesired resonance characteristics due to energy storage at the resonance frequencies. This is manifested in terms of ringing and more oscillations in the antenna impulse response and limits the performance of the system considerably. Figure 2.13a,b shows the performance of the measured phase response (unwrapped phase) and group delay of a UWB printed monopole antenna.

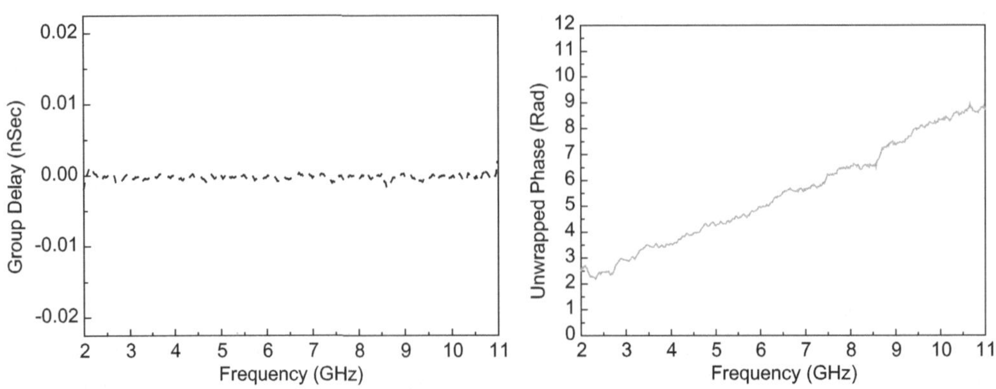

FIGURE 2.13
Measured (a) phase response and (b) group delay of an UWB printed monopole antenna.

2.3.2.5 Fidelity

Fidelity is a measure of the quality of the received UWB pulse compared to the transmitted pulse. It is a good indication of the spatial distortion of the radiated UWB pulse. Since decaying of the signal due to path loss is unavoidable, fidelity only focuses on the pulse shape without considering the amplitudes. This is done by normalizing both the transmitted and received pulse by a proper normalizing factor to have unit energy. Fidelity is then defined as the maximum value of the cross-correlation between transmitted pulse $x(t)$ and the received pulse $y(t)$. Thus, fidelity is defined as,

$$F = \max \frac{\int_{-\infty}^{\infty} x(t)y(t-\tau)dt}{\sqrt{\int_{-\infty}^{\infty} |x(t)|^2 dt \int_{-\infty}^{\infty} |y(t)|^2 dt}} \tag{2.40}$$

The minimum spatial distortion between transmitted and received pulse is achieved when fidelity F attains the value of 1. It must be noted that fidelity is also a directional quantity $F(\theta,\phi)$.

2.3.3 Experimental Setup for Time-Domain Characterization

Ultrawideband antennas when measured in frequency domain over the whole frequency band could take an excessively long time. This is particularly true for electrically large antennas [3]. Due to this shortcoming, time-domain measurement technique has received considerable attention [1, 2]. As shown in Figure 2.14, the time-domain measurement method consists of standard measurement equipment such as the pulse generator, sampling unit, and the sampling oscilloscope. The pulse generator generates pulses, the sampling unit samples, the received waveform, and the sampling oscilloscope processes the data and displays it.

2.4 Wideband and UWB Antenna: A Brief Review

In the previous section, we had a brief overview of fundamental antenna parameters followed by basic requirement and characterization of UWB antennas. Before we deal with recent trends on UWB antenna design, specifically the planar versions, which is the main

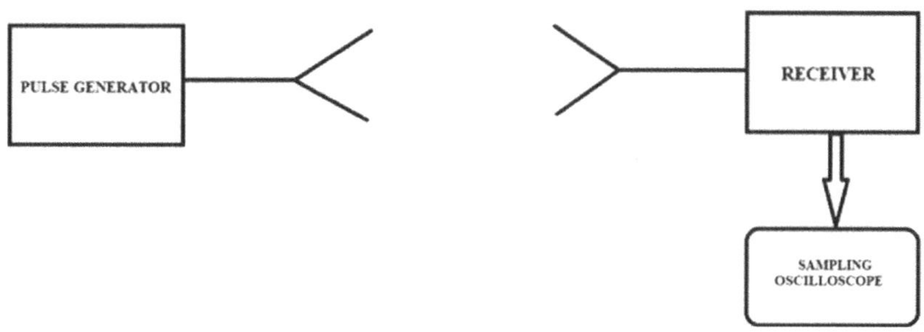

FIGURE 2.14
Schematic diagram of the time domain measurement setup.

focus of this book, in this section, we will review classic WB and UWB antennas. The discussion will be qualitative in nature and therefore would provide more stress on the fundamental principles. This brief overview is extremely useful, in particular for beginners new to the vast areas of UWB antennas, to understand the evolution of modern UWB printed antennas.

2.4.1 Frequency-Independent Antennas

Frequency-independent antennas (FIA), which are designed based on Rumsey's principle [4], can, depending on the physical size and if the particular design allows it, provide a very wide bandwidth. However, most of the frequency-independent antennas are highly dispersive which limits their use in practical UWB applications. The fundamental design principles of FIA are based on "angle invariant scaling from a small-scale portion" to a "large scale portion" [1]. The small-scale and large-scale portion defines the high-frequency and low-frequency limit of the antenna. FIA antenna can be designed to achieve a decade or even more bandwidth. However, FIAs suffer from high dispersion and are therefore not a good candidate for UWB applications. FIA can be broadly classified into equiangular antenna and log-periodic antennas.

2.4.1.1 Equiangular Antenna

Some of the most popular equiangular antennas are the log-spiral antenna, conical log-spiral antenna and Archimedean spiral antenna. Figure 2.15a–c shows the structures of these three antennas. The basic design criteria for these antennas is that the entire geometry of the antenna should be defined by angles. These antennas are self-scaling antennas as, for these antennas, the geometry is invariant when multiplied by a scaling factor. These antennas consist of two spirals wound in opposite directions on a dielectric sample. Basic geometry of these antennas is described by,

$$r = Ae^{a\phi} \tag{2.41}$$

where:

A is a constant that is equal to the radius at the origin ($\phi = 0$)

"a" indicates the expansion coefficient of the spiral.

The log-spiral and Archimedean spiral are 2-D antennas as the spirals are printed on a plane dielectric sheet. In case of the Archimedean spiral antenna, unlike the log-spiral case,

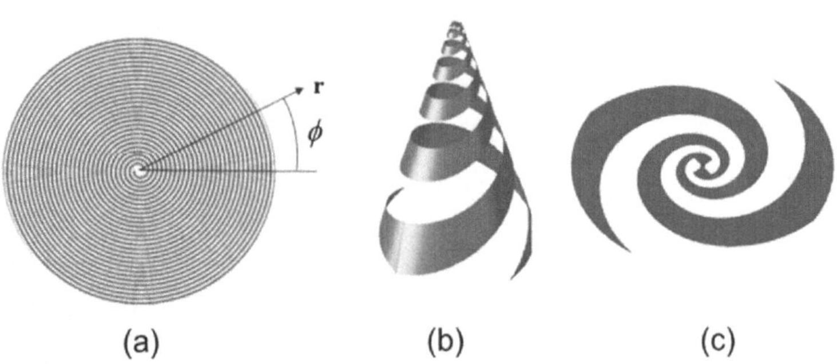

FIGURE 2.15
Geometries of the (a) log-spiral antenna, (b) conical log-spiral antenna, (c) Archimedes spiral antenna.

the width of the spirals is fixed. However, Archimedean spiral antennas have phase center stability and improved axial ratio. Both log-spiral and Archimedean spiral antennas have poor directivity as they have a bidirectional radiation pattern. For these antennas, lower frequency radiation is contributed by the part of the spiral far away from its center, while high-frequency radiation is contributed by the innermost regions and hence restricted by spacing of the feeding points. One of the major design challenges in realizing these antennas is designing the proper feeding network. As the antenna is balanced type, a wideband balanced feed is necessary to feed the antenna. A co-axial feeding cable can't be readily used as it is an unbalanced line. Thus, a wideband balun (balanced to unbalanced) must be used between the antenna feed-port and co-axial cable. A very wideband balun with extremely low loss is quite useful for this purpose. These antennas radiate CP signal over most of its bandwidth. When radiation from regions is far away from center, i.e. toward the lower frequency band, the polarization of the radiated signal is elliptic and tends to be linear.

Conical log-spiral antenna is a 3-D version of the log-spiral antenna in which the spirals are printed on a dielectric cone. It has a unidirectional radiation pattern with maximum radiation toward the apex of the cone. The angle of the cone is the most critical design parameter and has great impact on the impedance and radiation pattern of the antenna. Conical spiral antennas radiate circularly polarized signal.

2.4.1.2 Log-Periodic Antenna

Log periodic is a popular frequency-independent antenna, first introduced by Dwight Isbell and Raymond DuHamel at the University of Illinois in 1958. The design is based on a group of self-similar elements where a basic unit cell is repeated with a progressive change by scaling factor. There are various variants of log-periodic antenna. Figure 2.16 shows a circular log-periodic antenna. This antenna is designed in a way such that the ratio of two consecutive radii given by,

$$\rho = \frac{R_{n+1}}{R_n} \tag{2.42}$$

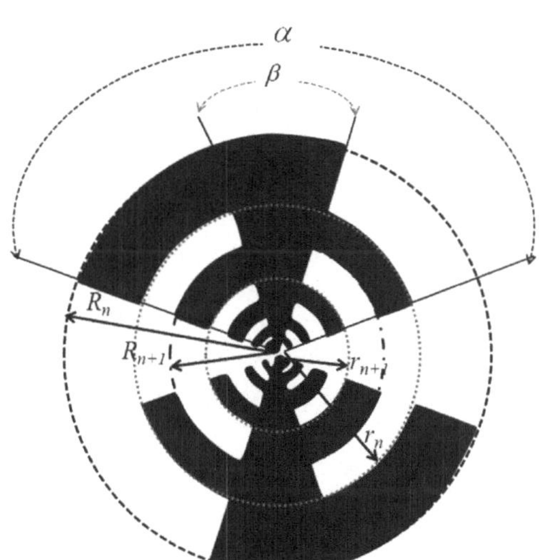

FIGURE 2.16
Basic geometry of the log-periodic antenna.

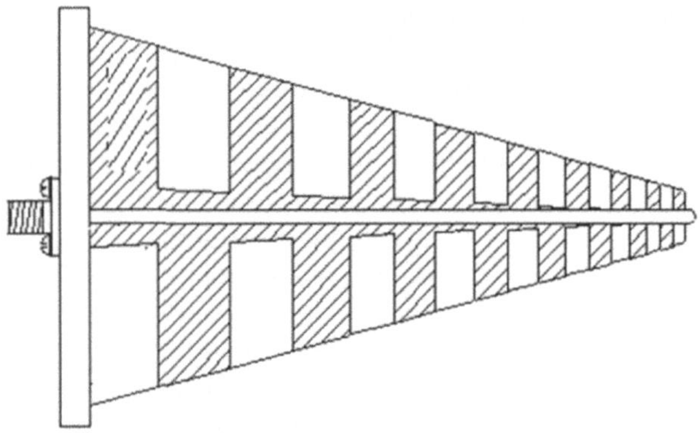

FIGURE 2.17
Geometry of the printed log-periodic antenna.

is related with the scaling factor, $k = \rho^m$ where $m = \pm 1, \pm 2, \pm 3$ etc. This ensures that the antenna exhibits similar properties at f_0 and f_m, where,

$$f_m = \rho^m f_0 \tag{2.43}$$

$$\log \frac{f_m}{f_0} = m \log \rho \tag{2.44}$$

Due to the relation of Equation 2.44, the antenna is named "log-periodic". The lowest and highest operating frequency of this antenna is governed by the tooth length of the largest and smallest teeth in the structure. The circular log-periodic antenna provides symmetrical bidirectional linearly polarized radiation. Figure 2.17 shows another version of log-periodic antenna consisting of dipoles of gradually changing size. The length and separation between successive elements are scaled by a factor to obtain a wide impedance bandwidth from the antenna. At any particular frequency, only one dipole, having resonant length, is directly excited while all other dipoles are parasitically excited by the field of the resonant dipole. The dipole which is larger compared to the resonant dipole acts as the reflector while all other elements smaller than the resonant dipole act as the directors. Thus, like Yagi antennas, dipole-based log-periodic antennas provide directional pattern having maximum directivity toward the shorter elements. The maximum and minimum operating frequency of the antenna is governed by the dipoles of shortest and longest length. As various resonating dipoles located at different positions contribute to the overall wideband radiation, the phase center of the antenna continuously varies with frequency. Because of this, this antenna is highly dispersive and naturally is not an attractive choice for UWB applications. Peak directivity of this antenna varies typically from 5 to 11 dBi [5].

2.4.1.3 Biconical Antenna

Biconical antenna, invented by Lodge in the 1890s [6] and extensively studied by Schelkunoff in the 1930s [7], is a popular wideband antenna which is capable of providing 6:1 impedance bandwidth. As shown in Figure 2.18, it consists of two metallic cones

FIGURE 2.18
Basic geometry of the biconical antenna.

oriented in opposite directions. Theoretically, if the metallic cones are of finite length, this acts as a frequency-independent antenna having an input impedance,

$$Z_{in} = Z_0 = 120\ln\left[\cot\left(\frac{\alpha}{4}\right)\right] \tag{2.45}$$

where α is angle of the cone. Practical realization limits the size of the antenna when the cones are truncated. This in turn limits the bandwidth of the antenna. Radiation pattern of this antenna is similar to a dipole, i.e. omnidirectional in a plane perpendicular to the axis of the cones having null along the axis of the cone.

2.4.1.4 Discone Antenna

The discone antenna is a modified variant of a biconical antenna. Figure 2.19 shows a discone antenna. As the name suggests, the antenna consists of a disc and a cone. The circular disc placed just above the apex of the cone acts as the ground plane of the antenna. The antenna is fed by connecting the inner pin of the co-ax cable to the disc while the outer body of the co-ax is connected to the apex of the cone. The oblique height of the cone is typically ~0.25λ and the disc diameter ~0.18λ. Discone antennas provide a dipole-like radiation pattern. Trimming the apex of the cone of the discone antenna provides a better impedance matching. An optimized design of discone antenna can provide an impedance bandwidth higher than 150% [8].

2.4.1.5 Bow Tie Antenna

A bow tie antenna is a planar version of a biconical antenna that can be printed on a dielectric substrate. As shown in Figure 2.20, it consists of two metal triangles in inverted configuration. Antenna is fed between the two vertices of the triangle. Spacing of the feed

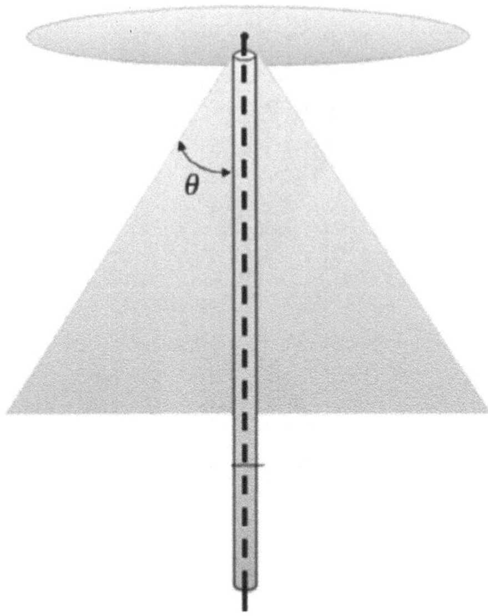

FIGURE 2.19
Basic geometry of the discone antenna.

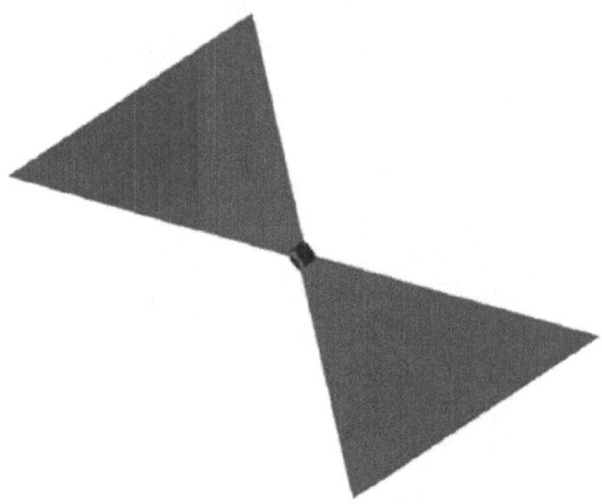

FIGURE 2.20
Basic geometry of the bowtie antenna.

region and the apex angle of the triangle in a bow tie antenna are the most critical design parameters that influence the impedance bandwidth of the antenna.

2.4.2 UWB Printed Monopole Antenna

Among the various types of antennas that are currently being used for UWB applications, the printed monopole antenna is the most popular due to features such as its simple structure, easy fabrication, wide frequency bandwidth, and good radiation patterns. It has been shown that if a fundamental thin wire monopole antenna is replaced with a conducting disc, the resultant monopole antenna exhibits a bandwidth of 1:8 as defined by return loss less than –10 dB and maintains omnidirectional coverage [9, 10].

As described by Honda et al. [9] and derived with multipole expansion [11], it was inferred that any current source whose maximum dimension is much smaller than wavelength is

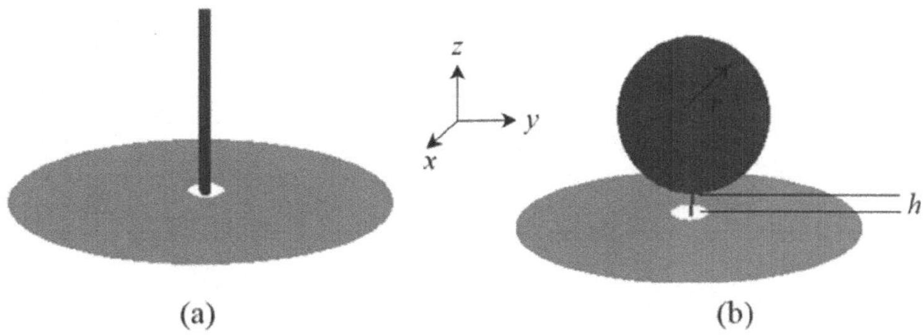

FIGURE 2.21
(a) A vertical thin wire monopole antenna, (b) vertical disc monopole antenna.

equivalent to an electric dipole in the far field which yields in an omnidirectional radiation pattern of all monopole antennas. Moreover, the symmetries in current distribution provide a broadband response. The impedance matching can be further improved by removing stray capacitance between the monopole and the ground plane [9, 12]. In this section, the printed monopole antennas will be studied in order to understand their operation.

2.4.2.1 Vertical Disc Monopole

The vertical disc monopole is so called because the disc radiator is vertically placed on the ground plane. As described by Honda et al. [9], it evolved from a conventional straight wire monopole which is replaced with a disc copper plate as shown in Figure 2.21.

The geometry of the vertical disc monopole and its coordinate system is illustrated in Figure 2.21b. A conducting circular disc with radius r is mounted vertically above a circular ground plane. The antenna is fed using a probe connecting the bottom of the disc through the ground plane via a connector. The height h between the feed gap and the ground controls the impedance bandwidth [13]. The performance of the antenna is mainly dependent on the feed gap h and also on the dimension of the ground plane. The finite size of the ground plane affects the image current which in turn has an effect on the impedance bandwidth. The widening of the wire into a disc yields a broadband response. As mentioned by Stratton [1], that widening of the conductor is better for broadband antenna design, which was understood as early as the 1940s by J.C. Slater, R.W.P. King, and others [1]. The widened or the "fat" conductor minimizes the stored reactive energy.

The printed disc monopole is a version of the vertical disc monopole but printed on a substrate. The radiator as well as the ground planes are uniplanar if the antenna is coplanar waveguide (CPW) fed or biplanar if the antenna is microstrip fed as shown in Figure 2.22a–b.

A disc monopole antenna with a radius of r and CPW line is printed on the same side of a dielectric substrate as shown in Figure 2.22a. The coplanar waveguide is formed using two ground planes and a signal line separated by two slot lines which control the characteristic impedance of the coplanar waveguide. The disc monopole is connected to the signal line and separated by a distance h which forms the feed gap between the disc and the ground plane.

Other than a CPW-feeding structure, a planar circular disc monopole can also be realized by using a microstrip feedline, as illustrated in Figure 2.22b. For a microstrip-fed monopole, the radiator is connected to the stripline which is separated from the ground plane by the thickness of the substrate. However, the impedance bandwidth of the antenna is also controlled by a critical separation, δ, between the conductor and the ground plane.

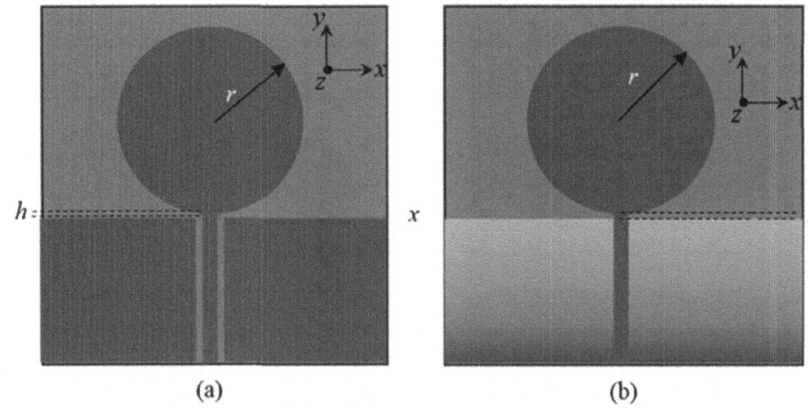

FIGURE 2.22
(a) A CPW-fed printed monopole, (b) a microstrip-fed printed monopole.

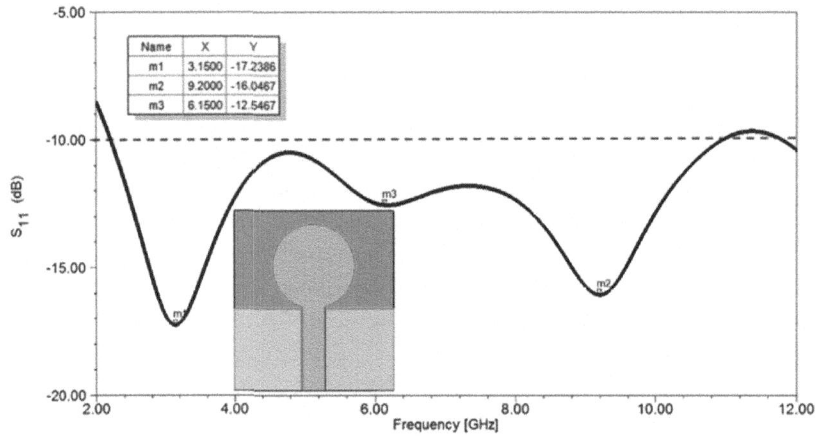

FIGURE 2.23
Magnitude of the reflection coefficient versus frequency of an UWB CPW-fed printed monopole antenna.

For a CPW-fed antenna having a radius of 12 mm and printed on a substrate having dielectric constant of 2.33 and thickness of 1.575 mm, the optimal feed gap, *h*, was found to be between 0.3–0.7mm which yielded an UWB response [13]. Figure 2.23 shows the UWB return loss (<10 dB) over the frequency range of 2.25 GHz to 11 GHz, where the input impedance is matched to 50 Ω, i.e. the input resistance *R* is close to 50 Ω while the input reactance *X* is nearly zero. For higher values of *h*, the resistance and reactance vary significantly resulting in impedance mismatch at the antenna which decreases the operating bandwidth [13].

The diameter of the disc corresponds to the quarter wavelength of the first resonant dip of a circular monopole disc. The resonant frequency is determined by the circular disc and the size of the ground plane. The disc monopole supports multiple resonance modes, with the higher order modes being the harmonics of the fundamental mode of the disc. Moreover, the higher order modes are closely spaced which leads to the UWB characteristic due to the overlapping of the resonance modes. The current is mainly distributed along the edge of the circular disc which is why the resonant frequency is associated with the

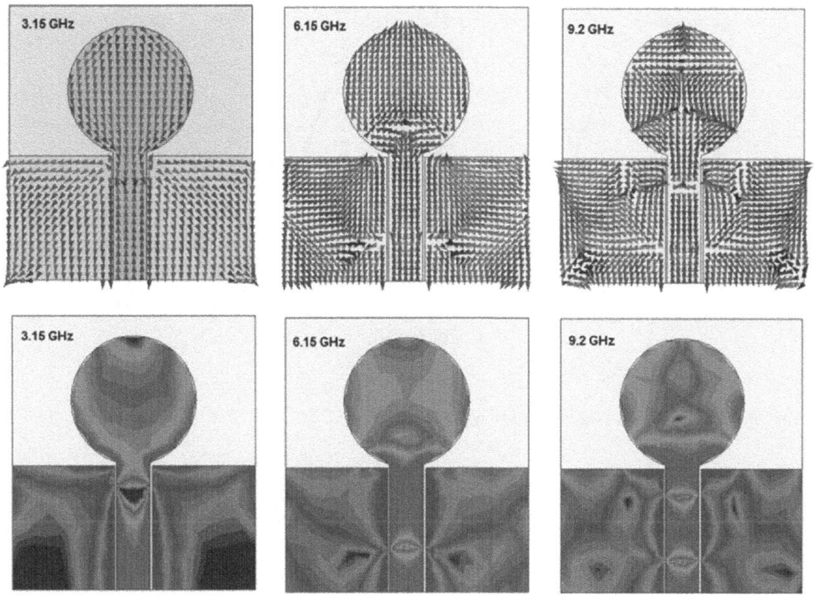

FIGURE 2.24
Surface current distribution of UWB CPW-fed printed monopole antenna.

dimension of the disc. It has been demonstrated that the overlapping of closely distributed resonance modes in both vertical and CPW-fed disc monopoles is responsible for an ultra-wide bandwidth. The surface current distribution of a CPW-fed UWB monopole antenna for three different frequencies (3.15 GHz, 6.15 GHz, and 9.2 GHz) covering the UWB spectrum is shown in Figure 2.24.

At the low-frequency end (i.e. the first resonance) when the wavelength is bigger than the antenna dimension, the EM wave can easily "couple" into the antenna structure so that it operates in an oscillating mode, i.e. a standing wave. With the increase of the frequency, the antenna starts to operate in a hybrid mode of standing and traveling waves as shown in Figure 2.25 [13]. At the high-frequency end, the traveling wave becomes more critical to the antenna operation since the EM wave needs to travel down to the antenna structure which is big in terms of the wavelength. For the CPW-fed antenna, the slots formed by the lower edge of the disc and the ground plane with a proper dimension can support the traveling wave very well. Thus, an optimally designed CPW-fed circular disc monopole can exhibit an extremely wide –10 dB bandwidth. The UWB characteristic of the antenna can again be attributed to the overlapping of the first and higher order resonances which are closely distributed across the spectrum. However, the –10 dB bandwidth of this antenna is always limited at the higher end around 10 GHz.

The 3-D radiation patterns of a CPW-fed printed monopole is shown in Figure 2.26 for three different frequencies of operation covering high, center, and low frequencies of the entire UWB spectrum. With respect to the coordinate system shown in Figure 2.22, the x-y and the y-z planes form the E-planes in a rectangular coordinate system, whereas the x-z plane forms the H-plane. The pattern is directive in the x-y and y-z planes with a null in the y direction for the low as well as the higher frequencies. The patterns also show the omnidirectionality of the antenna in the H-plane, i.e. in the x-z plane. At low frequency, the pattern is circular in the x-z plane. With increasing frequency, the splitting of the beam

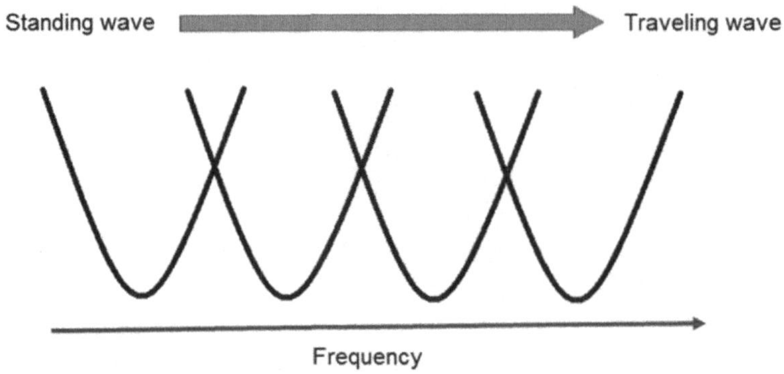

FIGURE 2.25
Operating principle of a UWB monopole antenna.

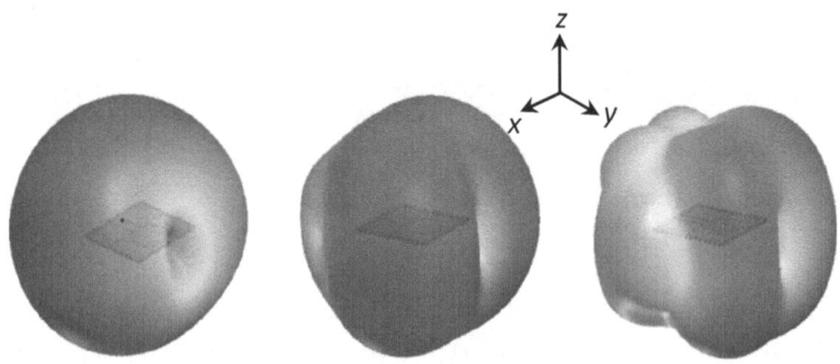

FIGURE 2.26
3-D radiation patterns of a UWB CPW-fed printed monopole antenna at three different frequencies: (a) 3.15 GHz, (b) 6.15 GHz, (c) 9.2 GHz.

in the *x*-*y* plane is more prominent and the beam is directed more toward an oblique angle between +*x*,+*y* and −*x*,-*y*. For the E-planes, the patterns have large back lobes and with increasing frequency they become smaller, splitting into many minor ones. The ground planes play a major role for the back lobe as the back lobe is bigger at low frequency than at high frequency because the ground plane is electrically small compared to the wavelength at low frequency.

2.4.2.2 Printed Slotted Monopole

As seen in Figure 2.24, in printed monopole antennas (MPA) the current is mainly distributed along the edge of the disc for both CPW-fed and microstrip-line-fed disc monopoles. This implies that the performance of the antenna is independent of the central part of the disc, and hence cutting this part to form a slotted monopole can still achieve an ultrawide bandwidth. For a CPW-fed disc monopole, retaining the optimal parameters, a slot in the inner region is etched from the disc at the disc center. Figure 2.27 illustrates the schematic diagram of a slotted UWB printed disc monopole having inner and outer radii, R_2 and R_1, respectively. An inner slot radius of 0.3 times the outer radius provides a similar return loss almost to that of a conventional circular disc monopole.

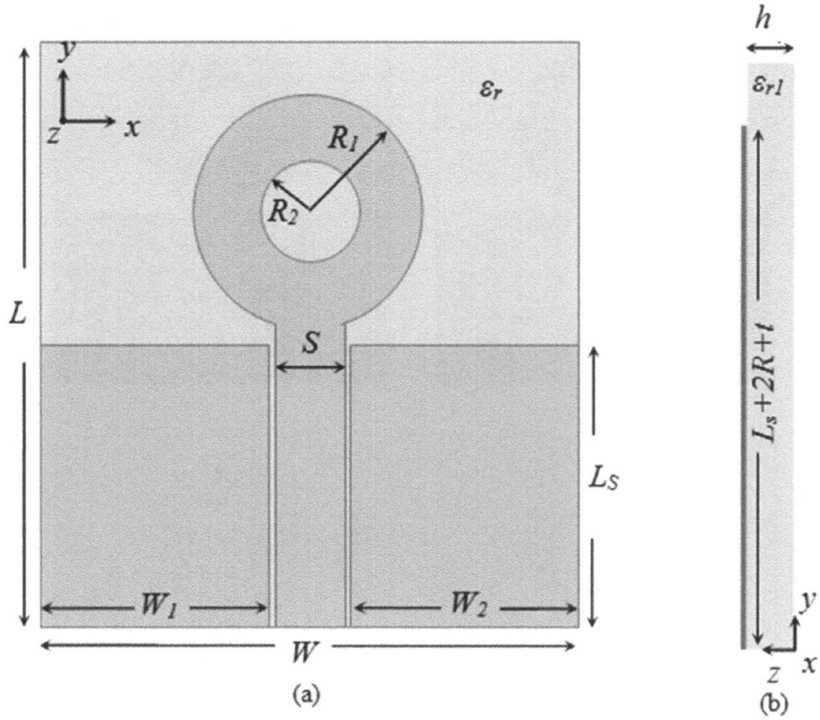

FIGURE 2.27
Schematic of a printed slotted monopole antenna.

The surface current distribution of a CPW-fed slotted UWB disc monopole antenna for a lower frequency at 3.12 GHz and at higher frequency at 9.2 GHz is shown in Figure 2.28. It is evident from the surface current plot that very low current resides at the center of the radiator and hence the slot does not have a large impact on the impedance bandwidth of the antenna.

2.4.3 UWB Tapered Slot Antenna

In 1979, Gibson proposed an application incorporating a Vivaldi aerial on an alumina substrate, operating in a frequency range of 8–40 GHz [14]. In the same year, Prasad and Mahapatra [15] introduced a similar antenna, namely the linearly tapered slot antenna (LTSA). Tapered slot antennas (TSAs) are a class of end-fire radiating antennas derived from Vivaldi aerials [14] which are extensively used as wideband antennas for radar and ground-penetrating radar applications. The tapered slot is produced by the gradual widening of a slotline [15, 16] to form a flare whose width increases with increasing distance from the feed. The planar geometry of a TSA allows it to be easily integrated with other planar devices such as filters and Schottky diode mixers [17]. There are different classes of tapered slot antennas (TSAs) which are classified according to their tapering, namely exponentially tapered slot antennas (ETSAs), also known as Vivaldi antennas, linearly tapered slot antennas (LTSAs), constant width slot antennas (CWSAs), etc. Figure 2.29 shows the different forms of tapering in TSAs.

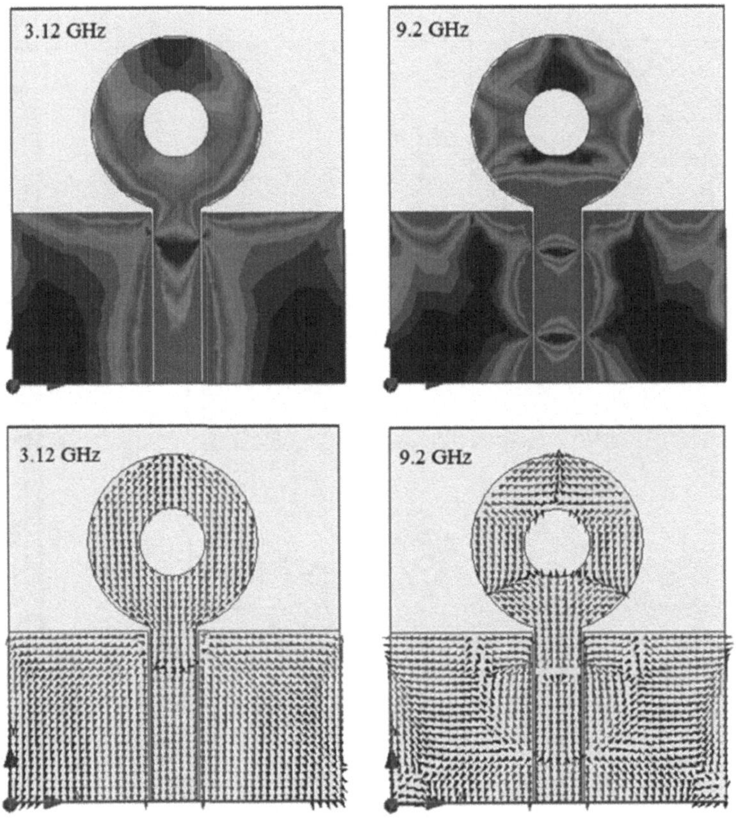

FIGURE 2.28
Surface current distribution of a UWB printed CPW-fed slotted disc monopole antenna.

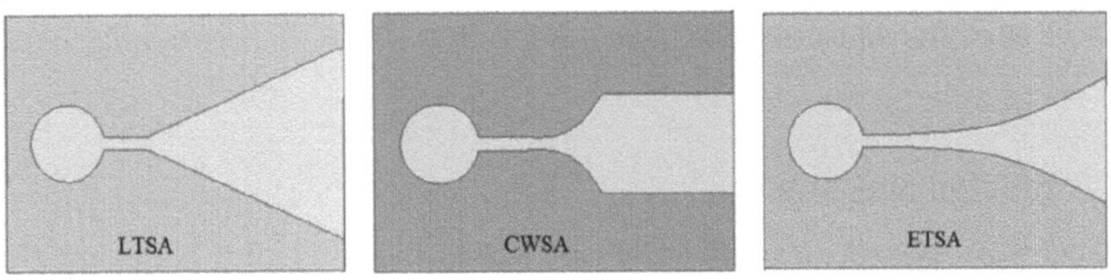

FIGURE 2.29
Different types of tapering in TSAs.

In a tapered slot antenna (TSA), the traveling wave is non-resonant and the radiation is produced by higher order Hankel function modes traveling along the antenna. The energy in the traveling wave is tightly bound to the conductors when the separation between them is very small and the bond becomes progressively weaker as the separation is increased which results in radiation. All tapered slot antennas are surface wave antennas which use a traveling wave structure and produce end-fire radiation [18] (Figure 2.30).

The performance of a TSA depends on both the geometry of the flare and on the thickness and dielectric constant of the substrate used. If the substrate is too thick, surface modes are launched that degrade the antenna performance. If the substrate is too thin,

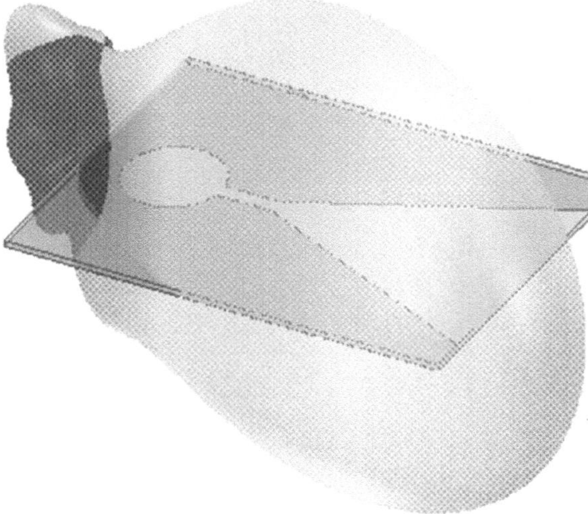

FIGURE 2.30
The 3-D end-fire radiation pattern of the LTSA.

less symmetrical beam patterns are formed due to higher cross-polarization effects. As a result, it is important to choose an optimum substrate thickness during the design of the antenna.

The main advantages of TSAs are that they are linearly polarized antennas. They exhibit broadband operation, moderate gain, low side lobes, almost symmetric E- and H-plane patterns, and a planar footprint. Tapered slot antennas are traveling wave structures and are thus well suited to high operating frequencies (greater than 10 GHz). TSAs can also be designed with variations in non-linear tapering as shown in Figure 2.29. However, in TSAs with non-linear tapers, the variation in beamwidth with opening angle is more pronounced with fewer nulls in the E-plane than what occurs in antennas with linear tapers.

A tapered slot antenna (TSA) operates in a traveling-wave mode and can have a large bandwidth if it has a perfect impedance match at both the feed transitions. The frequency bandwidth of a TSA for which VSWR is less than 2 generally decreases with a decrease in its radius of curvature.

A typical exponential TSA or Vivaldi antenna is shown in Figure 2.31. It consists of a metal-plated dielectric substrate that has symmetrical slots etched on one or both its sides. One end of the slot ends in a cavity and the other end is flared to form an exponential curve. The function of the cavity is to act as an "open circuit" and minimize any reflections at the point of termination of the slotline (the stripline-slotline junction). The shape of the flare and the slotline cavity determine most of the characteristics of the antenna. The straight section of the slot is fed by a stripline which ends in a stub. At a given wavelength, only a section of the exponential region will radiate efficiently. As the wavelength is scaled, the radiation will occur from a different section which is scaled in proportion to the wavelength and has the same relative shape [19]. The bandwidth of a Vivaldi antenna is limited by the physical size of the antenna and the fabrication capabilities available. The feed of the antenna generally determines the high-frequency limit, and the aperture size determines the low-frequency limit [20].

The Vivaldi antenna has a modestly wide beamwidth which is approximately the same in both the E- and H-planes, that is, the antenna can produce a symmetrical pattern though

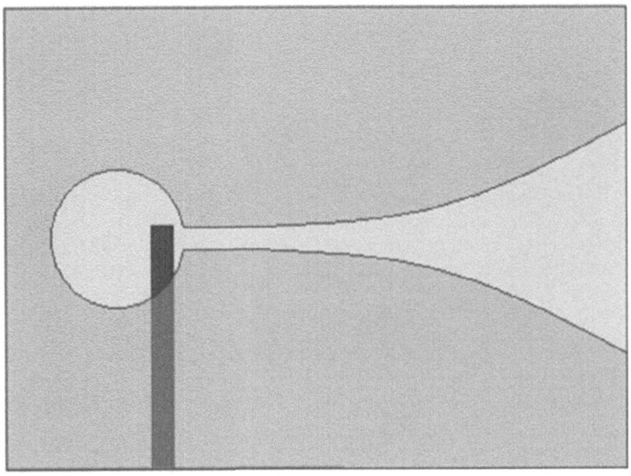

FIGURE 2.31
ETSA or the Vivaldi antenna with a microstrip line feed.

it has a planar structure. The gain of the antenna is proportional to its length as is the case with most traveling wave antennas. The aperture size of the antenna in wavelengths increases with frequency, so the gain also increases with frequency.

2.4.3.1 Feeding Mechanism of Tapered Slot Antennas

Tapered slot antennas are extensions of a slotline, therefore feeding a TSA often requires designing a transition between a slotline and some other transmission media. Thus, the slotline must be coupled to the actual antenna feed with an appropriate transition. Such a transition should be compact, low loss, and easy to fabricate. In addition, the transition should have small parasitics over a wide bandwidth. One way of exciting a Vivaldi slotline is by using a microstrip to slotline transition as shown in Figure 2.31. The simplest wideband transition is employing microstrip and slotline crossing each other at a right angle. A simple microstrip-slotline transition can be formed by etching a microstrip line on one side of the substrate and extending it past the center of a slotline on the opposite side of the substrate. Usually, the microstrip extends about one quarter of a wavelength beyond the slot, and the slot extends about a quarter of a wavelength beyond the microstrip, as shown in Figure 2.31. The microstrip must be terminated after a finite length, and the termination used may be a soldered short or a virtual short, such as a microstrip stub. The bandwidth of a transition with a virtual short is always smaller than the bandwidth of a transition with a soldered microstrip short, but this bandwidth reduction is rather low, especially for larger stub diameters. A circular slot stub may be used to increase the achievable bandwidth as shown in Figure 2.31. This can be done by increasing the circular slot stub diameter which in turn increases the characteristic impedance of the stub.

Impedance matching of the slotline is difficult to achieve in a conventional TSA. This is because a slotline has relatively high impedance if it is fabricated on a low dielectric constant substrate, and the microstrip-to-slotline or the stripline-to-slotline transition limits the operating bandwidth of the TSA. The antipodal TSA overcomes this problem, since it is fed directly by a microstrip feed as proposed by Gazit [21]. An antipodal TSA, shown in Figure 2.32, can be formed by gradually flaring the strip conductors of a microstrip on opposite sides of the dielectric substrate.

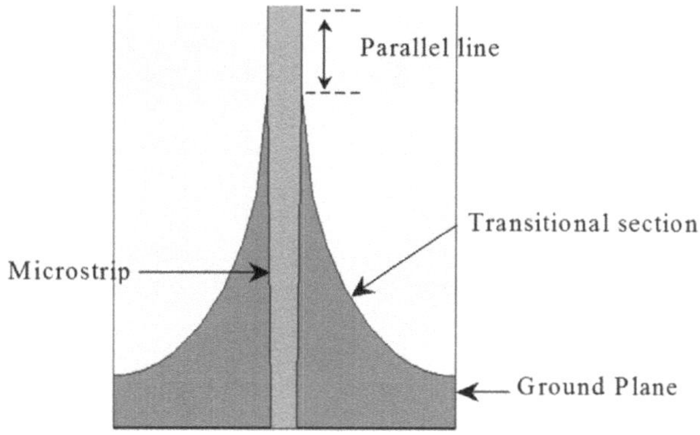

FIGURE 2.32
Schematic of an antipodal feed section for a TSA.

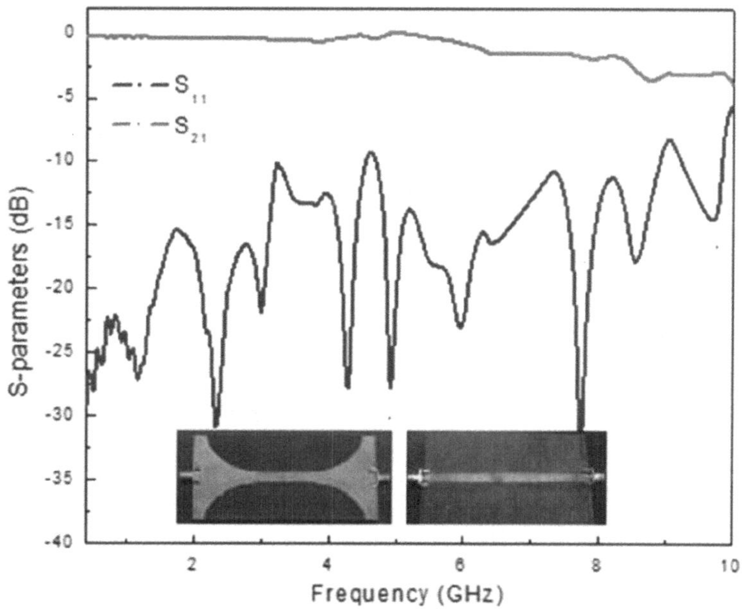

FIGURE 2.33
S-parameters response of antipodal feed balun.

Figure 2.33 shows the S_{11} and S_{21} response as a function of frequency of a back-to-back antipodal feed balun. The fabricated prototype (top and bottom) is shown in the inset.

Figure 2.34a shows a schematic of an antipodal tapered slot antenna designed with elliptical conductors [2]. The elliptical conductors printed on two sides of the substrate form the tapered slot as shown in Figure 2.34b, and is fed by a balun. Figure 2.35 shows the simulated and measured S_{11} response of the prototype shown in Figure 2.34. The antenna operates from 400 MHz to 9.8 GHz due to excellent matching of the balun feed. The measured E-plane radiation patterns are shown in Figure 2.36 and the simulated time-domain response of the antenna is shown in Figure 2.37.

As seen from Figure 2.36, the antipodal tapered slot antenna has a poor cross-polarization performance as the frequency of operation increases. To overcome this problem,

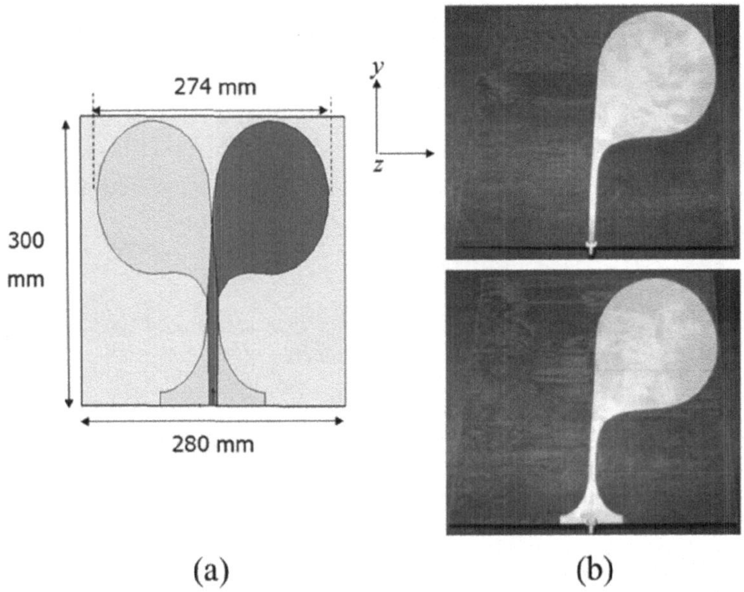

FIGURE 2.34
Antipodal tapered slot antenna with elliptical conductors: (a) schematic view, (b) fabricated prototype (top view and bottom view).

Reproduced with permission @2011 IEEE [2].

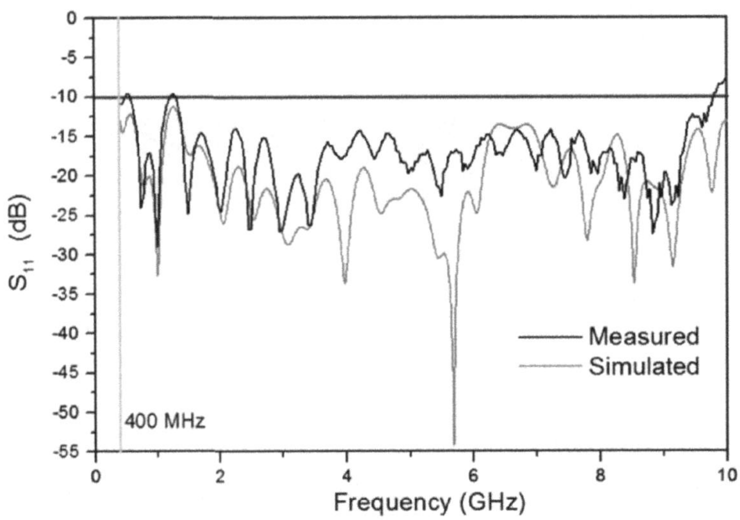

FIGURE 2.35
Simulated and measured S_{11} of the antipodal tapered slot antenna with elliptical conductors shown in Figure 2.34.

Reproduced with permission @2011 IEEE [2].

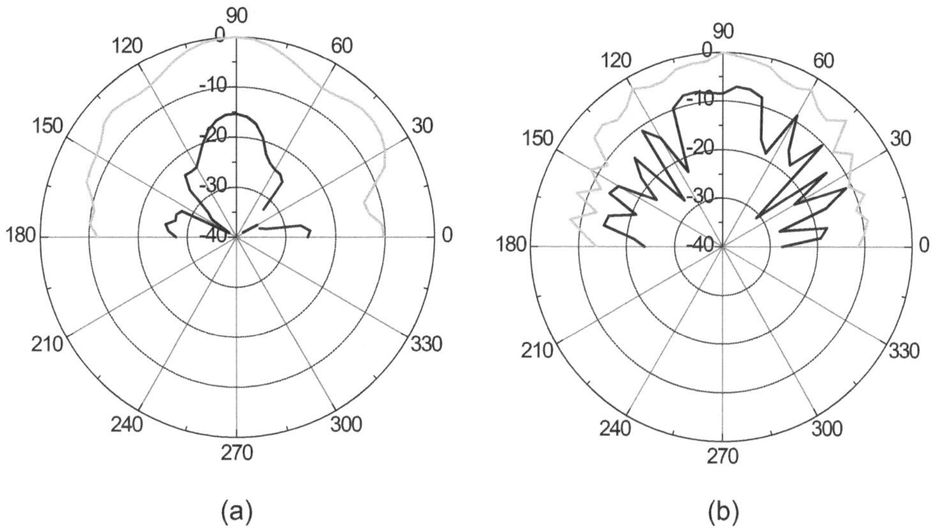

FIGURE 2.36
Measured E-plane radiation pattern of the antipodal tapered slot antenna with elliptical conductors shown in Figure 2.34: (a) 2 GHz (b) 8 GHz.

Reproduced with permission @2011 IEEE [2].

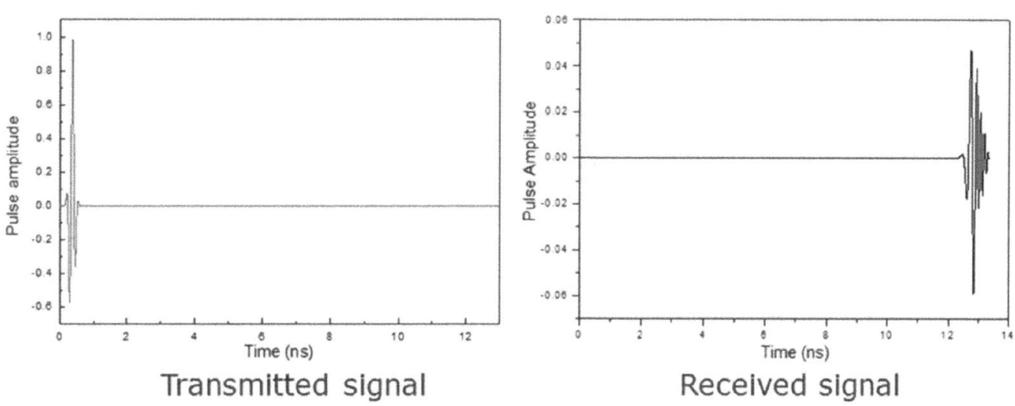

FIGURE 2.37
Simulated time domain response of the antipodal tapered slot antenna shown in Figure 2.34.

Langley [22] proposed the balanced antipodal TSA. The cross-polarization performance of an antipodal TSA is improved by converting it to a triplate structure with the addition of another dielectric and metallization layer which balances the E-field distribution in the flared slot. A balanced antipodal TSA (BATSA) starts in a stripline and the input track on one side of the board is flared to produce one half of a conventional TSA, as shown in Figure 2.37 [23]. The ground planes on the other side are reduced to form what will be a balanced set of lines. The "ground lines" are then flared out in the opposite direction to form the overall antenna, as shown in Figure 2.38.

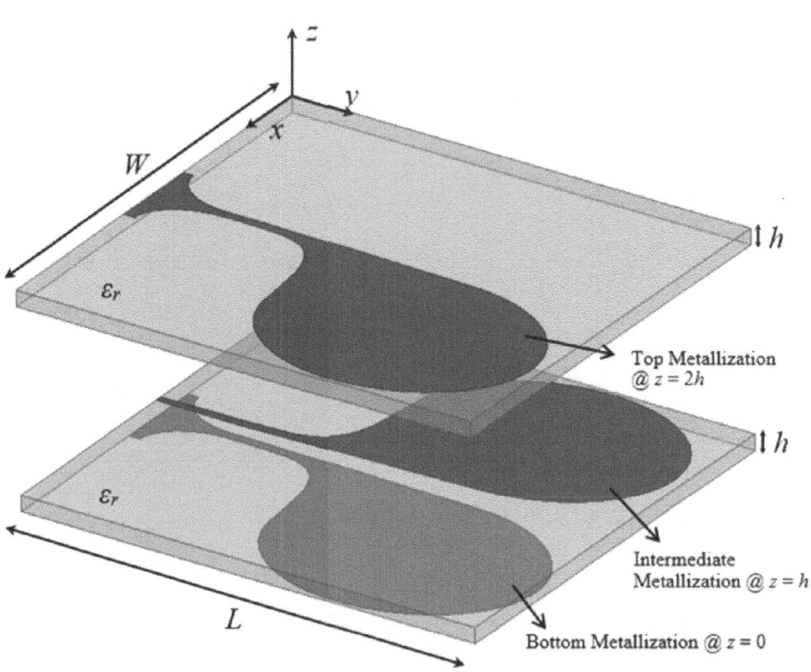

FIGURE 2.38
Schematic of a balanced antipodal TSA (BATSA).

Reproduced by permission @2018 IET [23].

Figure 2.39 shows the simulated and measured S_{11} response of the conventional antipodal tapered slot antenna and the balanced antipodal tapered slot antenna. The S_{11} response of both the antennas is similar with slight deterioration in the higher frequencies for BATSA.

2.4.4 UWB Fractal Antenna

Fractal antenna design is inspired by nature and was originally used by Mandelbrot [24] to describe complex shapes having an inherent self-similarity in their geometrical structure [25]. This section deals with the study of fractal-shaped antenna elements. The most common fractal geometry is the Sierpinski gasket shown in Figure 2.40.

Other useful fractals commonly used are the Koch snowflake and the Hilbert curve. The common fractal geometries have applications in antenna engineering. The Koch snowflakes are used to develop new designs for miniaturized loop as well as microstrip patch antennas [26, 27]. The Sierpinski gasket has been used to develop multiband and broadband antenna elements [28–31]. Fractal antennas are multi-resonant antennas. Variation of fractal parameters has an impact on the primary resonant frequency of the antenna. A fractal antenna is capable of operating with excellent performance at many different frequencies simultaneously. This makes the fractal antenna an excellent design for wideband and multiband applications [32].

2.4.5 UWB Dielectric Resonator Antennas

A commonly used antenna for UWB operations is the dielectric resonator antenna (DRA). Low-loss dielectrics, resonators with high Q elements for circuit applications such as filters

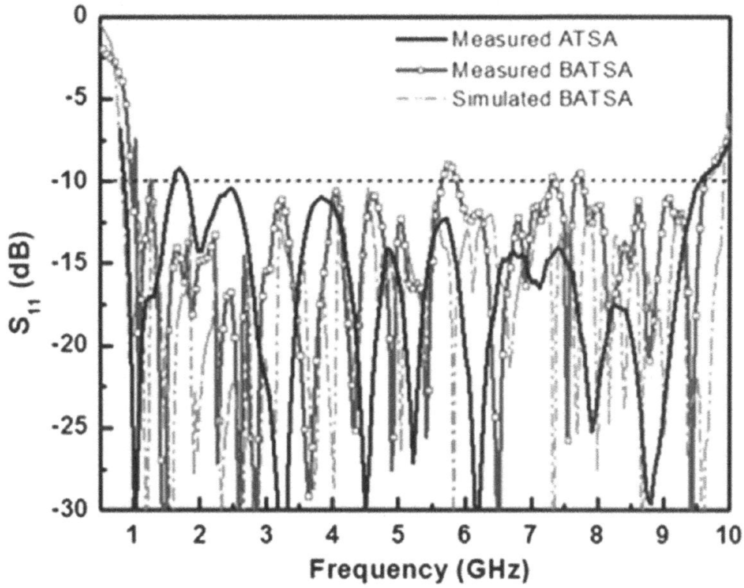

FIGURE 2.39
Simulated and measured S_{11} of the ATSA and BATSA.

Reproduced by permission @2018 IET [23].

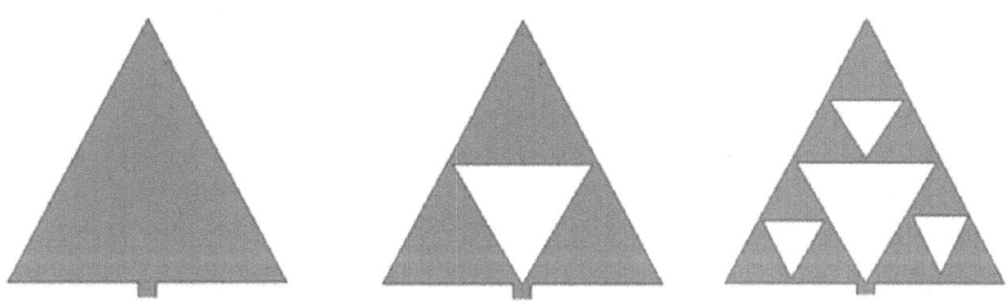

FIGURE 2.40
The first three stages of Sierpinski gasket fractal geometry.

and oscillators, were developed which increased the possibilities for developing more compact integrated microwave circuits. With proper excitation of certain modes, the resonators could be converted to efficient radiators as reported in 1983 by Long et al. [33]. In this work, cylindrical dielectric resonator antennas (DRAs) were used which subsequently led to the investigation of rectangular [34] and hemispherical [35] DRAs. DRA antennas have multiple advantages as DRAs offer a high degree of flexibility to an antenna designer. The DRA size is proportional to $\lambda_0/\sqrt{(\varepsilon_r)}$, where λ_0 is the wavelength at resonant frequency and ε_r is the dielectric constant of the DR, hence for the same frequency there is a natural reduction in size, compared to other printed antennas. Moreover, different values of ε_r (ranging from 4 to 100) can be used, thus allowing the designer the flexibility in controlling the size and bandwidth. Depending on the resonator shape, various modes can be excited within the DRA element. These modes can produce different radiation patterns for various coverage requirements. Many of the existing feeding schemes can be used (slots, probes, microstrip, coplanar waveguides, dielectric image guide, etc.). Compared to the microstrip

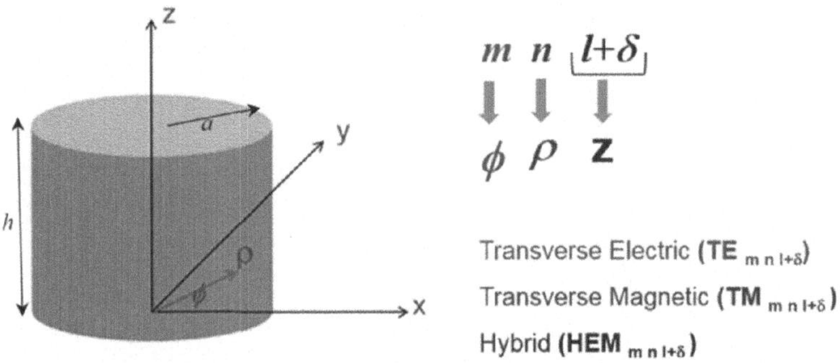

FIGURE 2.41
Different modes of a cylindrical DRA.

antenna, a DRA has a much wider impedance bandwidth. Moreover, the operating bandwidth of a DRA can be varied by suitably choosing the dielectric constant of the resonator material and its dimensions. DRAs have a high dielectric strength and hence higher power-handling capacity. In addition, the temperature-stable ceramics enable the antenna to operate in a wide temperature range. One of the limitations of DRAs is that they have restricted application as space antennas.

2.4.5.1 Modes and Radiation Mechanisms of DRAs

A microwave resonator has an infinite number of resonant modes, each corresponding to a particular resonant frequency at which the stored electric energy is equal to the magnetic energy. For a cylindrical resonator, the excited modes can be classified into three types: TE, TM, and hybrid. The fields for TE and TM modes are axisymmetric, whereas hybrid modes are azimuthally dependent. The TE, TM, and hybrid modes are classified as $TE_{mnl+\delta}$, $TM_{mnl+\delta}$, and $HEM_{mnl+\delta}$, respectively, as shown in Figure 2.41. The mode nomenclature makes possible the accurate prediction of far-field radiation of dielectric resonators in their application as antennas.

The modes that are most commonly used for radiating applications are $TE_{01\delta}$, $TM_{01\delta}$, and $HE_{01\delta}$. The corresponding field distributions for cylindrical DRs are shown in Figure 2.42.

2.4.5.2 Broadband DRA Techniques

Broadband DRAs can be broadly classified as single element methods, multiple DRAs, hybrid antennas, and DRA arrays. The single element method uses optimization of the DRA parameters and the feed geometries (stub matching), shaping the DRA or using parasitic coupling with different resonators. A wide range of bandwidths has been reported based on the feeding techniques and modification of DRA parameters [36–48], DRAs in stacked configurations [49–52] and multilayered DRAs [53, 54]. Array configuration of DRA radiators for improved bandwidth was shown in Loos and Antar, Petosa et al., and Fan et al. [55–57]. One of the recent techniques developed to have an ultrawideband hybrid antenna [58] is shown in Figure 2.43. Here a dielectric ring resonator surrounds a monopole antenna. Both are centered at the same axis and the monopole acts as a quarter-wavelength radiator as well as a feed for the DRA. The DRA is designed to excite in the dominant $TM_{01\delta}$ mode, which has a circular symmetric field distribution similar to that of a monopole. For wideband operation, the monopole is designed to operate at the lower end of the frequency

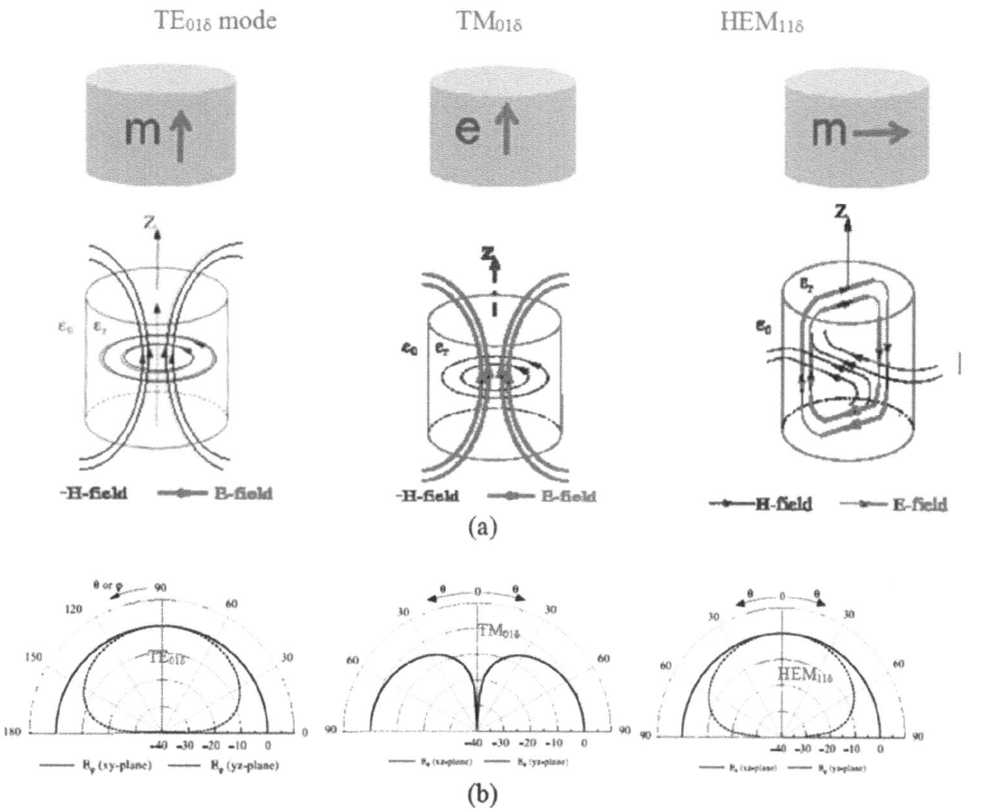

FIGURE 2.42
Different modes of a cylindrical DRA: (a) field distribution of the $TE_{01\delta}$ mode, $TM_{01\delta}$ mode, and $HEM_{11\delta}$ mode, (b) far-field radiation pattern of the $TE_{01\delta}$ mode, $TM_{01\delta}$ mode, and $HEM_{11\delta}$ mode.

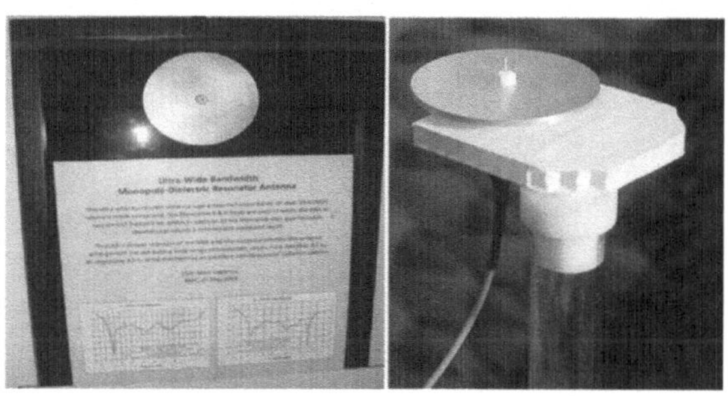

FIGURE 2.43
Photograph of a monopole-fed ultrawideband cylindrical DRA.

Reproduced by permission @2005 IEEE [58].

spectrum, while the DRA operates toward the upper end. The design parameters can be selected to have separate frequencies, so that dual band operation occurs. The improved design guidelines are given in Guha et al. [59].

Figure 2.44a shows the cross-sectional view of a monopole-fed cylindrical DRA and the measured return loss of the fabricated prototype compared with the simulation is shown in Figure 2.44b. The measured and simulated return loss exhibits excellent correspondence.

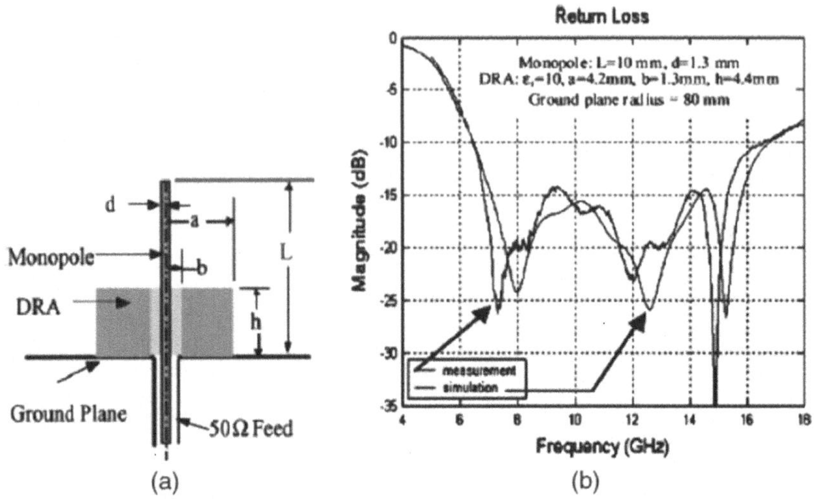

FIGURE 2.44
(a) A monopole-fed cylindrical DRA, and (b) its simulated and measured S_{11}.

Reproduced with permission @IEEE [58].

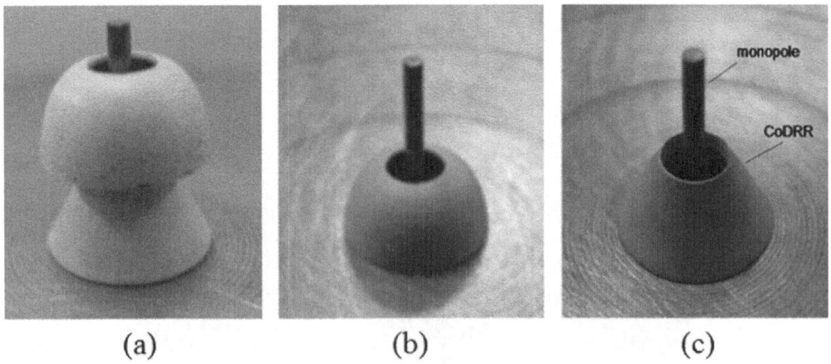

FIGURE 2.45
(a) photograph of a pawn-shaped DRA, (b) photograph of a hemispherical ring DRA, (c) photograph of a conical ring DRA.

Reprinted with permission @IEEE [60].

Other widebanding techniques include changing the shapes of the DRA or stacking them as shown in Figure 2.45. The conical DRA stacked with a hemispherical DRA in Figure 2.45a and fed by the monopole effectively reduces the length of the monopole and resonates at a higher frequency and the collective frequencies offer a wider bandwidth [60]. Similar design and shaping of DRAs using segmented hemispherical and half-hemispherical shapes are described in Guha et al., Guha, Gupta, and Antar, and Guha and Antar [61–63].

2.4.6 UWB Antennas for Special Applications

Ultrawideband antennas are used for various special applications such as radar, see through wall imaging, ground-penetrating radar, etc. UWB antennas are also used for electronic counter measure (ECM), electronic surveillance monitoring, etc. Depending on the precision of application, the designer determines the type of UWB antenna to be used. Some applications such as impulse radar require minimum distortion to the transmitted pulse which requires antenna with a stable phase center. Ground-penetrating radars have been used for demining. Antennas such as the biconical and bow tie, horn, and Vivaldi/tapered slot are some of the common antennas used for GPRs.

References

1. H. Schantz, *The Art and Science of Ultrawideband Antennas*, Boston, MA, Atrech House, 2005.
2. J.Y. Siddiqui, Y.M.M. Antar, A.P. Freundorfer, E.C. Smith, G.A. Morin, and T. Thayaparan, "Design of an Ultrawideband Antipodal Tapered Slot Antenna using Elliptical Strip Conductors," *IEEE Antennas and Wireless Propagation Letters*, vol. 10, 251–254, 2011.
3. R.K. de Jongh, M. Hajian, and L.P. Ligthart, "Antenna Time-Domain Measurement Techniques", *IEEE Antennas and Propagation Magazine*, vol. 39, no. 5, 7–12, 1997.
4. V.H. Rumsey, "Frequency independent antennas", 1957 IRE National Convention Record, pp. 119–128.
5. C.A. Balanis, *Antenna Theory: Analysis & Design*, John Wiley & Sons. Inc., New York, 1997.
6. F.J. Jucker, and H. Jasik, *Antenna Engineering Handbook*, McGraw-Hill, New York, 1961.
7. S.A. Schelkunoff and H.T. Friis, *Antennas Theory and Practice*, John Wiley & Sons, New York, NY, 1952.
8. J. Kim and S-O. Park, "Novel ultra-wideband discone antenna", *Microwave and Optical Technology Letters*, vol. 42, no. 2, 113–115, 2004.
9. S. Honda, M. Ito, H. Seki, and Y. Jingo, "A disc monopole antenna with 1:8 impedance bandwidth and omnidirectional radiation pattern", *In Proceeding of International Symposium on Antennas and Propagation*, Sapporo, Japan, 1145–1148, September 1992.
10. N.P. Agrawall, G. Kumar, and K.P. Ray, "Wide-Band Planar Monopole Antennas", *IEEE Transactions on Antennas and Propagation*, vol. 46, no. 2, 294–295, 1998.
11. J.A. Stratton, *Electromagnetic Theory*, McGraw Hill, New York, A31–A438, 1941.
12. M. Hammoud, P. Poey, and F. Colombel, "Matching the Input Impedance of a Broadband Disc Monopole", *Electronics Letters*, vol. 29, no. 4, 406–407, 1993.
13. J. Liang, "Antenna Study and Design for Ultra Wideband Communication Applications", Ph.D. Dissertation, Queen Mary University of London, 2006.
14. P.J. Gibson, "The Vivaldi Aerial", *Proceeding of 9th European Microwave Conference*, Brighton, U.K., 101–105, 1979.
15. S.N. Prasad and S. Mahapatra, "A Novel MIC Slotline Aerial", *Proceeding of 9th European Microwave Conference*, Brighton, U.K., 120–124, 1979.
16. R.Q. Lee, "Notch Antennas", NASA Technical Report, July 2004.
17. P.R. Acharya, H. Ekstrom, S.S. Gearhart, S. Jacobsson, J.F. Johansson, E.L. Kollberg, and G.M. Rebeiz, "Tapered Slotline Antennas at 802 GHz", *IEEE Transactions on Microwave Theory and Techniques*, vol. 41, no. 10, 1715–1719, 1993.
18. H. Jaisk, "Surface and Leaky Wave Antennas", *Antenna Engineering Handbook*, McGraw Hill, New York, 1961.
19. D. Schaubert, E. Kollberg, T. Korzeniowski, T. Thungren, J. Johansson, K. Yngvesson, "Endfire Tapered Slot Antennas on Dielectric Substrates", *IEEE Transactions on Antennas and Propagation*, vol. AP-33, no. 12, 1392–1400, 1985.
20. R.Q. Lee and R.N. Simons, "Overview of Tapered-Slot Antennas", NASA Technical Report, May 1998.
21. E. Gazit, "Improved Design of the Vivaldi Antenna", *IEE Proceedings H (Microwaves, Antennas and Propagation)*, vol. 135, no. 2, 89–92, 1988.

22. J.D.S. Langley, P.S. Hall, and P. Newham, "Novel Ultrawide-Bandwidth Vivaldi Antenna with Low Cross Polarization", *Electronic Letters*, vol. 29, 2004–2005, 1993.
23. C. Sarkar, C. Saha, L.A. Shaik, J.Y. Siddiqui, and Y.M.M. Antar, "Frequency notched balanced antipodal tapered slot antenna with very low cross-polarised radiation", *IET Microwaves, Antennas and Propagation*, 2018.
24. B.B. Mandelbrot, *The Fractal Geometry of Nature*, W. H. Freeman, New York, 1983.
25. D. Werner, J. Petko, and T.G. Spence, "Fractal Antennas", *Antenna Engineering Handbook*, John L. Volakis (Ed.), McGraw Hill, New York, 2007.
26. J.P. Gianvittorio and Y. Rahmat-Samii, "Fractal Antennas: A Novel Antenna Miniaturization Technique, and Applications," *IEEE Antennas and Propagation Magazine*, vol. 44, no. 1, 20–36, 2002.
27. D.H. Werner and S. Ganguly, "An Overview of Fractal Antenna Engineering Research," *IEEE Antennas and Propagation Magazine*, vol. 45, no. 1, 38–57, 2003.
28. C. Puente, J. Romeu, R. Pous, X. Garcia, and F. Benitez, "Fractal Multiband Antenna Based on the Sierpinski Gasket", *IEE Electronics Letters*, vol. 32, no. 1, 1–2, 1996.
29. C. Puente, J. Romeu, R. Pous, and A. Cardama, "On the Behavior of the Sierpinski Multiband Fractal Antenna", *IEEE Transactions on Antennas and Propagation*, vol. 46, no. 4, 517–524, 1998.
30. G.J. Walker and J.R. James, "Fractal Volume Antennas", *IEE Electronics Letters*, vol. 34, no. 16, 1536–1537, 1998.
31. C.T.P. Song, P.S. Hall, H. Ghafouri-Shiraz, and D. Wake, "Fractal Stacked Monopole with Very Wide Bandwidth", *IEE Electronics Letters*, vol. 35, no. 12, 945–946, 1999.
32. K.J. Vinoy, "Fractal shaped antenna elements for wide and multi-band wireless applications", Thesis, Pennsylvania State University, Aug. 2002.
33. S.A. Long, M.W. McAllister, and L.C. Shen, "The Resonant Cylindrical Dielectric Cavity Antenna", *IEEE Transactions on Antennas and Propagation*, vol. AP-31, 406–412, 1983.
34. M.W. McAllister, S.A. Long, and G.L. Conway, "Rectangular Dielectric Resonator Antenna", *Electronics Letters*, vol. 19, 218–219, 1983.
35. M.W. McAllister and S.A. Long, "Resonant Hemispherical Dielectric Antenna", *Electronics Letters*, vol. 20, 657–659, 1984.
36. A. Ittipiboon, A. Petosa, D. Roscoe, and M. Cuhaci, "An Investigation of a Novel Broadband Dielectric Resonator Antenna", *IEEE International Symposium on Antennas and Propagation*, 2038–2041, 1996.
37. R. Chair, A.A. Kishk, K.F. Lee, "Wideband Simple Cylindrical Dielectric Resonator Antennas", *IEEE Microwave and Wireless Components Letters*, vol. 15, no. 4, 241–243, 2005.
38. K.W. Leung and C.K. Leung, "Wideband Dielectric Resonator Antenna Excited by Cavity-Backed Circular Aperture with Microstrip Tuning Fork", *IEE Electronics Letters*, vol. 39, no. 14, 1033–1035, 2003.
39. K.W. Leung, K.M. Luk, K.Y. Chow, and E.K.N. Yung, "Bandwidth Enhancement of Dielectric Resonator Antenna by Loading a Low-Profile Dielectric Disk of Very High Permittivity", *IEE Electronics Letters*, vol. 33, no. 9, 725–726, 1997.
40. A.A. Kishk, "Experimental Study of Broadband Embedded Dielectric Resonator Antennas Excited by a Narrow Slot", *IEEE Antennas and Wireless Propagation Letters*, vol. 4, 79–81, 2005.
41. Y. Coulibaly, T.A. Denidni, and L. Talbi, "Wideband Impedance Hybrid Dielectric Resonator Antenna for X-Band Application", *IEEE International Symposium on Antennas and Propagation*, vol. 3, 2429–2432, 2006.
42. A.A. Kishk, "Wide-Band Truncated Tetrahedron Dielectric Resonator Antenna Excited by a Coaxial Probe", *IEEE Transactions on Antennas and Propagation*, vol. AP-51, no. 10, 2913–2917, 2003.
43. A.A. Kishk, Y. Yin, and A.W. Glisson, "Conical Dielectric Resonator Antennas for Wideband Applications", *IEEE Transactions on Antennas and Propagation*, vol. AP-50, no. 4, 469–474, 2002.
44. X.L. Liang and T.A. Denidni, "Wideband Rectangular Dielectric Resonator Antenna with a Concave Ground Plane", *IEEE Antennas and Wireless Propagation Letters*, vol. 8, 367–370, 2009.

45. L.C.Y. Chu, D. Guha, and Y.M.M. Antar, "Conformal Strip-Fed Shaped Cylindrical Dielectric Resonator: Improved Design of a Wideband Wireless Antenna", *IEEE Antennas and Wireless Propagation Letters*, vol. 8, 482–485, 2009.
46. R. Chair, A.A. Kishk, K.F. Lee, and C.E. Smith, "Wideband Flipped Staired Pyramid Dielectric Resonator Antenna", *IEE Electronics Letters*, vol. 40, no. 10, 581–582, 2004.
47. A.S. Al-Zhoubi and A.A. Kishk, "Wideband Strip-Fed Rectangular Dielectric Resonator Antenna with Improved Radiation Patterns", *IEEE International Symposium on Antennas and Propagation*, 2009.
48. X.L. Liang, T.A. Denidni, and L.N. Zhang, "Wideband L Shaped Dielectric Resonator Antenna with a Conformal Inverted-Trapezoidal Patch Feed", *IEEE Transactions on Antennas and Propagation*, vol. AP-57, no. 1, 271–274, 2009.
49. A.A. Kishk, B. Ahn, and D. Kajfez, "Broadband Stacked Dielectric-Resonator Antennas", *IEE Electronics Letters*, vol. 25, no. 18, 1232–1233, 1989.
50. K.M. Luk, K.W. Leung, and K.W. Chow, "Bandwidth and Gain Enhancement of a Dielectric Resonator Antenna with the Use of a Stacking Element", *Microwave and Optical Technology Letters*, vol. 14, no. 4, 215–217, 1997.
51. Q. Rao, T.A. Denidni, and A.R. Sebak, "Broadband Compact Stacked T-Shaped DRA with Equilateral-Triangle Cross Sections", *IEEE Microwave and Wireless Components Letters*, vol. 16, no. 1, 7–9, 2006.
52. Y. Ge and K.P. Esselle, "A Wideband Probe-Fed Stacked Dielectric Resonator Antenna", *Microwave and Optical Technology Letters*, vol. 48, no. 8, 1630–1633, 2006.
53. W. Huang and A.A. Kishk, "Compact Wideband Multi-Layer Cylindrical Dielectric Resonator Antennas", *IET Microwaves Antennas and Propagation*, vol. 1, no. 5, 998–1005, 2007.
54. K.W. Leung and K.K. So, "Theory and Experiment of the Wideband Two-Layer Hemispherical Dielectric Resonator Antenna", *IEEE Transactions on Antennas and Propagation*, vol. AP-57, no. 4, 1280–1284, 2009.
55. G.D. Loos and Y.M.M. Antar, "A New Aperture-Coupled Rectangular Dielectric Resonator Antenna Array", *Microwave and Optical Technology Letters*, vol. 7, no. 14, 642–644, 1994.
56. A. Petosa, A. Ittipiboon, M. Cuhaci, and R. Larose, "Band width Improvement for a Microstrip-Fed Series Array of Dielectric Resonator Antennas", *IEE Electronics Letters*, vol. 32, no. 7, 608–609, 1996.
57. Z. Fan, Y.M.M. Antar, A. Ittipiboon, and A. Petosa, "Parasitic Coplanar Three-Element Dielectric Resonator Antenna Subarray", *IEE Electronics Letters*, vol. 32, no. 9, 789–790, 1996.
58. M. Lapierre, Y.M.M. Antar, A. Ittipiboon, and A. Petosa, "Ultra Wideband Onopole/Dielectric Resonator Antenna", *IEEE Microwave and Wireless Components Letters*, vol. 15, no. 1, 7–9, 2005.
59. D. Guha, Y.M.M. Antar, A. Ittipiboon, A. Petosa, and D. Lee, "Improved Design Guidelines for the Ultra Wideband Monopole-Dielectric Resonator Antenna", *IEEE Transactions on Antennas and Propagation*, vol. 5, 373–377, 2006.
60. D. Guha, B. Gupta, and Y.M.M. Antar, "New Pawn-Shaped Dielectric Ring Resonator Loaded Hybrid Monopole Antenna for Improved Ultrawide Bandwidth", *IEEE Antennas and Wireless Propagation Letters*, vol. 8, 1178–1181, 2009.
61. D. Guha, B. Gupta, C. Kumar, and Y.M.M. Antar, "Segmented Hemispherical DRA: New Geometry Characterized and Investigated in Multi-Element Composite Forms for Wideband Antenna Applications", *IEEE Transactions on Antennas and Propagation*, vol. 60, no. 3, 1605–1610, 2012.
62. D. Guha, B. Gupta, and Y.M.M. Antar, "Hybrid Monopole-DRAs Using Hemispherical/Conical-Shaped Dielectric Ring Resonators: Improved Ultrawideband Designs", *IEEE Transactions on Antennas and Propagation*, vol. 60, no. 1, 393–398, 2012.
63. D. Guha and Y.M.M. Antar, "New Half-Hemispherical Dielectric Resonator Antenna for Broadband Monopole-Type Radiation", *IEEE Transactions on Antennas and Propagation*, vol. 54, no. 12, 3621–3628, 2006.

3

Frequency-Notched UWB Antenna Design

3.1 Introduction

The design and development of various UWB printed antennas were discussed in the previous chapter. The most common UWB printed antennas, such as the printed monopole and the tapered slot antennas, were also discussed in detail in Chapter 2. This chapter deals with the frequency-notching techniques of various UWB antennas to minimize the interference of various narrowband radiations with the overlapping UWB radiation.

3.2 Spectrum Overlapping Between UWB and Narrowband Services

Ultrawideband (UWB) signals are suitable for underlay communications, over a frequency band with other active systems. Hence, coexistence of UWB and other systems is eminent, however, with minimal mutual interference impacting their respective performance [1].

3.2.1 RF Spectrum

RF spectrum spans from ultra-low frequencies (a few kilohertz) up to submillimeter waves corresponding to frequencies of 300 GHz and more. The low-frequency RF spectrum increases the operating range with large antenna dimension but reduced range of operating frequency. At higher frequencies, the operating range of frequency becomes wider with increased bandwidth and reduced antenna size but also with a reduced operating range [2]. The spectrum allocation is based on exclusive and non-exclusive usage based on spectrum property rights and on licensing or classes of users, respectively [2]. Different types of spectrum allocations such as primary allocations and secondary allocations and their compatibility in the same or adjacent spectrum is decided by different rules [2]. Figure 3.1 shows the "United States Frequency Allocations" chart which gives a pictorial view of the allocations of various frequency bands to different types of "service".

3.2.2 Interference Aspects

Ultrawideband (UWB) communications systems with very large relative and/or absolute bandwidth are mandated with low power spectral density (PSD) to allow their use as overlays over existing systems. They can be used in the same frequency range as existing systems without causing undue interference [3].

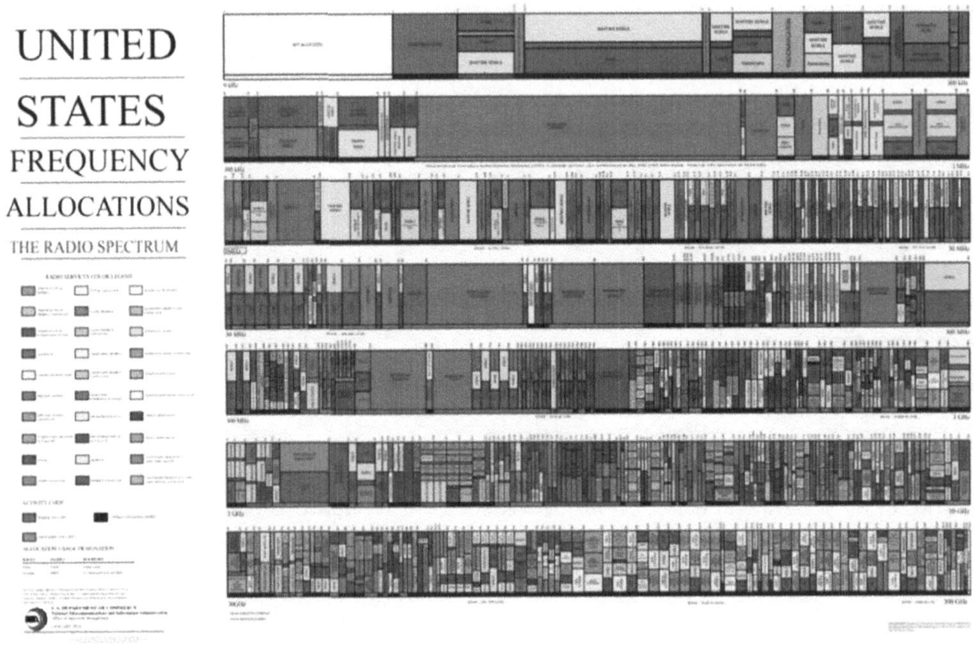

FIGURE 3.1
United States Frequency Allocations chart.

FIGURE 3.2
Coexistence between narrowband and ultrawideband communication.

As defined in Chapter 2, UWB communications systems are commonly defined as systems with large absolute and/or large relative bandwidth. Such a large bandwidth offers specific advantages with respect to signal robustness, information transfer speed, and/or implementation simplicity. However, it also leads to fundamental differences from conventional, narrowband, systems. The spark-gap generators used by Hertz and Marconi were the first practical UWB systems. Later, general interest turned to narrowband communications since the spectral efficiency of the signals generated by the spark-gap transmitters was low. Introduction of time-hopping impulse radio (TH-IR) in the early 1990s, with the help of pioneering work from Win and Scholtz [4–6], showed that impulse radio could sustain a large number of users by assigning pseudorandom transmission times to the pulses from the different users. The UWB systems, as shown in Figure 3.2, can be used as an overlay to existing systems and can be operated in parallel with existing legacy systems [3]. A UWB system will exhibit low PSD which is seen as noise to a narrowband receiver within its own system bandwidth as shown in Figure 3.3. The interference is determined by the PSD of the UWB radiation as seen by the victim device. Regulators impose limits on the PSD emitted by a UWB transmitter. Those limits or frequency masks are frequency dependent. For indoor communications, a PSD of −41.3 dBm/MHz is allowed in the frequency

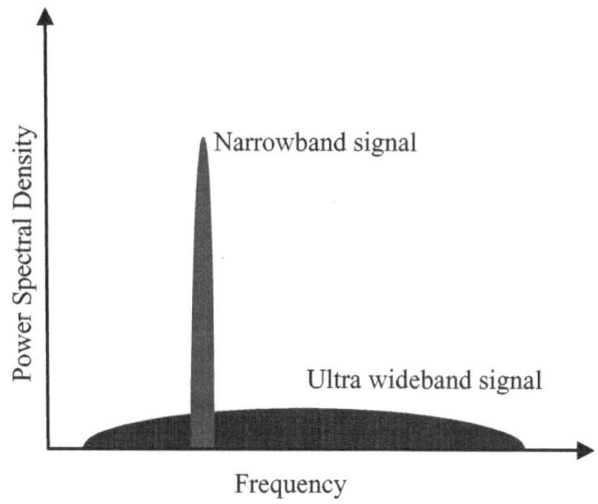

FIGURE 3.3
Power spectral density of UWB and NB systems.

band between 3.1 and 10.6 GHz. Outside of that band, no intentional emissions are allowed. For outdoor environments, only mobile, battery-operated transmitters are allowed. Again, intentional emissions are allowed in the 3.1–10.6 GHz range; the mask for spurious emissions is different from the indoor case [3]. However, there still is a concern that widespread use of UWB devices might lead to high interference, which could negatively impact operation of legacy systems [3].

3.3 Techniques of Frequency-Notched UWB Antennas

The interference between UWB devices and narrowband devices which hampers the overall performance of the communication systems is mitigated using some novel modifications on the antenna structures. Several techniques are proposed to design UWB antennas with frequency stop or frequency notch characteristics. Depending on the application and the nature and bandwidth of rejection band, researchers developed novel antennas with an integrated filtering mechanism which provides the desired response. Several antenna designs are reported which provide frequency notch response and the techniques are discussed in this chapter. Some common techniques used to obtain frequency notches are

- Modifications on the radiator
- Modifications on the ground plane or signal line
- Integrated filter techniques
- Metamaterial-inspired resonators

Modifications on the radiator and ground planes in the form of slots and cuts are the most common techniques to achieve frequency notches in a UWB-printed antenna. Several design configurations have been proposed in open literature using planar monopole antennas with modified radiator and/or ground plane to achieve this characteristic [7–25]. Notch characteristic with triple notch frequencies [7–11], dual, and single notch frequencies [12–25] are achieved using various design configurations employed with planar monopole printed antennas and are discussed in the following two subsections.

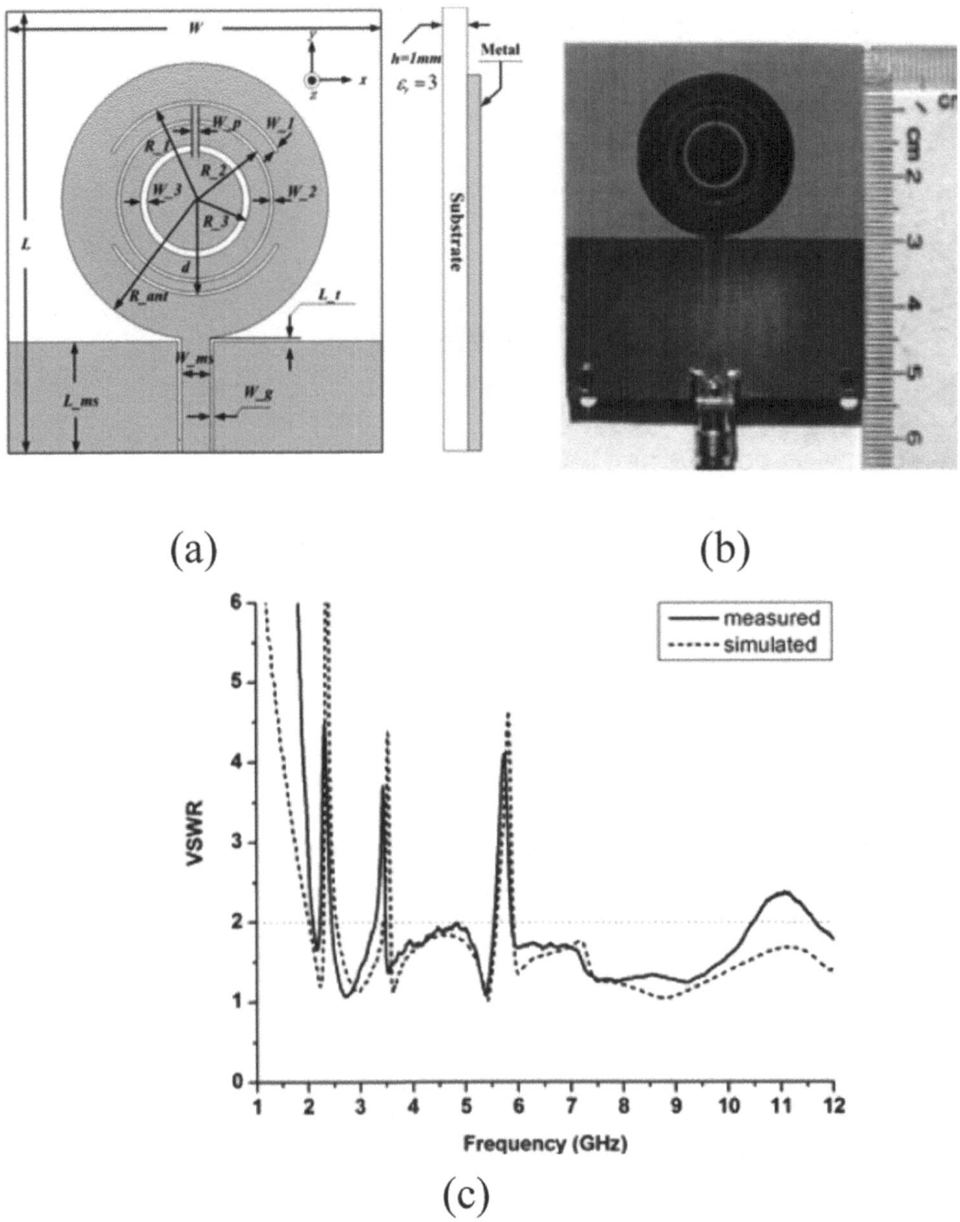

FIGURE 3.4
Circular monopole antenna with circular and arc-shaped slots [7]. (a) Antenna schematic, (b) fabricated prototype, and (c) VSWR plot as a function of frequency.

3.3.1 Modifications on the Radiator

Modifying the printed antenna with slot and cuts, embedded on the radiators, has been used as a common practice to alter the current path in printed antennas which in turn affects the input impedance of the antenna. Slots and cuts of different geometrical shapes and sizes yielded different results by changing the current distribution on the radiator. This perturbation of the currents on the radiator alters the input impedance at some frequencies based on the size of the slots and cuts along with the dielectric constant and height of the substrate. This technique has been employed in UWB antennas to filter out the interfering or the undesired frequency from the UWB spectrum. Several papers with design guidelines have been published to yield single and multiple notches [7–25]. Figure 3.4 shows an ultrawideband (UWB) antenna with triple notched band. The proposed antenna consists of a planar circular patch monopole UWB antenna and multiple etched split-ring slots and circular slots on the patch radiator. As shown in the figure, two split-ring slots and two arc slots have been etched on the patch to generate triple notched bands. Two split-ring slots are used to generate notched bands with a central frequency of 2.4 and 3.5 GHz, respectively, while the couple arc slots with the same radius correspond to the notched band centered on 5.8 GHz [7]. The technique is significant for designing UWB antennas with multiple narrow frequency-notched bands or for designing multiband antennas.

Similarly, a design using three open-ended quarter-wavelength slots as shown in Figure 3.5 [9], was used to obtain triple band-notched characteristics. The antenna uses three open-ended quarter-wavelength slots to create triple band-notched characteristics in

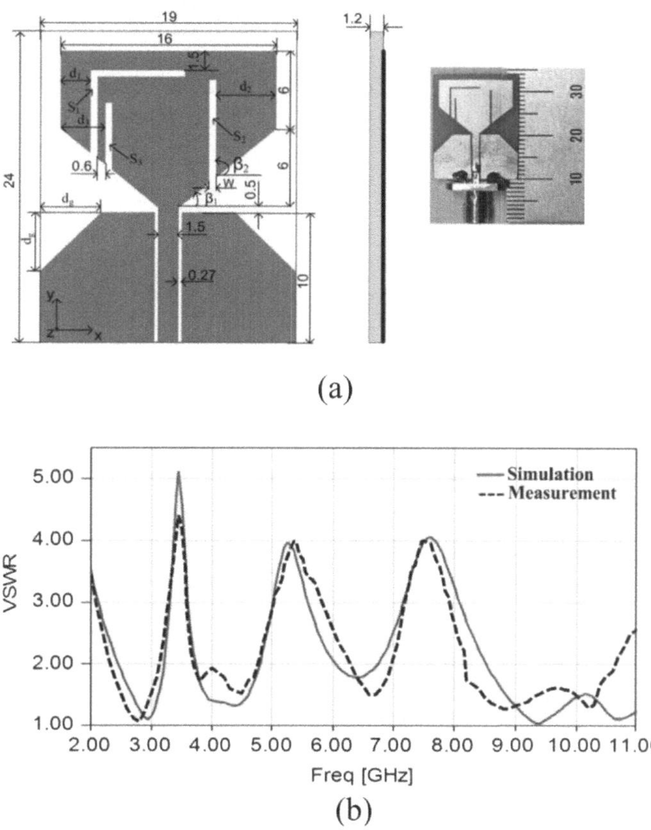

FIGURE 3.5
Geometry of the proposed antenna (units in millimeters) with top view, side view, and fabricated antenna, (b) VSWR plot as a function of frequency.

Reproduced by permission of © 2012 The IEEE [9].

3.3–3.7 GHz for WiMAX, 5.15–5.825 GHz for WLAN, and 7.25–7.75 GHz for downlink of X-band satellite communication systems, respectively.

Another design by embedding an omega-shaped slot on the radiating patch [10] as shown in Figure 3.6, was used to obtain triple notch. The measured result in Figure 3.6b shows that the intermediate frequency range between 5.35–5.8 GHz can be utilized with the proposed antenna, which is rejected by other WLAN band-notched antennas.

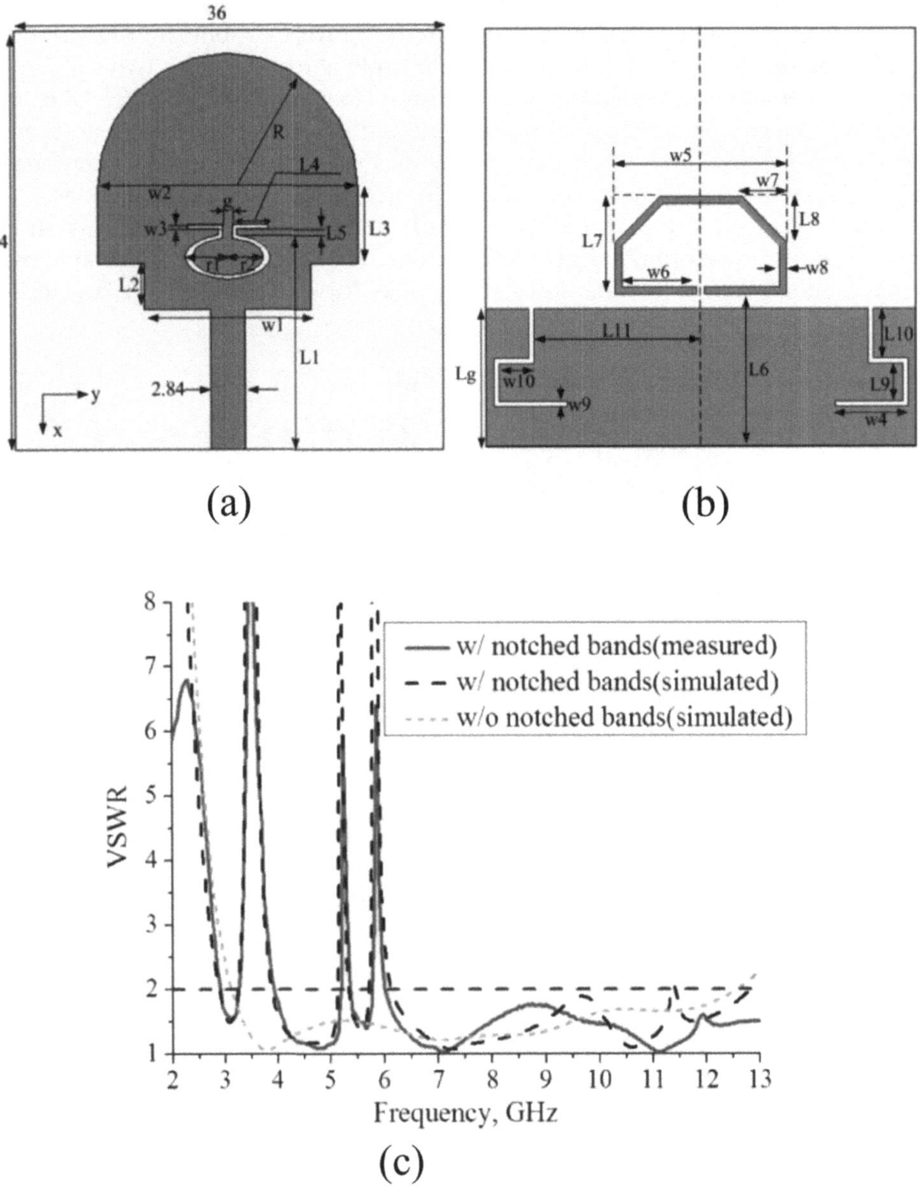

FIGURE 3.6
Geometry of the omega-shaped slot loaded monopole antenna: (a) top view, (b) back view, (c) VSWR plot as a function of frequency.

A pentagonal planar triple band-notched monopole antenna with two thin bent slots etched on the radiating patch is designed in Haykin [11]. Two slots with their length equal to half-guided wavelength at the center-rejected frequency [11] are optimally placed on the radiator. This yields a dual band-notched feature as shown in Figure 3.7. A similar resonant slot loaded monopole antenna was reported in Tang et al. [13] for band-notch characteristics. Multiple band notches featuring complementary co-directional split-ring

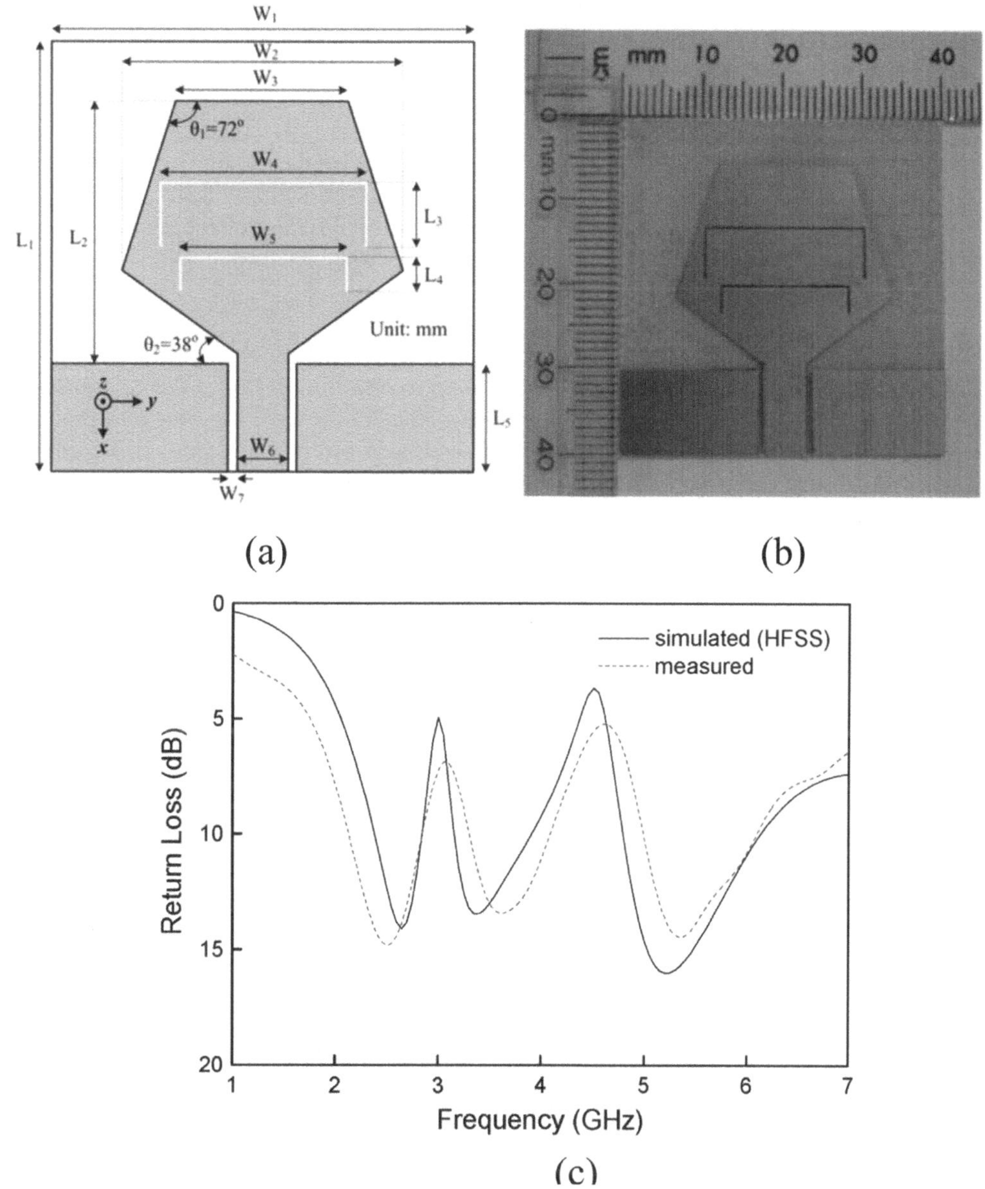

FIGURE 3.7
Geometry of the pentagon-shaped monopole antenna with slots: (a) schematic view, (b) prototype, (c) reflection coefficient plot.

resonator (CSRR) arranged in the middle of the radiating patch close to the feeding strip was demonstrated in Chu and Yang [14]. Similarly, a dual band-notch characteristic to reject undesired frequencies in UWB radiators was demonstrated in Li et al. [17] where two C-shaped slots were etched on a beveled rectangular patch.

A miniaturized elliptic-card UWB antenna with WLAN band (5.1–5.8 GHz) rejection capability was obtained by adding two slits within the elliptic element [15]. Figure 3.8 shows the miniaturized elliptic patch and the simulated and measured VSWR and gain plots as a function of frequency are shown in Figure 3.9a–b.

A UWB antenna with single or dual band notches for lower WLAN band and upper WLAN band was designed using a fork-shaped antenna. The band notches are realized by adding independent controllable strips in terms of the notch frequency and the width of the band notches to the fork shape of the UWB antenna [16]. The design provides the control of both the width of band notch and the rejected frequency. Figure 3.10 shows the schematic of the fork-shaped antenna with single and dual strips. Figures 3.11 and 3.12 show the VSWR response as a function of frequency for the single-frequency notch and dual-frequency notch, respectively.

3.3.2 Modifications on the Ground Plane or Signal Line

Modifications on the ground plane were also used to obtain notches in the operating frequency band of the UWB antennas. Like in Figure 3.6, along with the omega-shaped slot in the radiating patch, a semi-octagon-shaped resonator on the back side, and a defected grounded structure was also etched to obtain the desired characteristics. Another design by etching one quasi-complementary split-ring resonator (SRR) in the feedline was shown by Li et al. [17]. Several new designs, such as using a trapezoidal ground plane with a rectangular slot together with a modified complementary co-directional SRR etched on the radiating patch [18]; employing a U-slot (Figure 3.13) defected ground structure in the ground plane and etching a split-ring slot in the radiation patch [19]; and embedding an E-slot in the radiation patch and a U-slot defected ground structure in the feeding line [20], have also been used satisfactorily. Similarly, a single band-notch characteristic was proposed in an antenna consisting of a patch with an arc-shaped edge and a partially modified ground plane [21] as shown in Figure 3.14. The single frequency notch characteristic is also realized by introducing a microstrip feeder with a tuning stub [22], by utilizing a mushroom-type electromagnetic-bandgap (EBG) structure [23] as shown in Figure 3.14, and also by using a coplanar waveguide (CPW) with two asymmetrical ground planes [24].

FIGURE 3.8
Photograph of the miniaturized elliptic antenna with band-notched characteristics.

Reproduced by permission of © 2007 The IEEE [15].

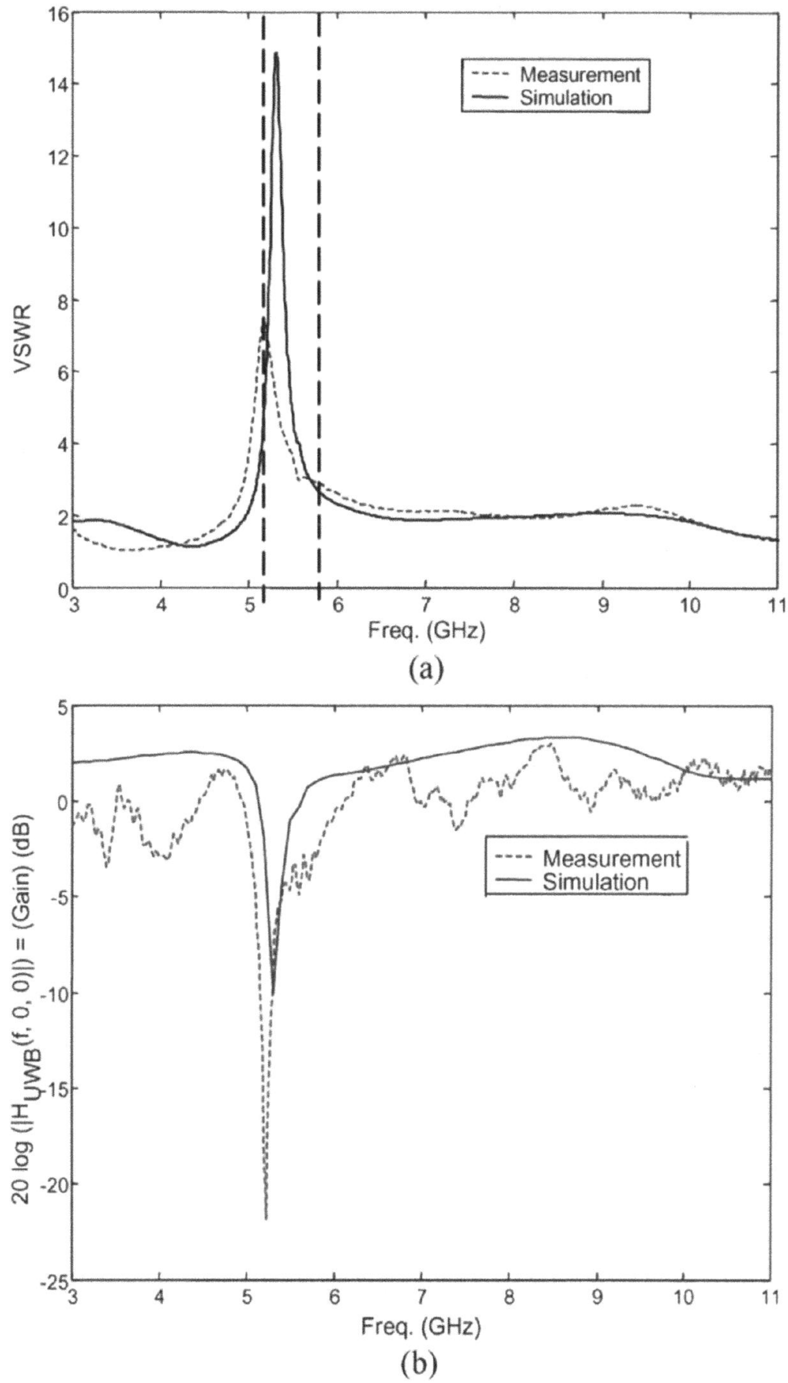

FIGURE 3.9
Simulated and measured (a) VSWR and (b) gain of the elliptic antenna.

Reproduced by permission of © 2007 The IEEE [15].

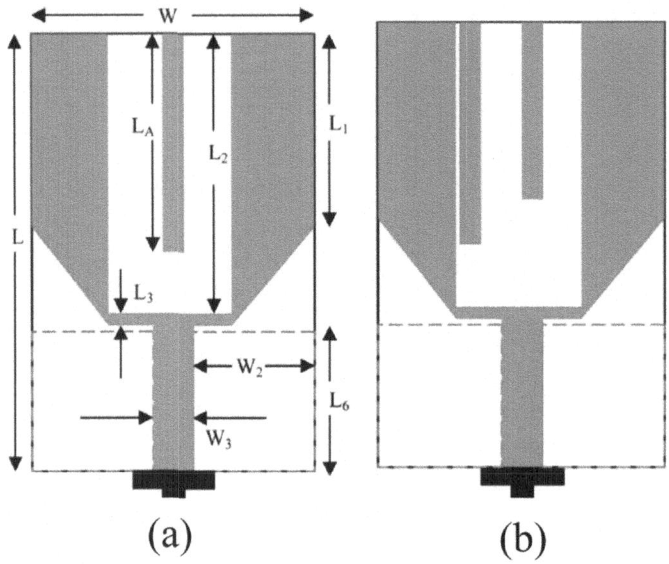

FIGURE 3.10
Schematic of the fork-shaped antenna with (a) single strip for single band-notched characteristic, (b) dual strip for dual band-notched characteristics.

Reproduced by permission of © 2009 The IEEE [16].

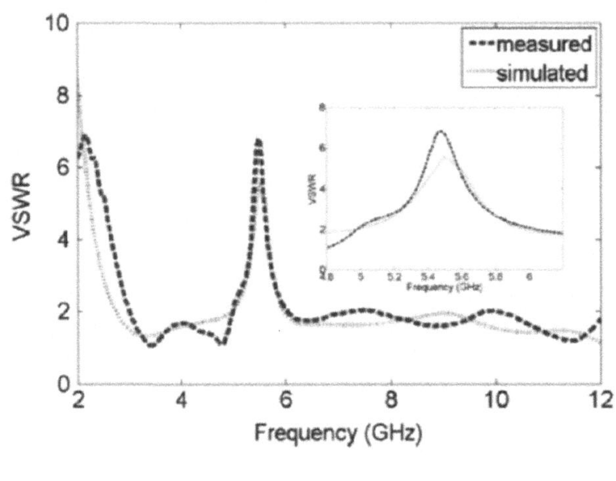

FIGURE 3.11
Simulated and measured VSWR plot of the single band-notched fork antenna.

Reproduced by permission of © 2009 The IEEE [16].

3.3.3 Integrated Filter Techniques

An ultrawideband planar monopole antenna with a tunable band-notch characteristic was also realized by loading an embedded resonant slot with a varactor as was demonstrated in Daviu et al. [25]. The UWB antenna with tunable band-notched behavior was implemented by loading the resonant slot of the antenna with a variable capacitance/varactor

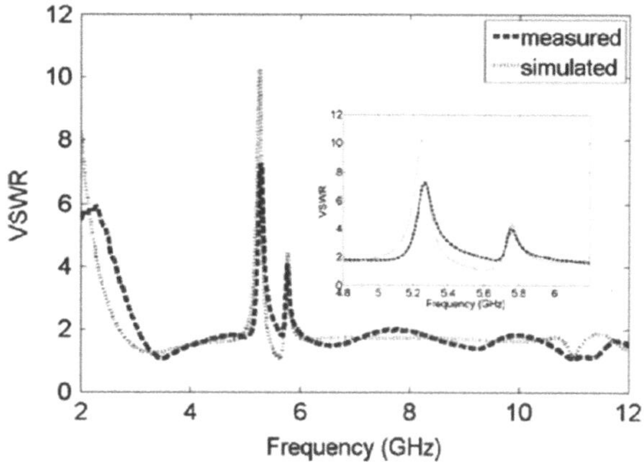

FIGURE 3.12
Simulated and measured VSWR plot of the dual band-notched fork antenna.

Reproduced by permission of © 2009 The IEEE [16].

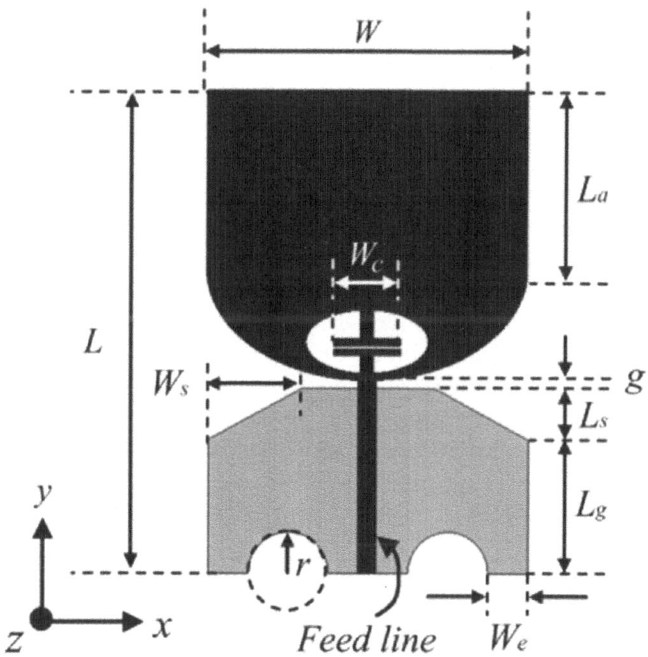

FIGURE 3.13
Schematic of the monopole antenna with arc-shaped edge and a partially modified ground plane.

Reproduced by permission of © 2007 The IEEE [21].

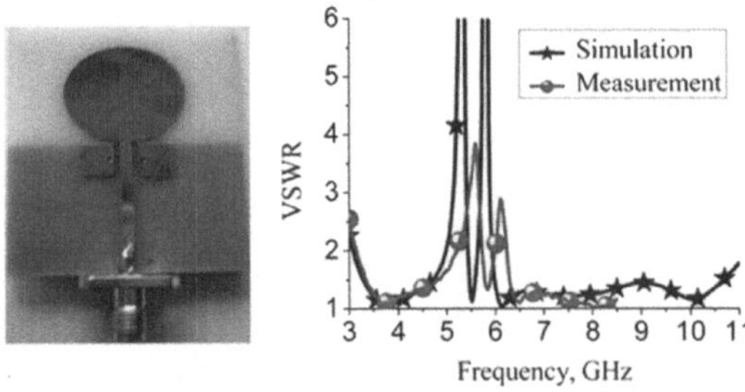

FIGURE 3.14
Prototype of UWB band-notched monopole antenna with electromagnetic-bandgap structures and VSWR plot.

Reproduced by permission of © 2011 The IEEE [23].

which provides the possibility of adjusting the rejected band within a wide range of frequencies. A control signal is used to tune the notch band.

3.3.4 Metamaterial-Inspired Resonators

Inspired by the work of Schelkunoff and Friis [26], in 1999 Pendry et al. [27] proposed that a pair of C-shaped open loops printed on a dielectric substrate can act as the "basis" for realization of negative magnetism because of its high magnetic polarizability [28] which was termed as the split-ring resonator (SRR).

The structure proposed by Pendry is shown in Figure 3.15a. The retrieved permeability (μ) curve against frequency *f* shown in Figure 3.15b, indicates a μ negative narrow frequency band around the magnetic resonance frequency of the structure. Such subwavelength resonant structures were used for passive planar circuit design [29, 30]. Recently, the split-ring resonator (SRR) has found applications in designing the narrow and wideband planar microwave filter [31–33], novel phase shifter [34], microwave power divider [35], etc. The split-ring resonator has also been considered as a potential candidate for designing compact low-frequency antennas [36]. Theoretical analysis suggests that a SRR-based leaky wave antenna gives higher radiated power, simultaneously reducing the reactance [37–39]. Associated research and development can be found from research publications which have appeared in open literature over the last decade [40–46]. For all such applications of SRR, accurate estimation of the resonance frequency is crucial. Other shapes of SRR that have been studied by a few groups [47–51] are the square SRR and hexagonal SRR. A comprehensive study of the SRR of various geometrical shapes is reported by Saha [52].

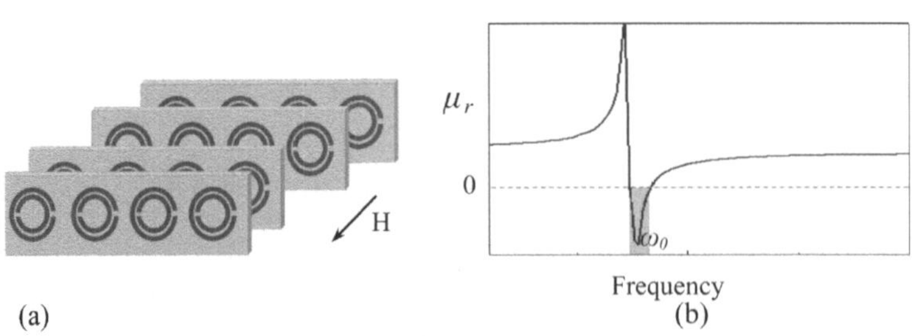

FIGURE 3.15
Negative μ medium: (a) array of split-ring resonator (SRR), (b) realized μ (real part) as a function of frequency.

Conventionally, a circular split-ring resonator consists of two concentric circular open loops printed on a dielectric substrate as shown in Figure 3.16a. The gaps of the two rings (g) are etched on diametrically opposite sides of the two edge-coupled rings and are very small compared to other physical dimension of the rings, namely the radius (r_{ext}) of the outer and inner rings (r_0), ring width (c), and separation between the two rings (d). The excitation mechanism can be easily understood from the diagrams shown in Figure 3.15a. The equivalent circuit is shown in Figure 3.17. When exposed to an electromagnetic wave with electric field vector in the plane of the SRR and magnetic field orthogonal to it, an electromotive force (EMF) is induced on the SRR which results in an induced current oscillating back and forth between the two rings and thus generates a resonance phenomenon. Determination of this resonance frequency has attracted many researchers in developing accurate models and proposing new formulations. Marquez et al. [53] was first to theoretically determine the resonance frequency of circular split-ring resonators using an equivalent circuit approach.

The most common SRR configurations that are used for frequency notch applications are

- Circular split-ring resonator (C-SRR)
- Square split-ring resonator (S-SRR)
- Hexagonal split-ring resonator (H-SRR)
- Rotational circular split-ring resonator (RC-SRR)

3.3.4.1 Calculation of SRR's Resonance Frequency

Figure 3.16 shows the schematic diagram of common SRR geometries, namely the circular, square, hexagonal, and rotational geometries as depicted in Figure 3.16a–d. Their equivalent circuit model is shown in Figure 3.17. Application of an external magnetic field along the z-axis of the SRR induces an electromotive force around the SRR with induced currents passing from one ring to the other through the inter ring spacing, d, and the structure behaves as a LC circuit. As shown in the equivalent circuit in Figure 3.17, the metallic rings contribute a total inductance L_T and distributed capacitances C_1 and C_2 forming at the two halves of the SRR structure above and below the split gaps. This new equivalent circuit also incorporates the gap capacitances C_{g1} and C_{g2} formed due to the split within the inner and outer rings, respectively.

3.3.4.2 Circular SRR

The resonant frequency f_0 of the circular SRR is given by,

$$f_0 = \frac{1}{2\pi}\sqrt{\frac{1}{L_T C_{eq}}} \tag{3.1}$$

where C_{eq} is the total equivalent capacitance of the structure.

Again, from the equivalent circuit of Figure 3.17, the total equivalent capacitance C_{eq} can be evaluated as,

$$C_{eq} = \frac{\left(C_1 + C_{g_1}\right)\left(C_2 + C_{g_2}\right)}{\left(C_1 + C_{g_1}\right) + \left(C_1 + C_{g_2}\right)} \tag{3.2}$$

Since the split gaps are of identical dimensions $g_1 = g_2 = g$, hence the gap capacitances $C_{g_1} = C_{g_2} = C_g$ and the series capacitances $C_1 = C_2 = C_0$ and therefore Equation 3.2 is modified as,

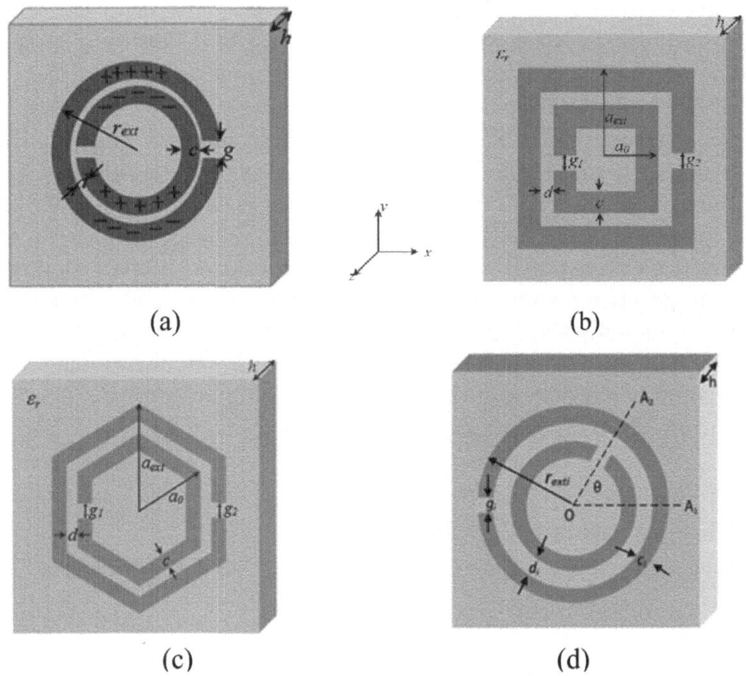

FIGURE 3.16
Schematic view of common SRR configurations: (a) circular, (b) square, (c) hexagonal, and (d) rotational.

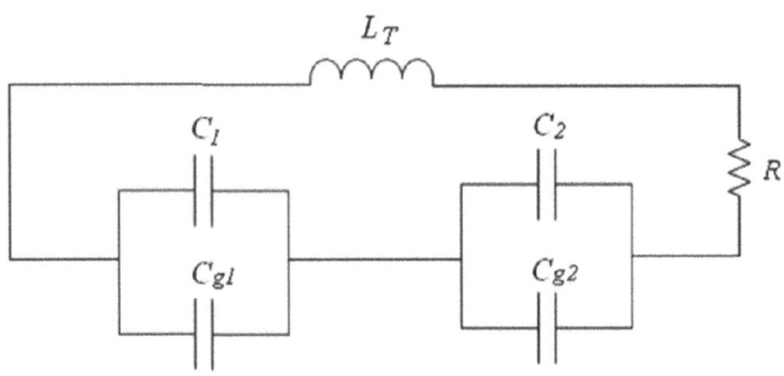

FIGURE 3.17
Equivalent circuit of the SRRs.

$$C_{eq} = \frac{(C_0 + C_g)}{2} \tag{3.3}$$

Considering a metal thickness t of the strip conductors, the gap capacitances C_{g1} and C_{g2} can be represented as,

$$C_{g_1} = C_{g_2} = C_g = \frac{\varepsilon_0 ct}{g} \tag{3.4}$$

where c and t are the width and thickness of the metallic rings, respectively and ε_0 is the free space permittivity. The distributed capacitances C_1 and C_2 are also a function of the split-gap dimensions $g_1 = g_2 = g$ and the average ring radius r_{avg}, and is given as,

$$C_0 = C_1 = C_2 = (\pi r_{\text{avg}} - g)C_{\text{pul}} \tag{3.5}$$

where,

$$r_{\text{avg}} = r_{\text{ext}} - c - \frac{d}{2} \tag{3.6}$$

and C_{pul} is the capacitance per unit length and is calculated as,

$$C_{\text{pul}} = \sqrt{\varepsilon_e} \Big/ c_0 Z_0 \tag{3.7}$$

where:

$c_0 = 3 \times 10^8 \text{m/s}$ is the velocity of light in free space
ε_e is the effective permittivity of the medium
Z_0 is the characteristic impedance of the line

The effective permittivity ε_e [54] can be calculated as,

$$\varepsilon_e = 1 + \frac{\varepsilon_r - 1}{2} \frac{K(k')K(k_1)}{K(k)K(k_1')} \tag{3.8}$$

where,

$$k = \frac{c/2}{c/2 + d} \tag{3.9}$$

$$k_1 = \frac{\sinh(\pi a / 2h)}{\sinh(\pi b / 2h)} \tag{3.10}$$

$$a = \frac{c}{2} \tag{3.11}$$

and,

$$b = \frac{c}{2} + d \tag{3.12}$$

$$k' = \sqrt{1 - k^2} \tag{3.13}$$

$K(k)$ is a complete elliptic function of the first kind and $K(k')$ is its complementary function. An approximate expression for $K(k)/K(k')$ is given as,

$$\frac{K(k)}{K(k')} = \left[\frac{1}{\pi}\ln\left(2\frac{1+\sqrt{k'}}{1-\sqrt{k'}}\right)\right]^{-1} \quad \text{for } 0 \le k \le 0.7 \tag{3.14}$$

$$\frac{K(k)}{K(k')} = \frac{1}{\pi}\ln\left(2\frac{1+\sqrt{k}}{1-\sqrt{k}}\right) \quad \text{for } 0.7 \le k \le 1 \tag{3.15}$$

The characteristic impedance Z_0 is given as,

$$Z_0 = \frac{120\pi}{\sqrt{\varepsilon_e}} \frac{K(k)}{K(k')} \tag{3.16}$$

Substituting the values of C_0 and C_g in Equation 3.3 we get,

$$C_{eq} = \frac{(\pi r_{\text{avg}} - g)C_{\text{pul}}}{2} + \frac{\varepsilon_0 ct}{2g} \tag{3.17}$$

Hence, the resonant frequency is computed as,

$$f_0 = \frac{1}{2\pi\sqrt{L_T C_{eq}}} = \frac{1}{2\pi\sqrt{L_T\left[\frac{(\pi r_{\text{avg}} - g)C_{\text{pul}}}{2} + \frac{\varepsilon_0 ct}{2g}\right]}} \tag{3.18}$$

A simplified formulation for the evaluation for the total equivalent inductance L_T for a wire of rectangular cross section having finite length l and thickness c [55] is proposed as,

$$L_T = 0.0002l(2.303\log_{10}\frac{4l}{c} - \gamma)\,\mu \cdot H \tag{3.19}$$

where the constant $\gamma = 2.451$ for a wire loop of circular geometry. The length l and thickness c are in mm. The evaluation of the wire length l is straightforward as,

$$l = 2\pi r_{\text{ext}} - g \tag{3.20}$$

For close proximity wires at high frequencies, the current is confined to the wire surfaces and effectively reduces the spacing between them [56]. The finite length l is calculated considering a single loop with r_{ext} as the radius.

3.3.4.3 Square SRR

The notch frequency f_0 which corresponds to the square SRR's resonance frequency $f_{\text{S-SRR}}$ is given by,

$$f_0 = f_{\text{S-SRR}} = \frac{\omega_0}{2\pi} = \frac{1}{2\pi}\sqrt{\frac{1}{L_T C_{eq}}} \tag{3.21}$$

where L_T is the total inductance of the SRR structure and C_{eq} is the total equivalent capacitance of the structure. This total equivalent capacitance C_{eq} can be evaluated as done in Equation 3.3 of the previous section. The distributed capacitances $C_1 = C_2 = C_0$ are also a function of the split-gap dimensions g and the average ring dimension a_{avg} and is given as,

$$C_0 = \left(4a_{\text{avg}} - g\right)C_{\text{pul}} \tag{3.22}$$

where,

$$a_{\text{avg}} = a_{\text{ext}} - c - \frac{d}{2} \tag{3.23}$$

and C_{pul} is the capacitance per unit length and as in Equation 3.7.

The resonant frequency of an S-SRR yielding the notch at its resonance can be expressed as,

$$f_{S-SRR} = \frac{1}{2\pi\sqrt{L_T C_{eq}}} = \frac{1}{2\pi\sqrt{L_T\left[\left(2a_{\text{avg}} - \frac{g}{2}\right)C_{\text{pul}} + \frac{\varepsilon_0 ct}{2g}\right]}} \tag{3.24}$$

The total equivalent inductance L_T is proposed as by Terman [55] as in Equation 3.19, where the constant $\gamma = 2.853$ for a square loop. The evaluation of the wire length l is given as,

$$l = 8a_{\text{ext}} - g \tag{3.25}$$

where length l and thickness c are in mm.

An example of an S-SRR resonance frequency calculation is given in Appendix A.

3.3.4.4 Hexagonal SSR

The notch frequency which corresponds to the hexagonal SRR's resonance frequency $f_{0\text{H-SRR}}$ is given by,

$$f_{0\text{H-SRR}} = \frac{1}{2\pi\sqrt{L_T C_{eq}}} = \frac{1}{2\pi\sqrt{L_T\left[\frac{(3a_{\text{avg}} - g)C_{\text{pul}}}{2} + \frac{\varepsilon_0 ct}{2g}\right]}} \tag{3.26}$$

3.3.4.5 Rotational Circular SRR

For a conventional SRR, the split gaps g_1 and g_2 lie on the same axis. Hence, C_1 and C_2, the capacitances of the upper and lower half of the rings, are equal. For rotational SRR, where the inner ring is rotated by an angle θ, the values of C_1 and C_2 become unequal. For an angle of rotation θ of the inner ring, the capacitance of the portion within OA and OB of the SRR as shown in Figure 3.18 is given as,

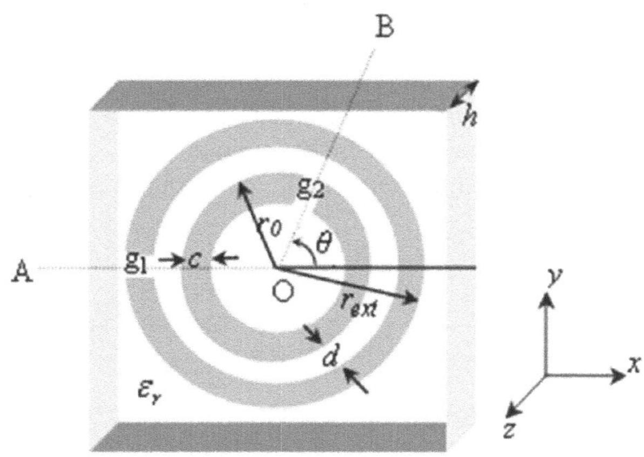

FIGURE 3.18
Schematic view of a circular split-ring resonator with rotated inner ring formed with metallic rings of width, c having radii, r_0 and r_{ext}.

$$C_1 = (\pi - \theta) r_{\text{avg}} C_{\text{pul}} \tag{3.27}$$

where C_{pul} is the per unit length capacitance between the inner and outer rings of the SRR and the uniform average dimension of the inner and outer rings of the SRR r_{avg} is calculated as

$$r_{\text{avg}} = r_{\text{ext}} - c - \frac{d}{2} \tag{3.28}$$

The capacitance of the remaining portion of the SRR is given by

$$C_2 = (\pi + \theta) r_{\text{avg}} C_{\text{pul}} \tag{3.29}$$

The gap capacitance C_g and per unit length capacitance C_{pul} between the inner and outer rings of the SRR is calculated as per the conventional circular SRR.

The equivalent capacitance is given as

$$C_{eq} = \frac{\left(C_1 + C_{g1}\right)\left(C_2 + C_{g2}\right)}{\left(C_1 + C_{g1}\right) + \left(C_2 + C_{g2}\right)} \tag{3.30}$$

Substituting C_1 and C_2 from Equations 3.27 and 3.29 in Equation 3.30 we get,

$$C_{eq} = \frac{[(\pi - \theta) r_{\text{avg}} C_{\text{pul}} + C_g][(\pi + \theta) r_{\text{avg}} C_{\text{pul}} + C_g]}{[(\pi - \theta) r_{\text{avg}} C_{\text{pul}} + C_g] + [(\pi + \theta) r_{\text{avg}} C_{\text{pul}} + C_g]} \tag{3.31}$$

$$C_{eq} = \frac{\left(\pi + q\right)^2 - \theta^2}{2\left(\pi + q\right)} r_{\text{avg}} C_{\text{pul}} \tag{3.32}$$

where,

$$q = \frac{C_g}{r_{\text{avg}} C_{\text{pul}}}. \tag{3.33}$$

As the angle of rotation increases and the angular separation between the split gap decreases, the area of the portion of rings between OA and OB decreases and the area of the remaining portion increases. This leads to decrease in C_1 and increase in C_2 which results in decrease of the equivalent capacitance C_{eq} of the structure, hence resulting in increase of the resonance frequency of the SRR structure.

So, the resonance frequency of the rotated SRR can be obtained as,

$$f_0' = \frac{1}{2\pi \sqrt{L_T \dfrac{\left(\pi + q\right)^2 - \theta^2}{2\left(\pi + q\right)} r_{\text{avg}} C_{\text{pul}}}} \tag{3.34}$$

For conventional SRR, resonance frequency is given by,

$$f_0 = \frac{1}{2\pi \sqrt{L_T \left[\dfrac{\pi r_{\text{avg}} C_{\text{pul}}}{2} + \dfrac{\varepsilon_0 c h}{2g}\right]}} \tag{3.35}$$

$$f_0 = \frac{1}{2\pi\sqrt{L_T\left(\frac{\pi}{2} + \frac{C_g}{2r_{\text{avg}}C_{\text{pul}}}\right)r_{\text{avg}}C_{\text{pul}}}} \quad (3.36)$$

Simplifying Equation 3.36 we get,

$$f_0 = \frac{1}{2\pi\sqrt{L_T\left(\frac{\pi + q}{2}\right)r_{\text{avg}}C_{\text{pul}}}} \quad (3.37)$$

A simplified formulation for the evaluation for the total equivalent inductance L_T is already derived in the previous section for a circular SRR.

From Equations 3.34 and 3.37 we get,

$$\frac{f_0'}{f_0} = \frac{\pi + q}{\sqrt{(\pi + q)^2 - \theta^2}} \quad (3.38)$$

If the gap dimension ($g < 0.2$ mm) is very small, we may neglect q. So, we get,

$$\frac{f_0'}{f_0} \approx \frac{\pi}{\sqrt{\pi^2 - \theta^2}} \quad (3.39)$$

where θ and π are in radian. Equation (3.39) is applicable for the rotation of both inner and outer rings.

3.3.4.6 Excitations of SRR for Notched Applications

The most convenient planar transmission lines that can be employed for synthesizing SRR modules are microstrip lines (ML) and coplanar waveguide (CPW). For microstrip-line-based characterization, the SRRs are to be placed laterally on the plane of the microstrip line. For linear or planar array, the SRRs are to be placed periodically along the microstrip line as shown in Figure 3.19. The associated fringing electric and magnetic field of the ML get coupled with the SRRs which excites them electromagnetically. This excitation becomes maximum at resonance corresponding to a dip in transmission.

In case of CPW-based synthesis, the SRRs are placed on the back side of the metallic strips (ground planes and central strip) of the CPW. The coplanar waveguide is designed with a 50 Ω impedance matching and a pair of SRRs is placed symmetrically on the other side in such a way that the centers of the SRRs coincide with the center of the two gaps of the CPW lines. Here the dominant factor exciting the SRRs is the coupling of the magnetic field associated with the CPW to the SRRs placed on the opposite side. Because of the positional advantage of the SRR in this method, the magnetic coupling is much stronger with CPW configuration compared to ML configuration. So, in this chapter and for subsequent work of the SRR loaded in planar transmission line, the CPW configuration is chosen. Another advantage of a planar transmission line, loaded with SRRs, is the need not to use any additional substrate for the SRR laminate. This is because the substrate for the CPW acts as the substrate of the SSRs as well. Figure 3.20 depicts the schematic diagram of a

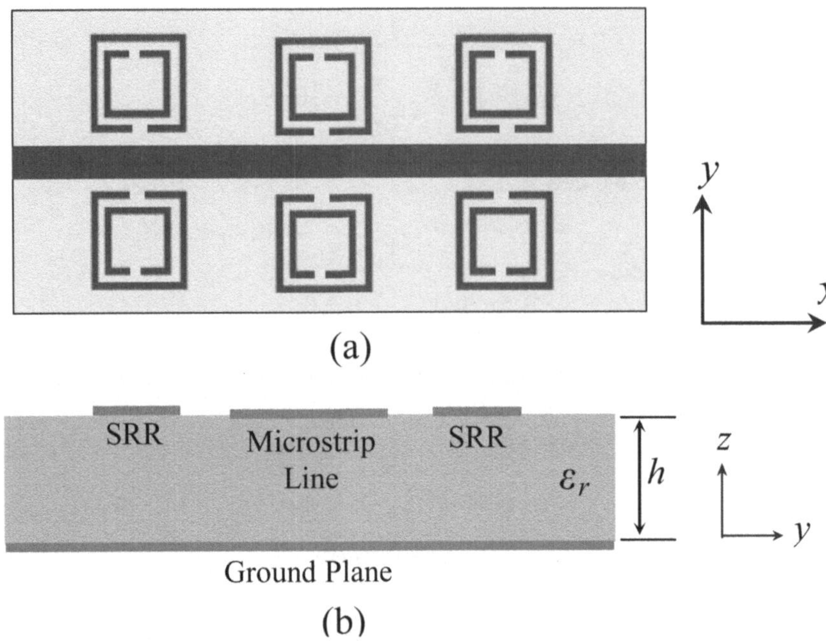

FIGURE 3.19
Schematic diagram of a 50 Ω microstrip line loaded with square split-ring resonator: (a) top view with the SRRs loaded on the same plane as the microstrip line, (b) cross-sectional view.

SRR loaded CPW printed on a dielectric substrate of height *h* and dielectric constant ε_r. The dark rectangular strips indicate copper lines (two ground planes and one central strip). The SRRs are on the other side of the substrate. However, in this method of characterizing SRRs in planar medium, proper impedance matching between transmission line (here CPW) and the exciting port is a necessary requirement. Since the SRRs are loaded symmetrically on the back side of the impedance matched CPW, we need to design the CPW in way such that it provides:

- the required 50 Ω impedance
- sufficient spacing between the two pairs of SRRs to be placed laterally along the CPW length.

The second condition arises since the centers of the SRRs are fixed (it is below the center of the slots of the CPW) and if sufficient spacing is not provided, the two pairs of SRRs cannot be accommodated.

Figure 3.20b shows the cross-sectional view of a coplanar waveguide. Here, *s* is the width of the central conductor of the CPW, *w* is the width of the slots between the central conductor and the two ground planes on either side of the central strip all having a thickness of *t*. They are printed on a dielectric substrate of relative permittivity ε_r and height *h*. The SRRs are loaded in the bottom surface of the designed CPW and are shown in Figure 3.20b. Figure 3.21 shows the fabricated prototype of a pair of SRRs printed on a substrate housing a CPW. The simulated and measured *S*-parameter responses are shown in Figure 3.22.

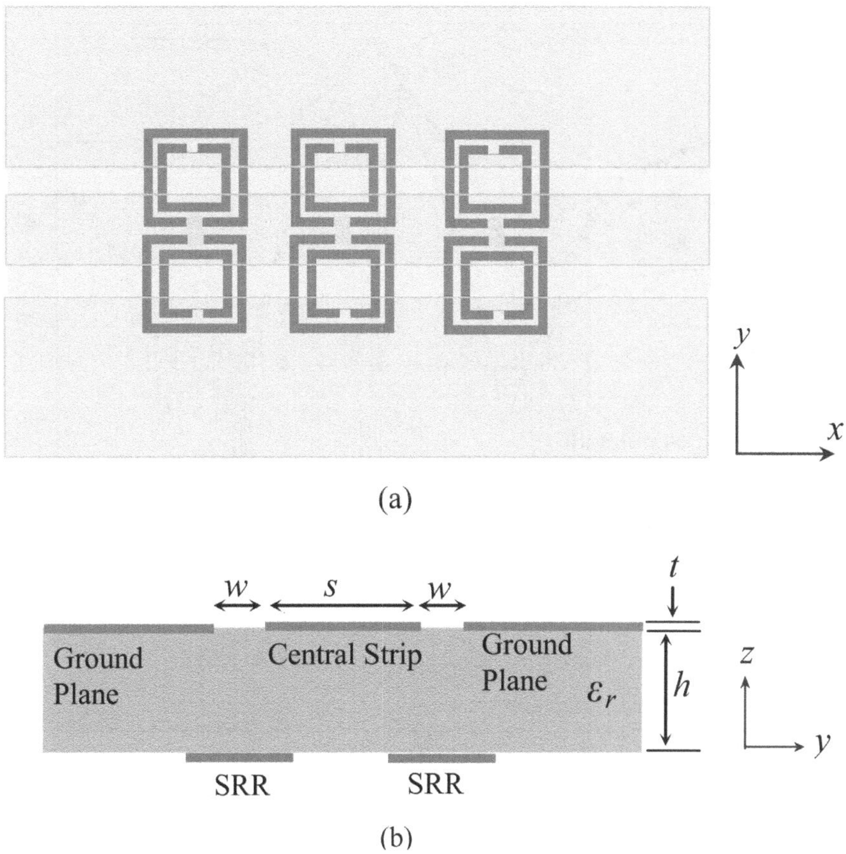

FIGURE 3.20
Schematic diagram of a coplanar waveguide loaded with square split-ring resonator on the back surface of the CPW: (a) top view, (b) cross-sectional view.

FIGURE 3.21
Fabricated prototype of a pair of S-SRR loaded coplanar waveguide: (a) top, (b) bottom.

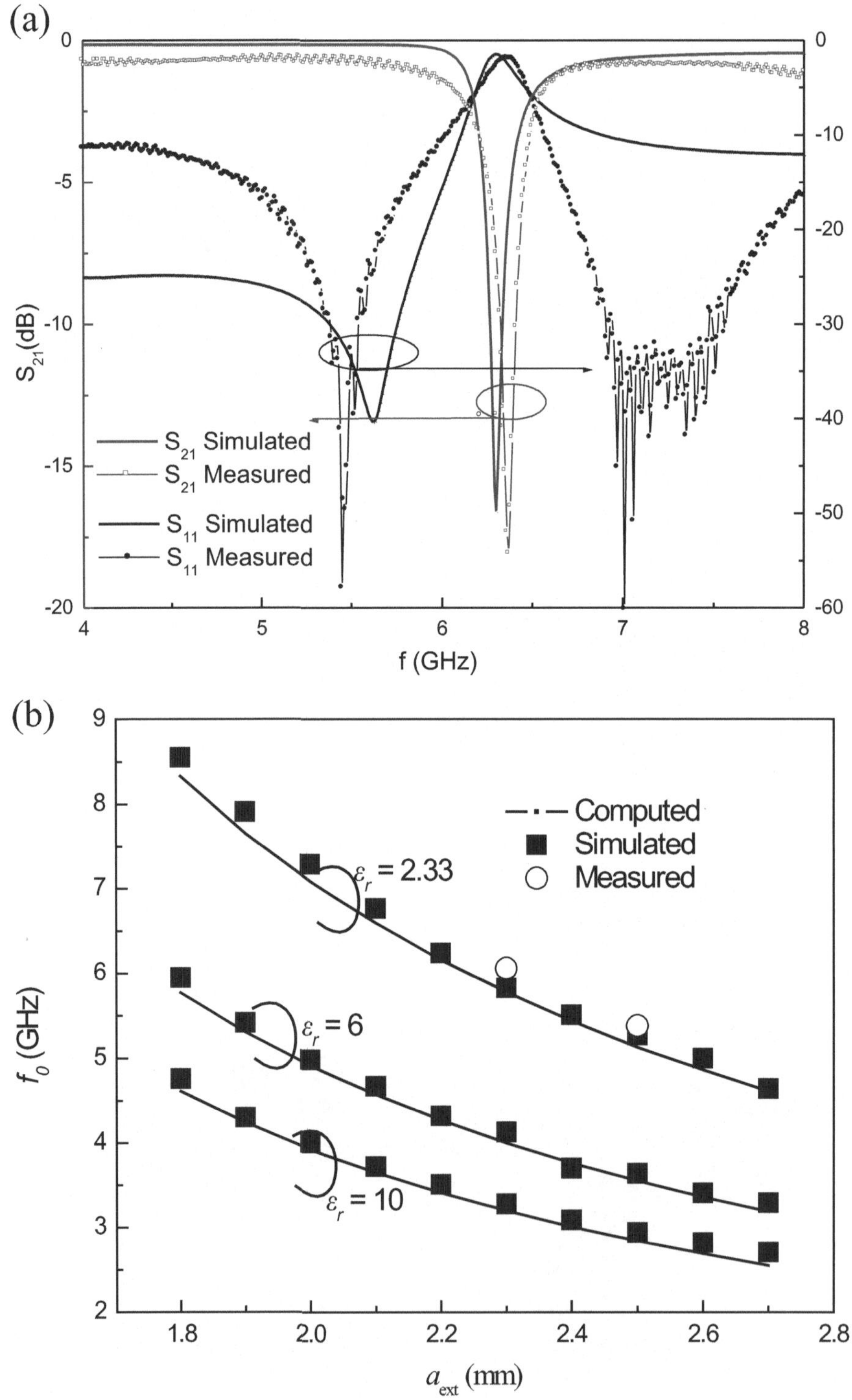

FIGURE 3.22
(a) Measured and simulated reflection and transmission coefficients of the S-SRR loaded CPW, (b) computed and simulated resonant frequency of square SRR as a function of a_{ext}. Parameters: $\varepsilon_r = 2.33$, $a_{ext} = 2.5$ mm, $c = 0.5$ mm, $g = 0.2$ mm, $h = 1.575$ mm, $t = 35$ μm, $d = 0.6$ mm.

3.4 Frequency-Notched UWB Monopole Antennas

3.4.1 Single Pair SRR Coupled CPW-Fed Antenna

The SRR loaded CPW is fed to excite a circular monopole antenna as shown in Figure 3.23 [57]. The SRRs are placed symmetrically on the back side of the printed monopole antenna which results in a notch frequency determined by the SRR's geometrical dimensions. The suppression of the radiation at the notch frequency is due to the effect of a strong magnetic coupling of the propagating EM signal with the SRR. This coupling between the SRR and the propagating EM signal can be used to filter out undesired frequencies and avoid possible interference within the UWB (3.1 GHz to 10.6 GHz). By loading multiple SRR pairs with varying dimensions, multiple resonances can also be achieved. Figure 3.23a,b shows the schematic of the proposed antenna. The circular monopole of radius R is fed by a CPW consisting of ground planes having widths W_1 and W_2, length Ls, and a signal line having width S and length $Ls+t$. The slots between the ground planes and signal line have width S_g. The antenna is printed on a substrate having thickness h and dielectric constant ε_r. Two square-shaped split-ring resonators have dimension a_{ext} which is half the dimension of the side-length of the SRR, conductor thickness c, separation between rings d and split gaps g_1 and g_2 as shown in Figure 3.23c, are printed on the other side of the substrate with their centers coinciding with the slot lines of the CPW feed. Unlike in most of the previously presented designs described earlier, where most of the inclusions and slots were arranged on the radiating patch itself or on the ground planes, the novelty in this design is that it can be employed on any CPW-fed planar monopole UWB antenna without tampering or changing the shape of the radiator or the ground plane.

The fabricated prototypes of the SRR loaded circular monopole antenna are shown in Figure 3.23d. The circular monopole, having a diameter of 25 mm, is fed with a coplanar waveguide with ground plane length $L_s = 22.5$ mm, width $W = 50$ mm, and feed gap dimension $t = 0.2$ mm. The prototype was fabricated on a Taconic substrate of thickness $h = 1.575$ mm, dielectric constant $\varepsilon_r = 2.33$, and a loss tangent $tan\delta = 0.0009$. The pair of SRRs are printed on the back side of the CPW with dimension $a_{ext} = 2.5$ mm. The propagating EM signal along the CPW with its magnetic field oriented along the axis of the SRR induces an electromotive force on the SRR which in turn induces currents oscillating between the two rings of the SSR. These oscillating currents between the two rings yield a resonance which is determined by the SRR's geometry and prohibit signal propagation at that frequency.

Figure 3.24 shows the measured and simulated magnitude of the reflection coefficient S_{11} of the SRR loaded circular monopole antenna. These SRRs are placed symmetrically on the back side of the CPW-fed planar monopole antenna which contributes to a notch in the reflection coefficient at around 6.38 GHz for simulated and 6.39 GHz for measured results as shown in the figure. This notch frequency corresponds to the resonance frequency of the printed SRR. The resonance frequency of the SRRs is determined and controlled by the geometrical dimension of the SRR and the material constitutive parameters of the host substrate.

The measured and simulated radiation patterns in the x-y-plane and x-z-plane for the CPW-fed planar monopole with SRR loading are presented in Figure 3.25 for 3.1 GHz, 6 GHz and 10 GHz. The radiation patterns show good directive patterns for the x-y-plane and omnidirectional pattern for the x-z-plane. The correspondence between the simulated and measured patterns in the x-y-plane from 0° to 180° is quite good.

The measured and simulated peak gain values versus frequency of the SRR loaded prototype are illustrated in Figure 3.26. The plot shows a reduction in gain at the notch frequency of operation whereas the gain for the rest of the band remains acceptable. The

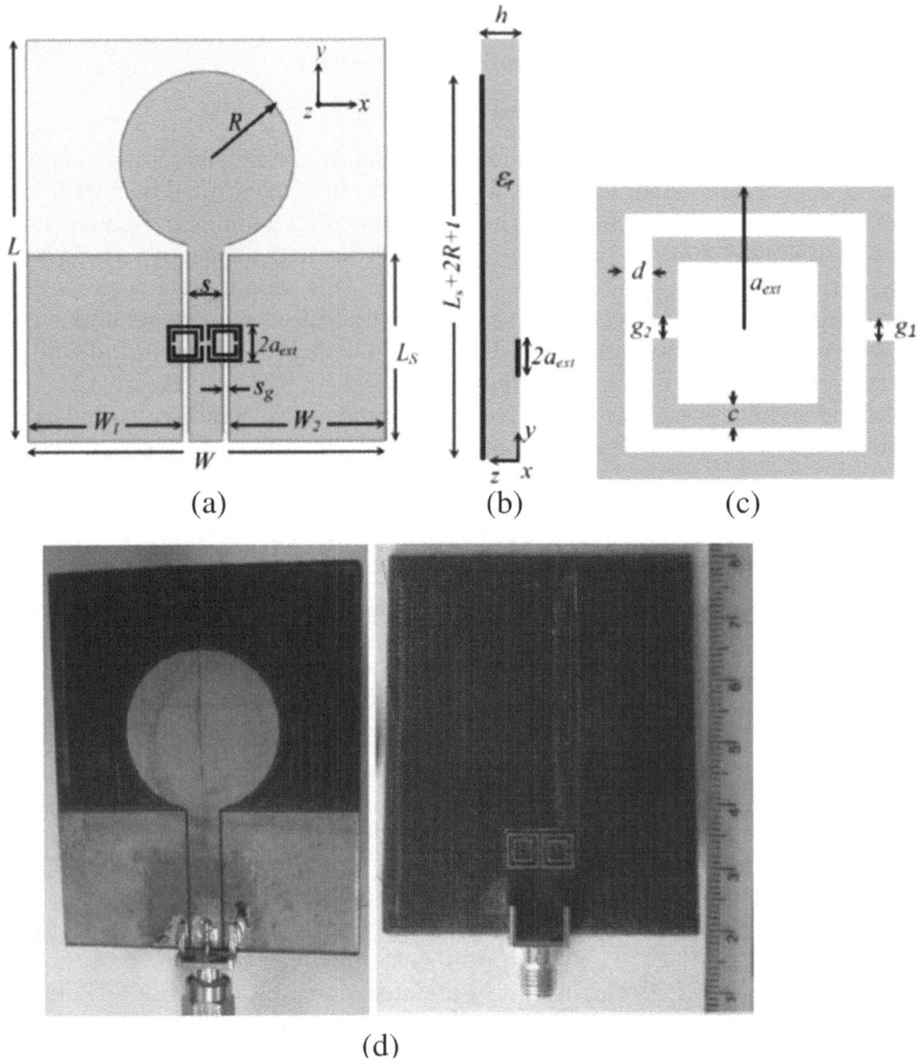

FIGURE 3.23
Schematic of a CPW-fed printed circular monopole antenna loaded with square SRR: (a) top view with a pair of SRRs (in darker shade) loaded on the back side, (b) side view showing the printed SRRs separated from the CPW-fed printed circular monopole, (c) enlarged view of the square SRR printed on the back side of the CPW-fed monopole antenna, (d) fabricated prototype.

Reproduced by permission © 2014 The IEEE [57].

measured results yield a gain of –7 dBi at 6.39 GHz, whereas the simulated result yields a reduced gain of –8 dBi at the notch frequency of 6.38 GHz.

Figure 3.27 shows the simulated S_{11} characteristics for two different a_{ext} values of 2.2 mm and 2.6 mm yielding notch frequencies of 6.35 GHz and 4.8 GHz, respectively. With the increase in a_{ext} values, the total inductance L_T and the capacitance of the structure increase which in turn decreases the resonance frequency of the structure. Varying a_{ext} can be used as a variation parameter for wide range tuning of the notch frequency.

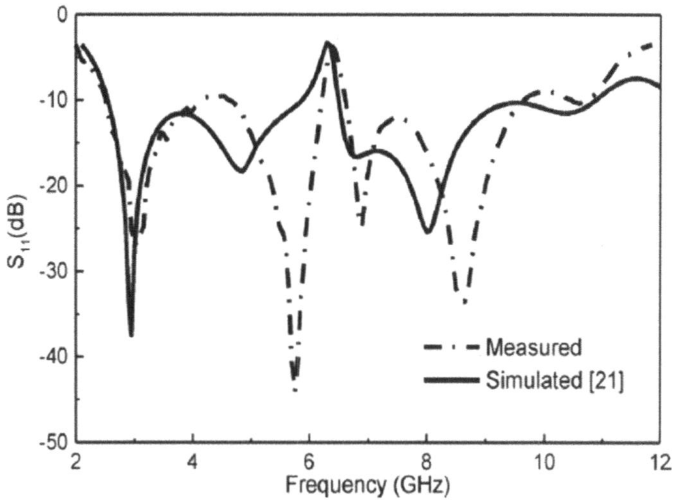

FIGURE 3.24
Simulated and measured S_{11} characteristics of the CPW-fed printed monopole antenna with SRR loading.

Reproduced by permission © 2014 The IEEE [57].

The simulated surface current distribution at three different frequencies, 6.38 GHz, 3.1 GHz and 10 GHz, of operation is shown in Figure 3.28. It is clearly visible that at the frequency of interest, i.e. at 6.38 GHz, corresponding to the resonance frequency of the SRR, the circular monopole is not excited resulting in suppression of radiation at that frequency.

3.4.2 Dual Pair SRR Coupled CPW-Fed Antenna

The design concept of frequency-notched UWB antenna using SRR loading can be excellently extended for multi-notch applications. For multiple notches, more than one pair of SRRs of varying dimensions may be placed along the transmission line, provided the length of the transmission line is long enough to accommodate them. Such a configuration is shown in Siddiqui et al. [58] where two pairs of square-shaped SRRs are printed on the back side of the CPW separated by the substrate thickness h as shown in Figure 3.29.

The dimensions of the pair of SRRs printed on the back side of the CPW are tailored to yield dual frequency notch operation as well as wideband notch operation [58].

Two separate prototypes for the dual frequency rejecting DF-DSRR and wideband rejecting WB-DSRR were fabricated on Taconic substrates with $\varepsilon_r = 2.33$, $tan\delta = 0.0009$, and thickness $h = 1.575$ mm. The circular monopole with radius $R = 12.5$ mm and fed with a CPW having ground plane length $Ls = 22.5$ mm, width $W = 50$ mm, and feed gap $t = 0.2$ mm was printed on one side of the substrate. The signal line S of the CPW was designed at 5 mm with slot gap dimension $S_g = 0.25$ mm to maintain an impedance close to 50 Ω. Figure 3.30 compares the measured and simulated VSWR of the dual frequency rejecting DF-DSRR antenna. Due to the dual SRR pairs of different geometrical parameters, the configuration contributes dual notches centered at 5.33 GHz and 7.9 GHz for the simulated results and 5.34 GHz and 7.95 GHz for the measured results. The first notch at 5.33 GHz/5.34 GHz is

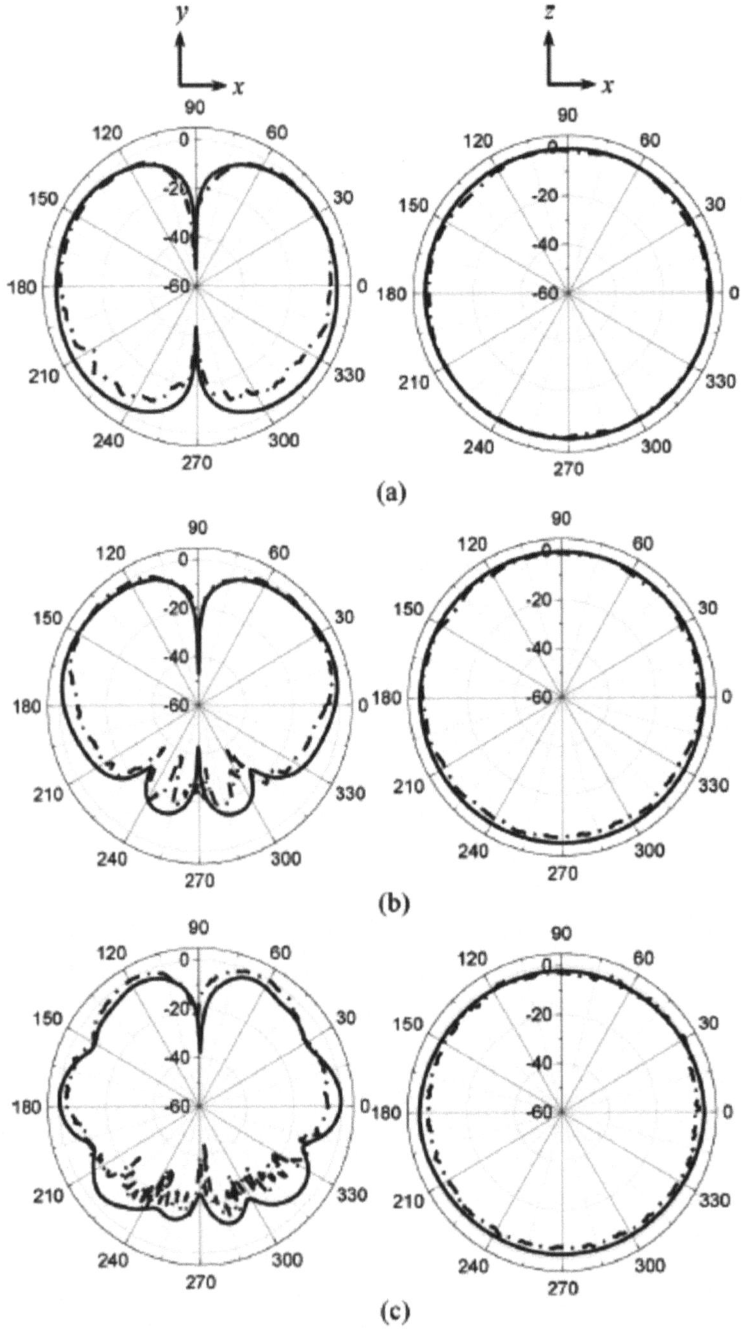

FIGURE 3.25
Simulated and measured E- and H-plane radiation patterns of the CPW-fed printed monopole antenna with SRR loading.

Reproduced by permission © 2014 The IEEE [57].

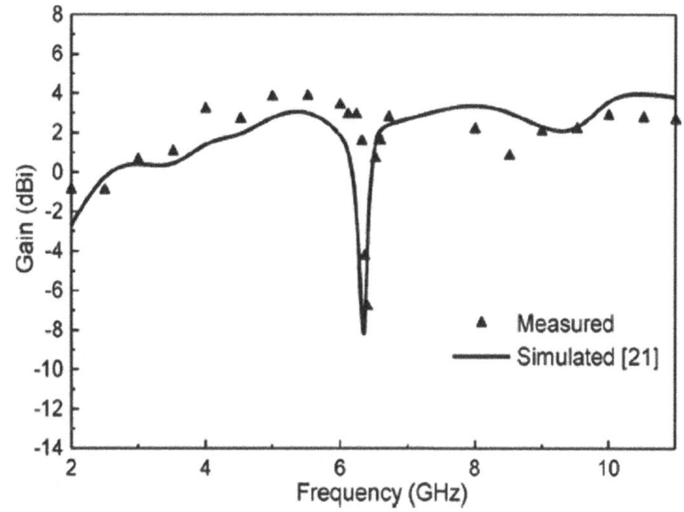

FIGURE 3.26
Simulated and measured peak gain of the CPW-fed printed monopole antenna with SRR loading.

Reproduced by permission © 2014 The IEEE [57].

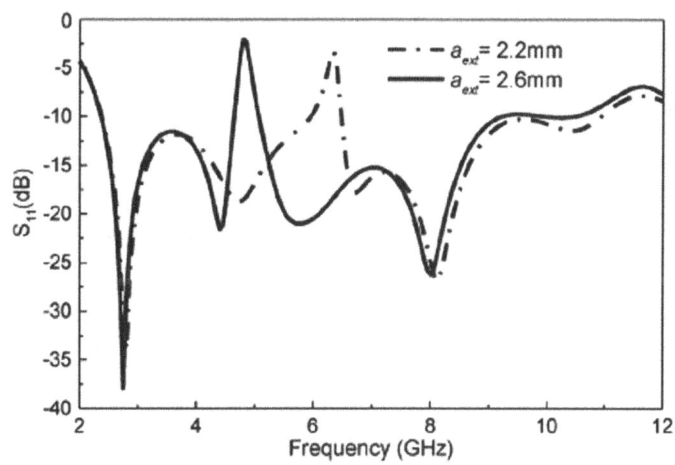

FIGURE 3.27
Simulated S_{11} of the CPW-fed printed monopole antenna with SRR loading for varying a_{ext}.

Reproduced by permission © 2014 The IEEE [57].

due to the resonance of the SRR pair with larger dimension ($aext_1 = 2.5$ mm), whereas the second notch at 7.9 GHz/7.95 GHz is contributed by the SRR pair with smaller dimension ($aext_2 = 2.1$ mm). Since the SRR's resonance frequency is determined and controlled by the geometrical dimension of the SRR and the material constitutive parameters of the host substrate, the notch frequencies can be controlled and designed by changing the SRR's dimension.

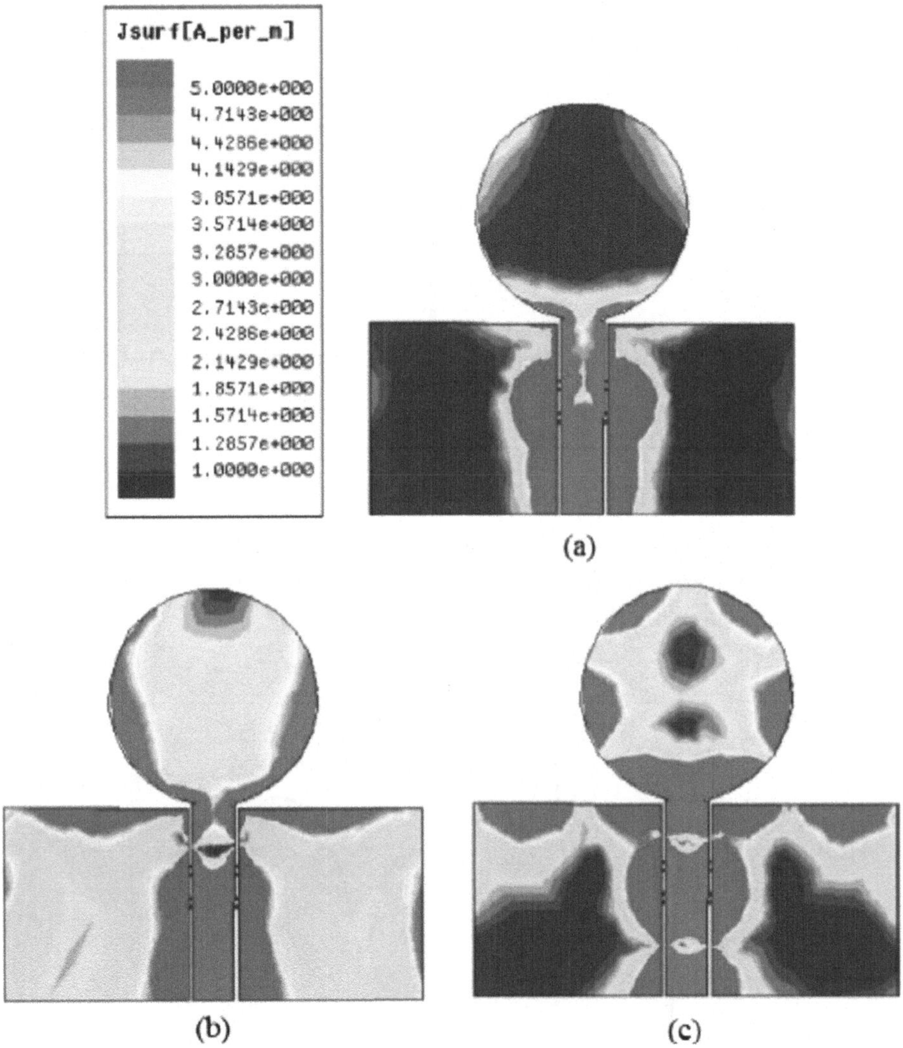

FIGURE 3.28
Simulated surface current distribution of the CPW-fed circular monopole with SRR loading: (a) 6.38 GHz, (b) 3.1 GHz, and (c) 10 GHz.

Figure 3.31 compares the measured and simulated VSWR of the wideband rejecting WB-DSRR antenna. The SRR pairs, with dimensions $a_{ext1} = 2.4$ mm and $a_{ext2} = 2.5$ mm, were tuned for proximity resonance to obtain wideband notch of 6.2–7.02 GHz for simulated results and 6.2–6.9 GHz for the measured results.

The maximum calibrated gain in the E(x-y) plane and the corresponding magnitude of the reflection coefficient as a function of frequency for dual frequency notches and wideband notch are plotted in Figures 3.32 and 3.33, respectively. As can be envisaged from the plots, the gain is suppressed at the notch frequencies where the reflection is maximum and hence this prohibits radiation.

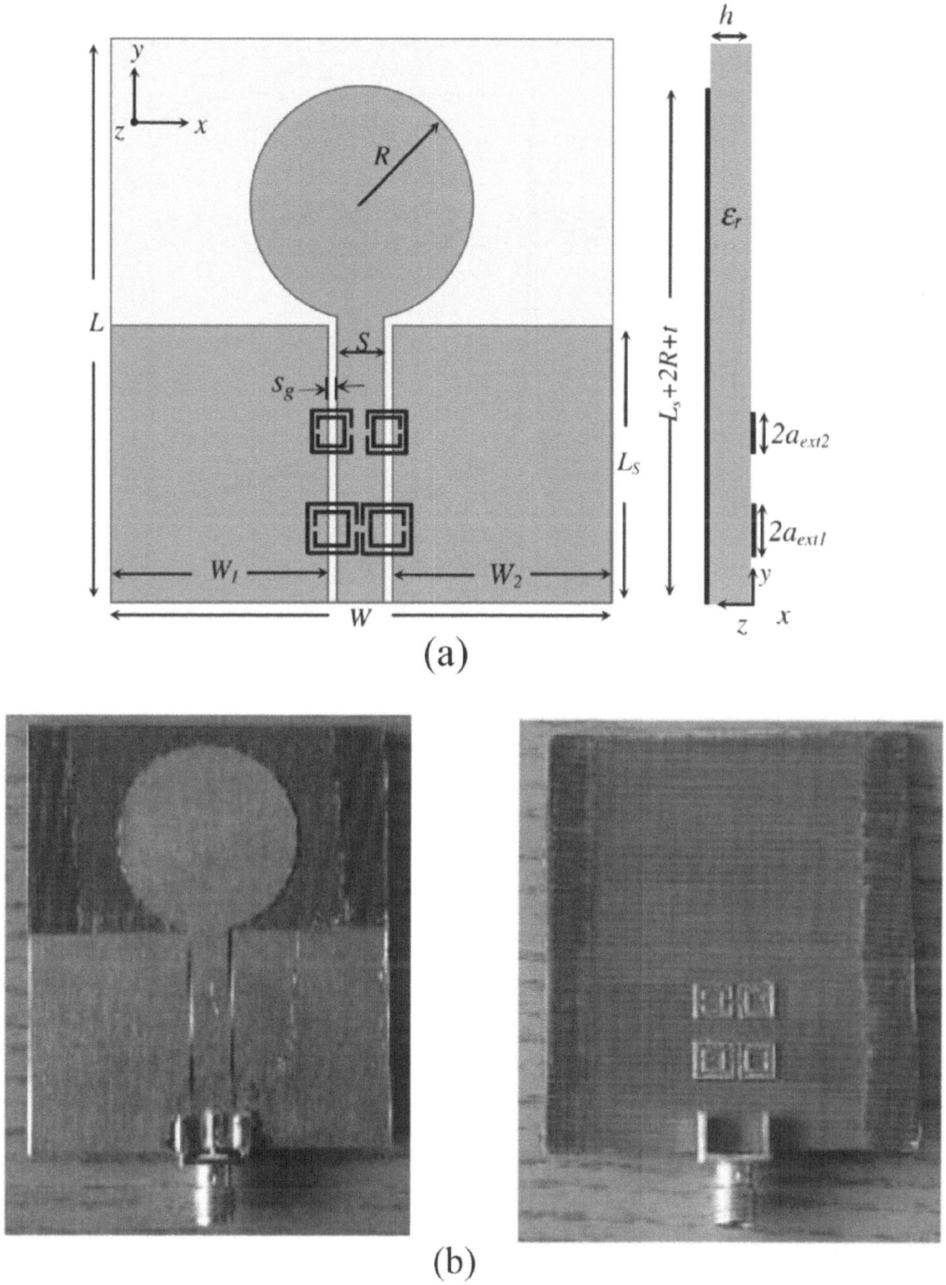

FIGURE 3.29
Schematic of a CPW-fed printed circular monopole antenna loaded with dual pair of square SRR: (a) schematic view, (b) fabricated prototype.

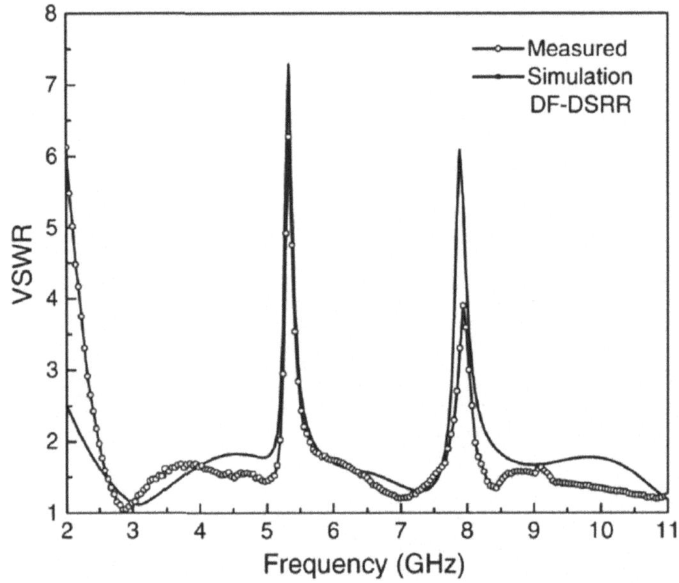

FIGURE 3.30
Measured and simulated VSWR plot of the CPW-fed printed monopole antenna with dual SRR loading for dual frequency notches.

Reproduced by permission © 2015 The IEEE [58].

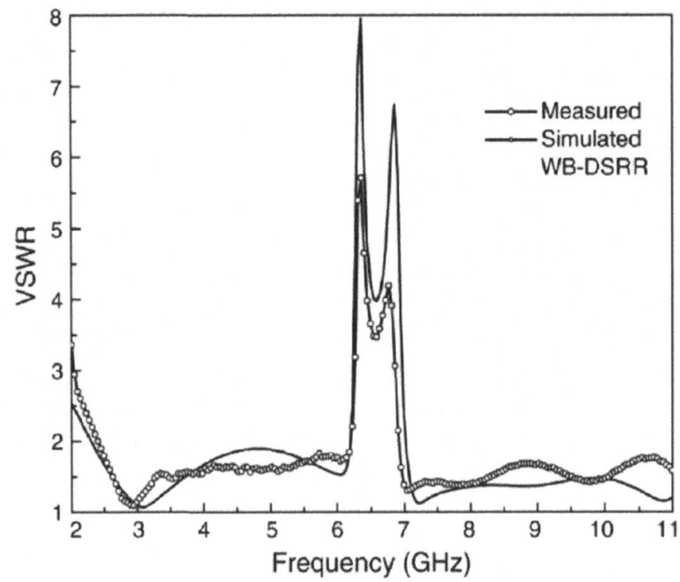

FIGURE 3.31
Measured and simulated VSWR plot of the CPW-fed printed monopole antenna with dual SRR loading for wideband notch.

Reproduced by permission © 2015The IEEE [58].

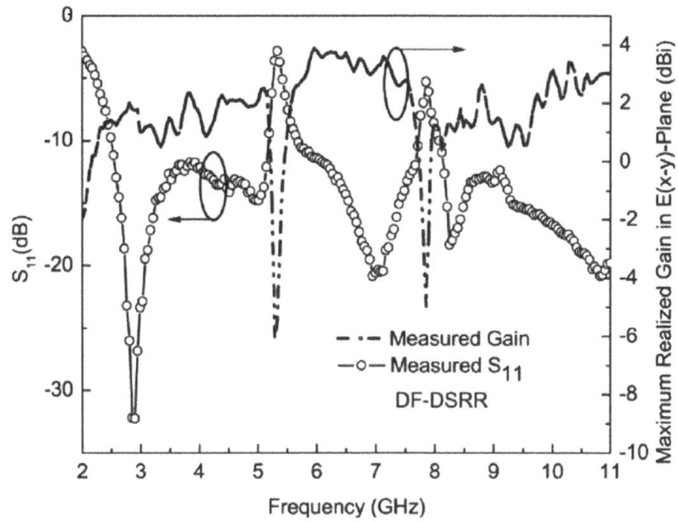

FIGURE 3.32
Measured reflection coefficient and realized gain of the CPW-fed printed monopole antenna with dual SRR loading for dual frequency notches.

Reproduced by permission © 2015 The IEEE [58].

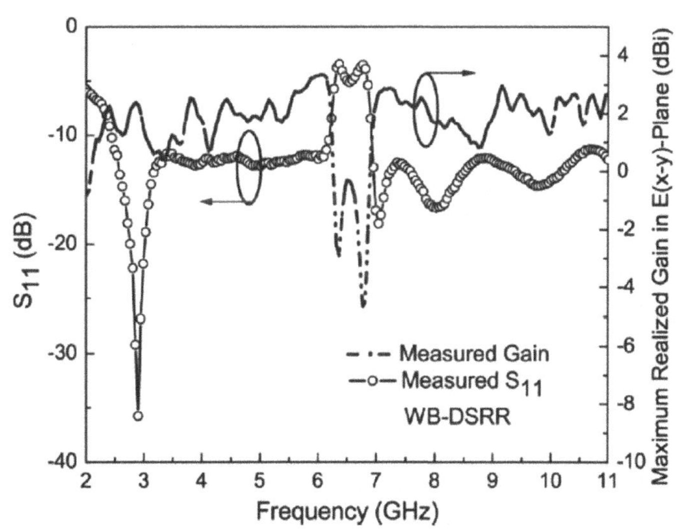

FIGURE 3.33
Measured reflection coefficient and realized gain of the CPW-fed printed monopole antenna with dual SRR loading for wideband notch.

Reproduced by permission © 2015 The IEEE [58].

3.4.3 Rotational SRR Coupled CP-fed Antenna

Tuning the resonant frequency of the SRRs by changing the SRR dimensions, which yields the notch frequencies, has been shown in the previous sections. However, the width of the signal line may limit the variation of the SRR. In Shaik et al. [59], a concept is proposed with a single printed circular monopole antenna which is loaded with multiple SRRs with different inner and outer ring angular orientations to yield multiple frequency notches in the UWB radiation.

Figure 3.34 shows the schematic diagram of the circular SRR, with external radius r_{exti}, conductor thickness c_i, separation between rings d_i, and split gap g_i. Along with these physical parameters of the SRR, one more parameter θ_i, indicating orientations of the inner ring of SRR, is chosen as an additional parameter to control the SRR's resonance, contributing to the notch frequency of the antenna. The resonance frequency f_0 of a rotational SRR being substantially greater than that of conventional SRR ($\theta = 0°$) of the same physical dimensions, angular orientation of the SRR helps in using a relatively big SRR (larger r_{ext}) to achieve the same resonance frequency. It becomes difficult to get a notch at higher frequency using a conventional SRR of reduced r_{ext} which yields a weak resonance phenomenon. This weak resonance, which is due to reduced excitation of the SRR for smaller r_{ext}, can be mitigated by optimally using rotational orientation of the SRR. For achieving triple notches, three pairs of circular SRRs with varying physical dimensions and orientation of the inner ring of the SRR are chosen, each providing a notch at their respective resonance frequency. The fabricated prototype of this configuration is shown in Figure 3.35a,b, with the expanded view of Figure 3.35b clearly indicating the gradually changing orientation of the inner rings of the SRR.

The measured and simulated plots for the S_{11} versus frequency of the proposed rotational circular SRRs loaded circular monopole antenna are shown in Figure 3.36a and compared with unloaded case. As revealed from the figure, the simple CPW-fed circular

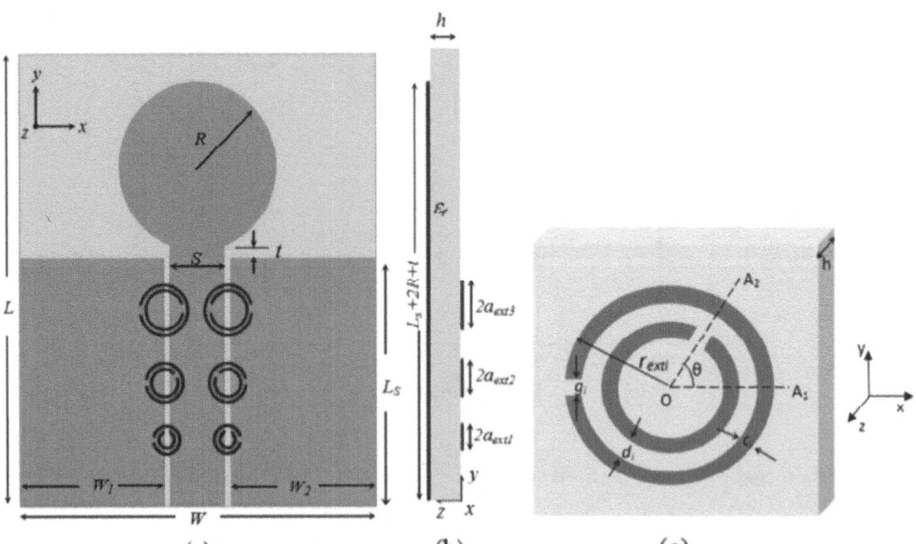

FIGURE 3.34
Schematic of a CPW-fed printed circular monopole antenna loaded with three rotational circular pairs of SRRs.

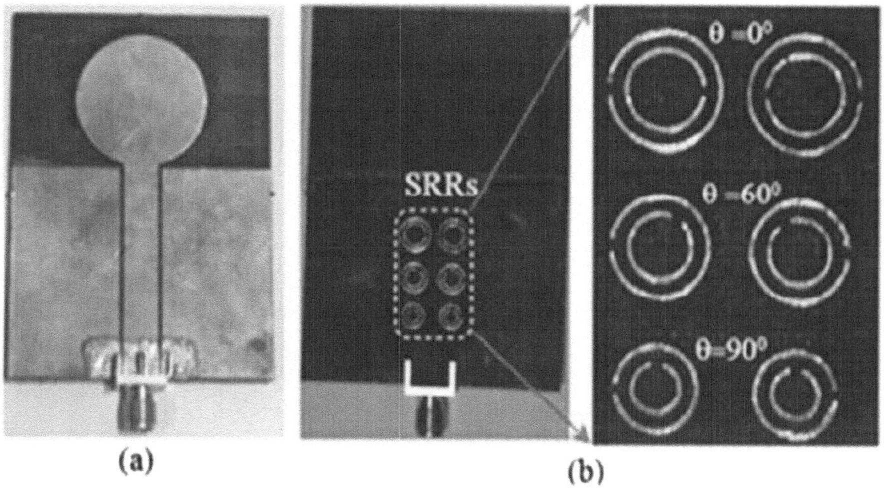

FIGURE 3.35
Fabricated prototype of a CPW-fed printed circular monopole antenna loaded with three rotational circular pairs of SRRs.

Reproduced by permission © 2016 The IET [59].

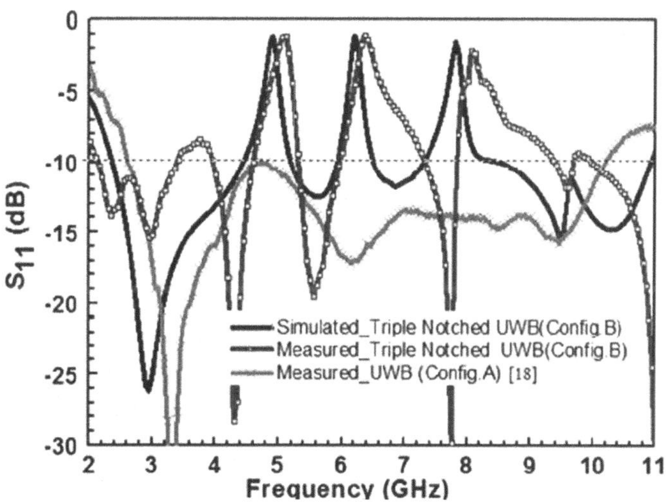

FIGURE 3.36
Measured S_{11} versus frequency of a CPW-fed printed circular monopole antenna loaded with three rotational circular pairs of SRRs.

Reproduced by permission © 2016 The IET [59].

monopole without any SRR loading operates from 2.6 GHz to 10.8 GHz with resonance dip around 3 GHz (corresponding to the quarter-wavelength resonance of the disc diameter) while the same antenna when loaded with rotational SRRs provides three distinct notches at 5.09 GHz, 6.34 GHz, and 8.04 GHz contributed by the corresponding SRR pair for measured results against 4.81 GHz, 6.19 GHz, and 7.8 GHz for simulation. The uniqueness of the present notching technique lies in its generality and versatility: in scaling for any other combination of notches by tailoring the SRR dimension without any change on the radiator and ground plane, and applying the proposed concept for any other CPW-/microstrip-fed UWB antenna.

Figure 3.37 shows the E-plane and H-planes measured normalized radiation patterns of the proposed antenna configuration at 4.2 GHz, 6 GHz, 7.2 GHz, and 9.4 GHz, exhibiting a monopole type radiation pattern with axial null for the E-plane and near omnidirectional

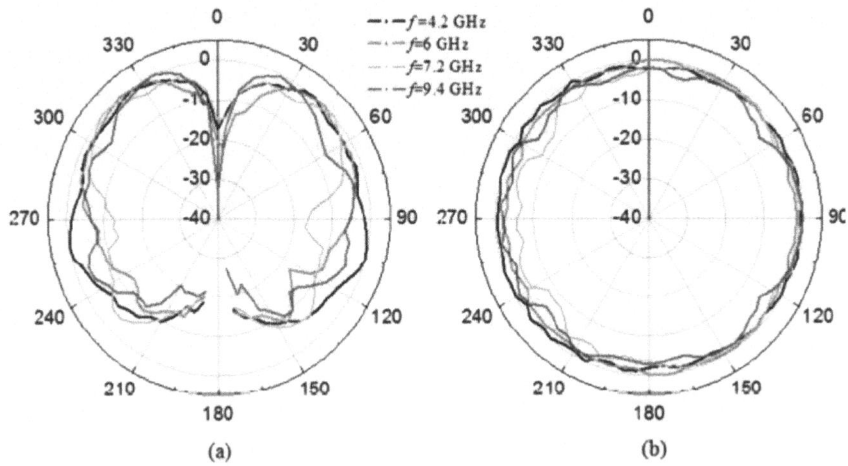

FIGURE 3.37
Measured radiation pattern of the E-plane and H-plane of a CPW-fed printed circular monopole antenna loaded with three rotational circular pairs of SRRs.

Reproduced by permission © 2016 The IET [59].

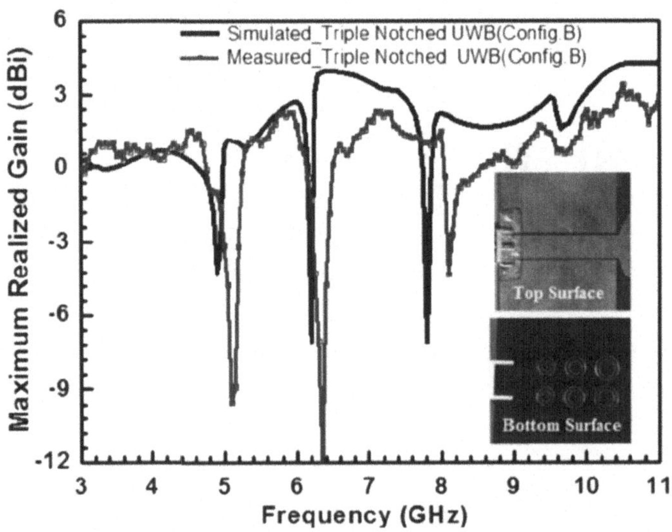

FIGURE 3.38
Measured peak gain of a CPW-fed printed circular monopole antenna loaded with three rotational circular pairs of SRRs.

Reproduced by permission © 2016 The IET [59].

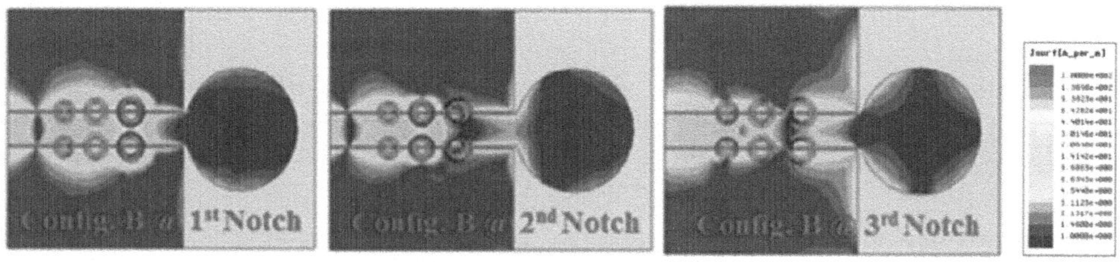

FIGURE 3.39
Simulated current distribution on a CPW-fed printed circular monopole antenna loaded with three rotational circular pairs of SRRs.

Reproduced by permission © 2016 The IET [59].

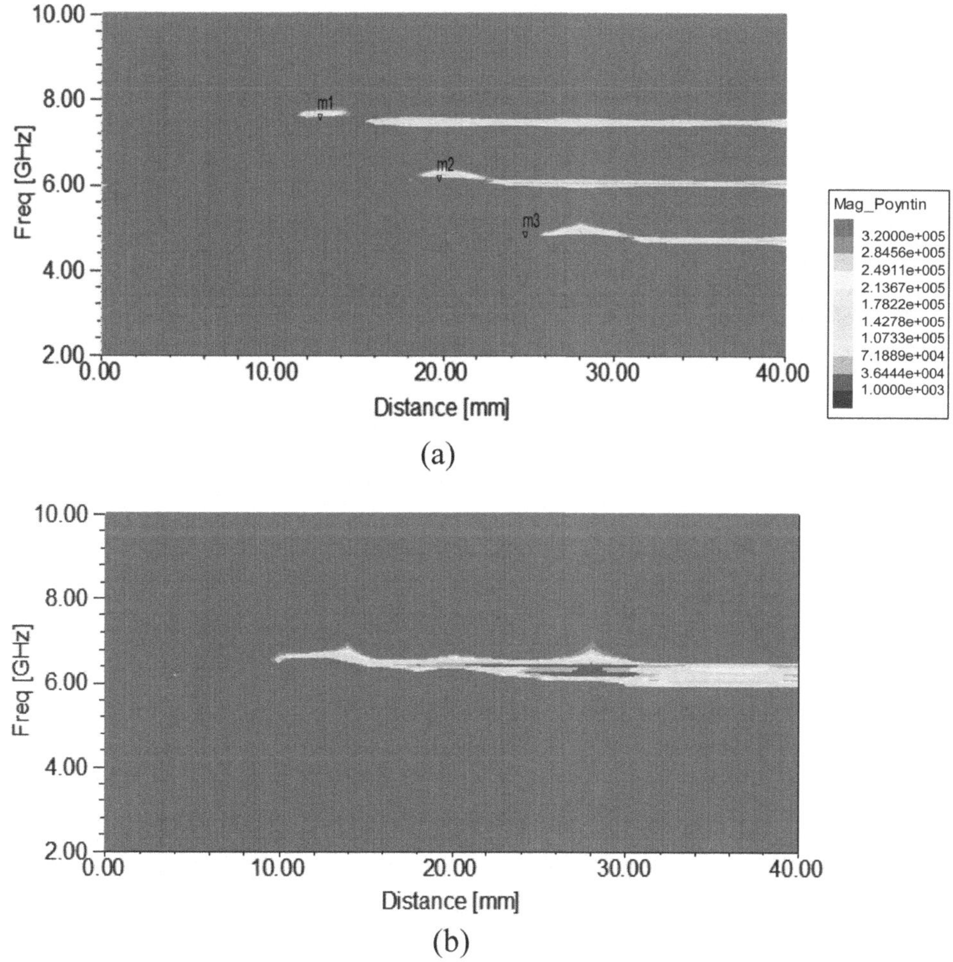

FIGURE 3.40
Simulated contour plots of Poynting vectors of the propagating electromagnetic energy through the longitudinal dimension of one of the slots as a function of frequency: (a) triple notched UWB antenna, indicating three notch frequencies corresponding to SRR's position, (b) wideband notched UWB antenna.

Reproduced by permission © 2016 The IET [59].

pattern for the H-planes. The frequencies are selected at different points over the entire UWB spectrum, excluding the notch frequencies where there is no effective radiation from the antenna. The measured and simulated maximum realized gain values versus frequency of the proposed multiple SRR loaded antenna, plotted in Figure 3.38, exhibits drastically reduced gain of the SRR loaded antenna (−9.6 dBi at 5.08 GHz, −12.7 dBi at 6.35 GHz, and −4.27 dBi at 8.1 GHz) reconfirming the presence of strong notches contributed by the SRR resonance. The maximum realized gain at other frequencies is within 0–4 dBi and is almost similar to simple CPW-fed UWB antenna.

Figure 3.39 shows the simulated surface current distribution of the antenna at three notch frequencies which clearly indicates non-excitation of the antenna due to strong magnetic resonance of the corresponding SRR pair which in turn inhibits the signal to excite the antenna.

Another unique and more insightful way to exhibit the antenna excitation/notching is to plot the magnitude of the Poynting vector along the feedline of the antenna as a function of distance from the feed port to the radiator and frequency in a rectangular contour plot as shown in Figure 3.40. Figure 3.40a clearly demonstrates existence of triple notches contributed by each SRR pair indicated by the three lighter shade (greenish in color) lines. It is interesting to note that the first notch (at 4.81 GHz) starts at a longer distance as this is contributed by the largest SRR which is farthest from the feed port of the antenna; similarly, the second notch (at 6.19 GHz) and third notch (at 7.8 GHz) start at a relatively lesser distance from the preceding notch due to their closer proximity to the feed port of the antenna.

The proposed multiple notched configuration can be converted to achieve a single wideband notched antenna by selecting the SRR size and angular orientation in such a way so that individual resonance is slightly shifted from the preceding SRR pair, resulting in a wider rejection band. The wider rejection band has been achieved for the same geometrical parameters of each SRR pair ($r_{ext}=2.8$ mm, $c=0.5$ mm, $d=0.4$ mm, and $g=0.4$ mm) and varying only in angular orientation θ = 0°, 60°, and 90° for the three SRR pairs. This also shows the generality and scalability of the proposed techniques of frequency notching. Figure 3.41 shows the simulated S_{11} and maximum realized gain of the proposed antenna

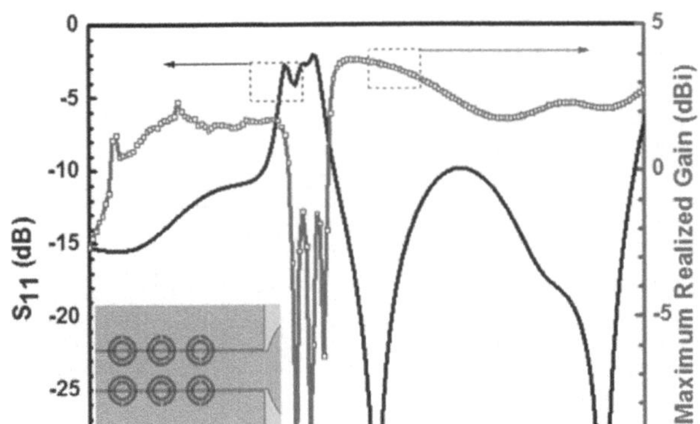

FIGURE 3.41
Simulated S_{11} and maximum realized gain (*x*-*y* plane) characteristics of the proposed CPW-fed circular monopole antenna with three pairs of SRRs printed on beneath the feedline for wideband notched UWB response.

Reproduced by permission © 2016 The IET [59].

versus frequency. The plot clearly indicates the presence of a notch band from 5.75 GHz to 6.4 GHz ($|S_{11}| < 5$ dB) providing a notch bandwidth of 625 MHz. The simulated gain of the antenna falls below –10 dB over this notch band and maintains a nearly flat profile over the other frequency band. As introduced previously, Figure 3.40b shows the simulated Poynting vector along the feedline of the antenna. This plot reveals a wider rejection band indicated by gradually increasing the width of the lighter (greenish and bluish in color) shade which is due to superposition of narrowly separated individual notches by different SRRs along the CPW line.

3.5 Frequency-Notched Horn Antennas

Exploiting the inherent narrow bandwidth of metamaterial-inspired resonators, a horn antenna with a resonant inclusion that stores/dissipates energy at a given frequency, leading to a band-notch in a narrow frequency range, was proposed by Barbuto et al. [60]. The structure, shown in Figure 3.42, consists of a WR-90 waveguide operating from 8.2–12.4 GHz, a regular pyramidal horn and the SRR etched on a substrate.

The notch frequency depends on the SRR dimensions and substrate parameters. Depending on the field distribution of the fundamental mode in the waveguide and horn, the distance between the center of the resonator and the throat of the horn influences the

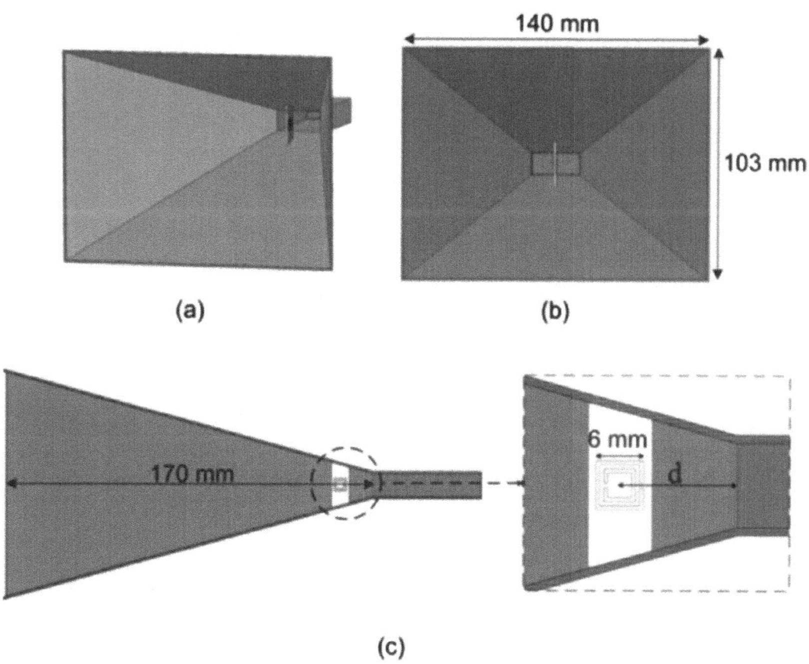

FIGURE 3.42
Geometrical sketch of the proposed horn antenna with the notched-band filter: (a) perspective view, (b) front view, (c) side view.

Reproduced by permission © 2015 The IEEE [60].

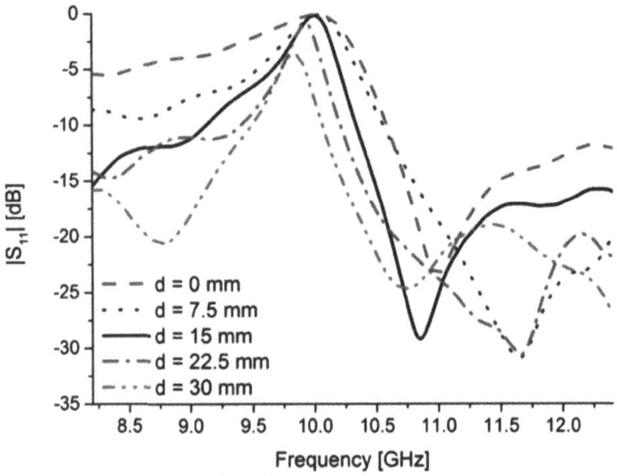

FIGURE 3.43
Reflection coefficient amplitude at the input port of the structure shown in Figure 3.36 for different positions of the SRR.

Reproduced by permission © 2015 The IEEE [60].

magnitude of the reflection/transmission coefficient. As the SRR is moved away from the throat of the horn, the reflection at resonance gradually lowers. Proper positioning of the SRR inside the horn antenna can yield the notched-band characteristic.

The SRR was printed on Rogers Duroid RT5870 dielectric substrate with a thickness of 0.787 mm and designed to operate at 10 GHz. Figure 3.43 shows the frequency variation of the reflection coefficient amplitude at the input port for different values of distance *d*. The amplitude of the peak is lower for larger values of *d*, so the optimum distance is determined from this plot [60].

Another design with two SRRs placed in the throat was also reported in order to have two almost independent resonant frequencies at 9.25 GHz and 10.75 GHz [60]. The schematic diagram is shown in Figure 3.44 and the measured and simulated S_{11} plot is shown in Figure 3.45. The realized gain as a function of frequency showing two notches is shown in Figure 3.46. The fabricated prototype of the horn antenna and the SRR loading is shown in Figure 3.47.

3.6 Frequency-Notched Tapered Slot Antennas

3.6.1 Printed Tapered Slot Antennas Loaded with SRRs

Following the design concepts reported by Siddiqui et al., Siddiqui et al., and Shaik et al. [57–59] and the works by Barbuto et al. [60], another work was proposed in Siddiqui et al. [61] where the concept was extended to a printed ultrawideband antipodal tapered slot antenna (ATSA). The proposed design techniques showed two different band notching techniques: by loading S-SRRs on the radiating aperture of the ATSA, and by placing the S-SRRs along the microstrip feedline of the ATSA. Unlike previously reported results, both the techniques yield sharp notches at the resonating frequencies of the S-SRRs which is validated experimentally in Siddiqui et al. [58]. The rejection bandwidth of the notches

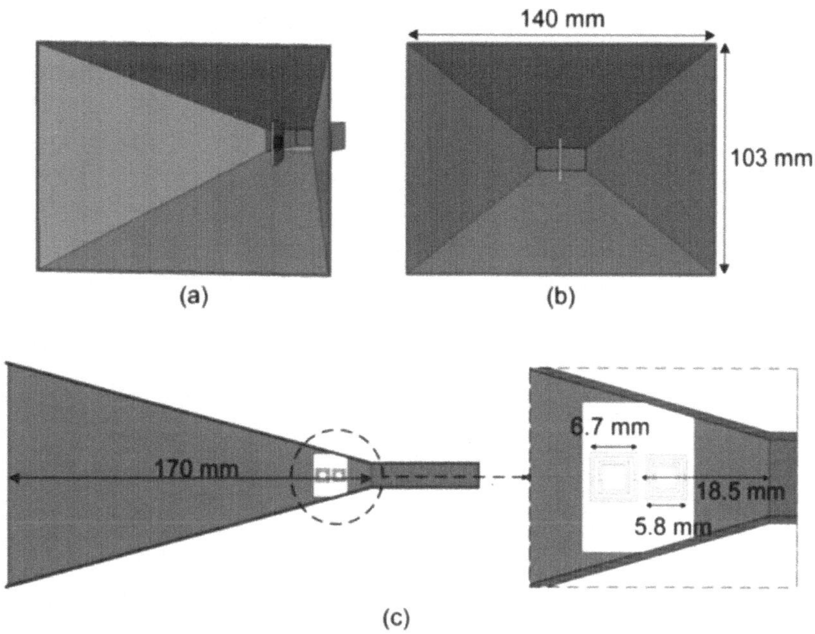

FIGURE 3.44
Geometrical sketch of the proposed horn antenna with dual band notch filter: (a) perspective view, (b) front view, (c) side view.

Reproduced by permission © 2015 The IEEE [60].

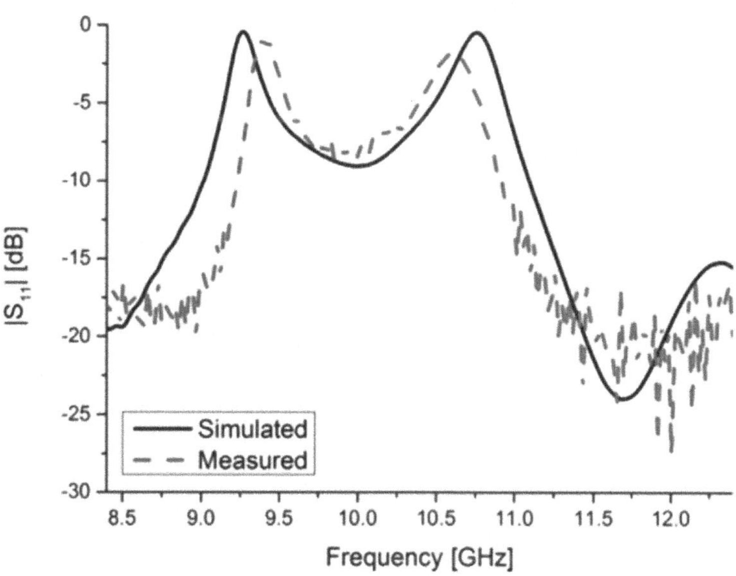

FIGURE 3.45
Simulated and measured reflection coefficient amplitude at the input port of the horn filtenna with a dual band-stop characteristic.

Reproduced by permission © 2015 The IEEE [60].

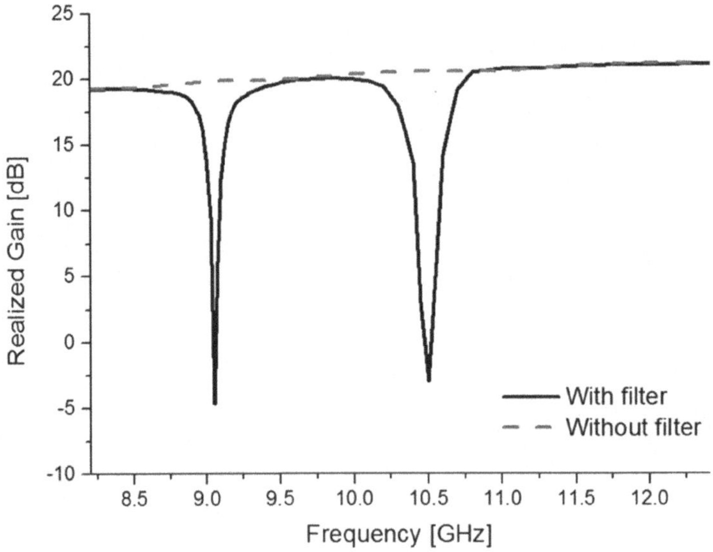

FIGURE 3.46
Simulated realized gain of the dual-notched antenna shown in Figure 3.38.

Reproduced by permission © 2015 The IEEE [60].

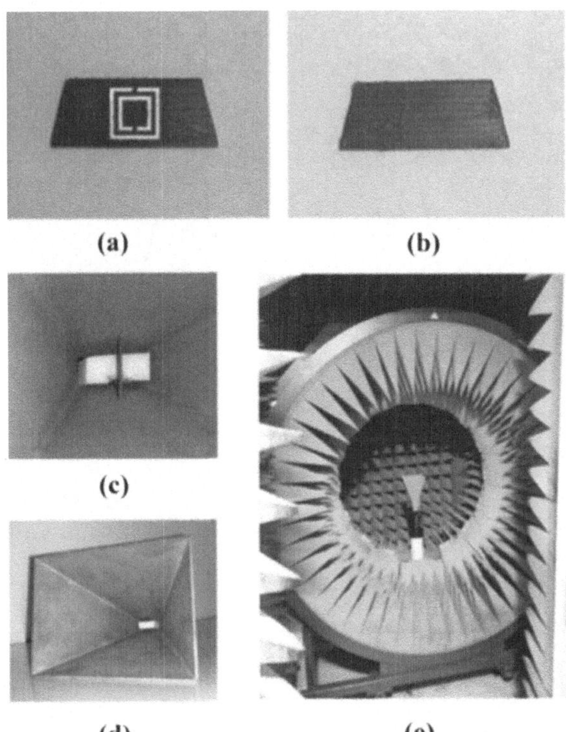

FIGURE 3.47
Fabricated prototype of the SRR-loaded horn antenna.

Reproduced by permission © 2015 The IEEE [60].

at $|S_{11}|$=10 dB is significantly narrower. In configuration A, S-SRRs are optimally placed between the flaring regions of elliptically shaped strip conductors which yields a sharp notch at the resonance frequency of the S-SRR. A similar notch is also yielded in configuration B in which the S-SRR is placed along the microstrip feedline of the antipodal configuration. The dimensions of the S-SRR can be easily designed to make the notched band centered at the frequency of the interfering signal without redesigning the radiator. Two or more S-SRRs may also be employed to suppress multiple interfering signals at different frequencies.

Figure 3.48 shows the schematic of the proposed S-SRR loaded antipodal tapered slot antenna consisting of elliptical radiators. Two elliptical strip conductors with similar dimensions are printed on each side of a substrate having dielectric constant ε_r and thickness h. The elliptical strip conductors with minor axis and major axis having radii R_1 and R_2 are designed with an ellipticity ratio of 1.3 for a wideband response. The ultrawideband ATSA operating from 400 MHz – 10 GHz is designed based on a work in Siddiqui et al. [62].

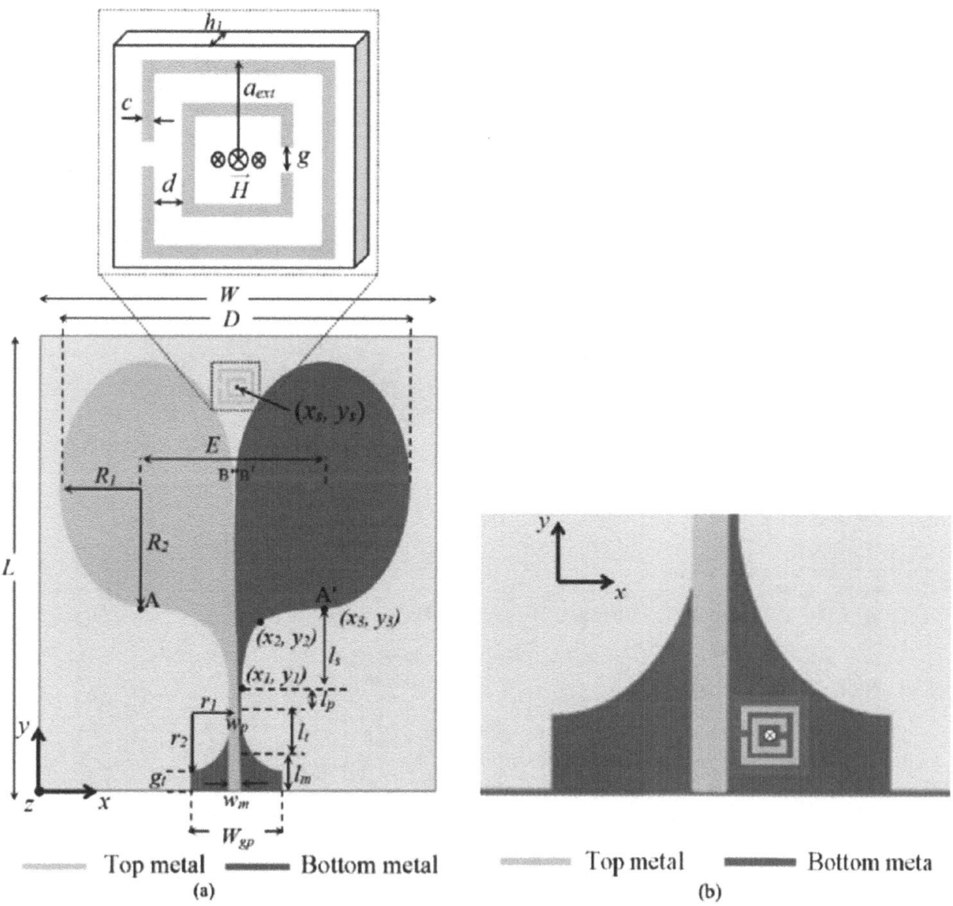

FIGURE 3.48
Schematic of a microstrip-fed ATSA loaded with S-SRR for two different configurations: (a) S-SRR loaded on the aperture of ATSA, (b) S-SRR loaded along the feeding microstrip line. Zoomed-in unit cell of S-SRR is indicated above.

Reproduced by permission of © 2018 The IEEE [61].

In both the configurations of Figure 3.48a,b, the magnetic fields of the propagating wave are axially oriented along the S-SRR for optimal excitation of the S-SRR. The axially oriented magnetic fields induce oscillating currents between the two rings of the S-SRR yielding a resonance determined by the S-SRR's geometry and prohibit signal propagation at that frequency yielding a notch.

The fabricated prototypes of the ATSA with S-SRR loadings in different configurations, configuration A, ATSA with aperture loaded S-SRR and configuration B, ATSA with SRR-coupled microstrip line as shown in Figure 3.49. Square SRRs of different dimensions were fabricated on PTFE substrate with dielectric constant 2.33 and thickness 0.787 mm and placed along the aperture and along the feedline of the antenna fabricated on the same substrate having dielectric constant 2.33 but a different thickness of 1.575 mm. The

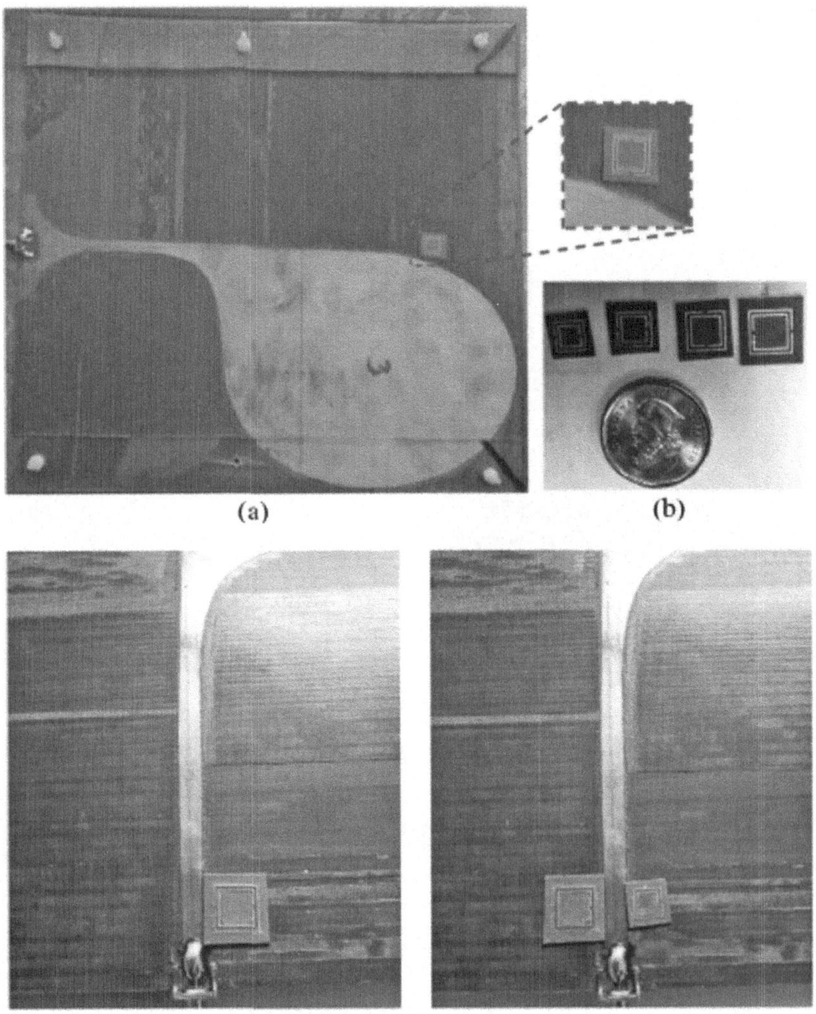

FIGURE 3.49
Photograph of the fabricated prototype of the S-SRR loaded ATSA for two different configurations: A and B.

Reproduced by permission of © 2018 The IEEE [61].

measured and simulated plots for the S_{11} versus frequency and maximum realized gain versus frequency of the standalone TSA are shown in Figure 3.50a,b.

In configuration A, for ATSA with aperture loaded S-SRR, the S-SRRs were placed on the aperture of the tapered slot antenna at a fixed *x*-coordinate of 141 mm from the origin and marginally varying *y*-coordinate. Four square SRRs, SRR1, SRR2, SRR3, and SRR4 with the same values of *c*, *d* and *g* but varying a_{ext} dimensions of 3.4 mm, 4 mm, 4.5 mm, and 5 mm, were used. Figure 3.51 shows the S_{11} versus frequency plot, and sharp and distinct measured notches were observed at 4.10 GHz, 3.18 GHz, 2.95 GHz, and 2.55 GHz which corresponds to the resonance frequencies of SRR1, SRR2, SRR3, and SRR4, respectively.

The measured maximum realized gain values versus frequency for two different SRRs, SRR1 and SRR2, loaded on the aperture of the TSA is plotted in Figure 3.52. The plot exhibits drastically reduced gain of −2.76 dBi at 4.10 GHz for SRR1 and −2.85 dBi at 3.18 GHz for SRR2 with acceptable gain above 2 dBi for the entire band and almost around 10 dBi for

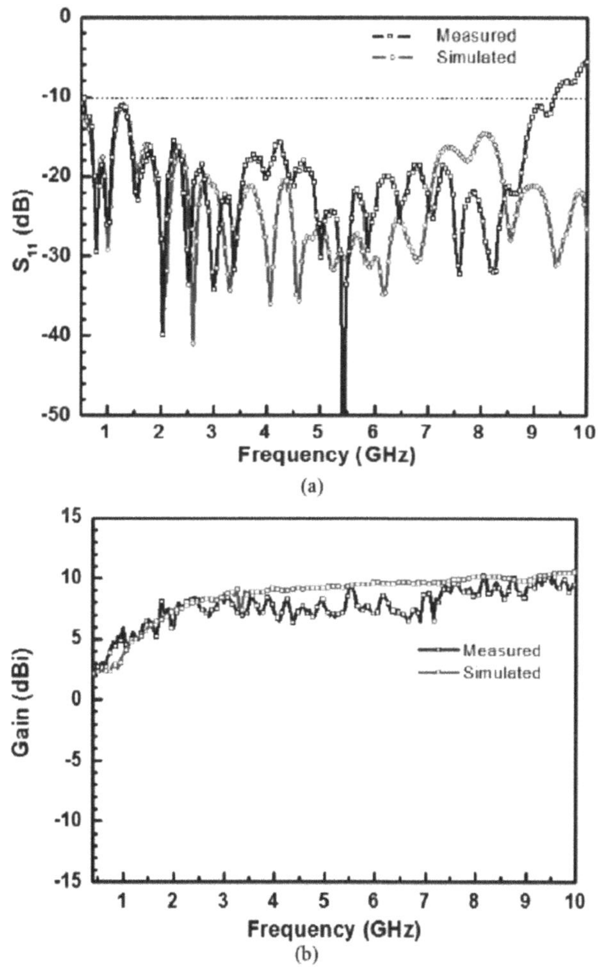

FIGURE 3.50
(a) Simulated and measured reflection coefficient as a function of frequency of the microstrip-fed ATSA, (b) simulated and measured realized gain of the microstrip-fed ATSA.

Reproduced by permission of © 2018 The IEEE [61].

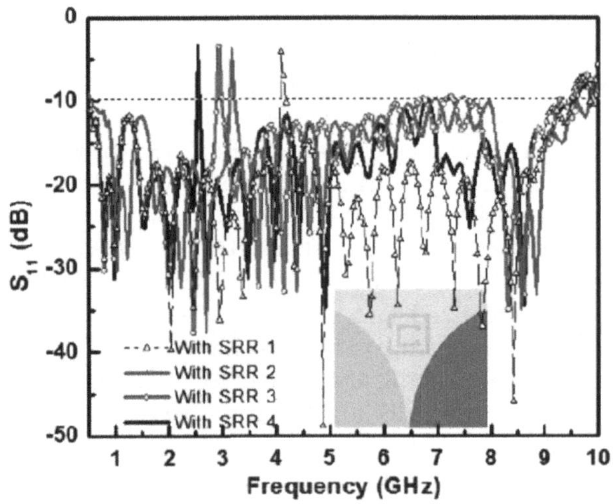

FIGURE 3.51
Measured reflection coefficient as a function of frequency of the various SRR loaded ATSA.

Reproduced by permission of © 2018 The IEEE [61].

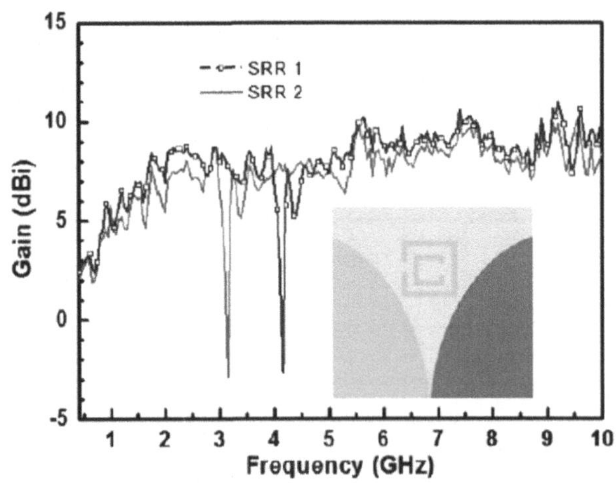

FIGURE 3.52
Measured realized gain as a function of frequency of the various SRR loaded ATSA.

Reproduced by permission of © 2018 The IEEE [61].

most of the spectrum from 400 MHz to 9.8 GHz reconfirming the presence of strong notch contributed by the SRRs resonance. The measured E-plane and H-plane radiation patterns for the SRR2 loaded TSA is shown Figure 3.53.

Since the aperture loaded SRRs are sub-wavelength resonators with $2a_{ext} < 10$, their presence should not impact the radiation performance of the antenna over the entire spectrum excepting the notch frequencies. To validate this point, the radiation pattern of the antenna is plotted at 3.8 GHz, 6.2 GHz, 8 GHz, and 9.4 GHz, covering the entire impedance bandwidth of the antenna. As revealed in the plots of Figure 3.53a,b, the proposed antenna exhibits directive patterns in both the principal planes similar to unloaded ATSA [61]. It

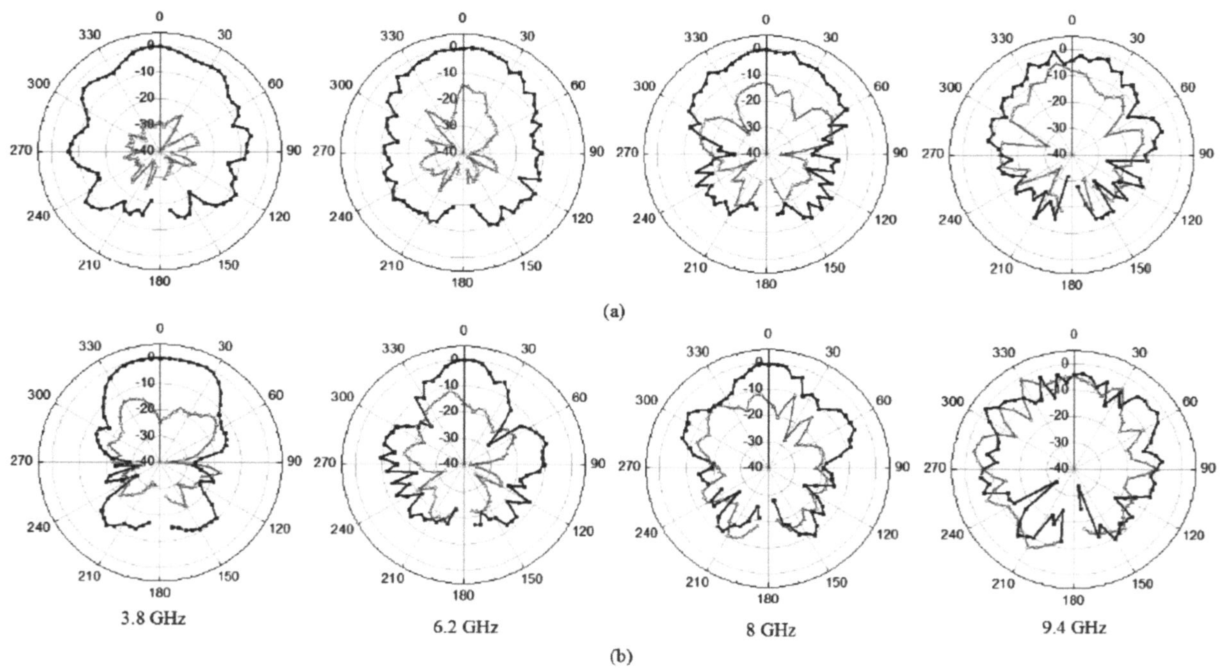

FIGURE 3.53
Measured radiation pattern: (a) E-plane, (b) H-plane of the SRR loaded ATSA.

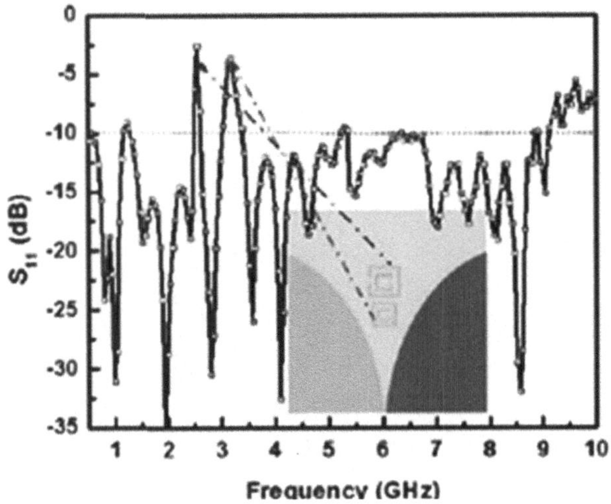

FIGURE 3.54
Measured reflection coefficient as a function of frequency of the various SRR loaded ATSA.

Reproduced by permission of © 2018 The IEEE [61].

can be noted that the S-SRR loaded ATSA exhibits good co- to cross-polar discrimination in the lower frequencies.

Multiple S-SRRs, SRR2 and SRR4, were placed simultaneously on the aperture to yield multiple notches in the reflection coefficient plot as shown in Figure 3.54. Dual frequency notches at 3.18 GHz and 2.55 GHz were obtained when SRR2 and SRR4 were optimally positioned on the aperture. This proves that proposed techniques can be easily tailored to achieve single (multiple) frequency notches of any arbitrary frequency (frequencies) by optimal design and placements of S-SRR (SRRs) along the aperture of ATSA.

In configuration B, the SRRs were placed along the microstrip line as shown in the inset in Figure 3.55. The plot in Figure 3.55 shows the measured maximum realized gain characteristics of the microstrip-fed ATSA with S-SRR loaded along the feeding microstrip line of the ATSA for two different cases (SRR3 and 4). Figure 3.56 shows the measured S_{11} plots for two different sets of SRRs.

3.6.2 Printed Tapered Slot Antennas Loaded with Spur Lines

In the previous section, the tapered slot antennas were loaded with SRRs to obtain notches. The same tapered slot antennas are also loaded with spur lines to obtain the desired notch frequency. The tapered slot antenna is also modified into a balanced antipodal tapered slot antenna (BATSA) which reduces the cross polarization. Figure 3.57 shows the 3-D perspective and cross-sectional view of the proposed frequency-notched BATSA [63].

The antenna as shown in Figure 3.57 is fed with a microstrip line connected with the antipodal coplanar stripline and forms the required balun to excite the slot [63]. The basic ATSA, discussed in the previous section, is modified to achieve electromagnetic balance of the radiated fields. The balancing of the radiated electric field is due to two elliptic conductors acting as the common ground placed on the opposite side of another elliptic conductor acting as the signal line of the antenna. This effectively forms two slots between the signal line and the ground planes. The skewed electric field at any point along the

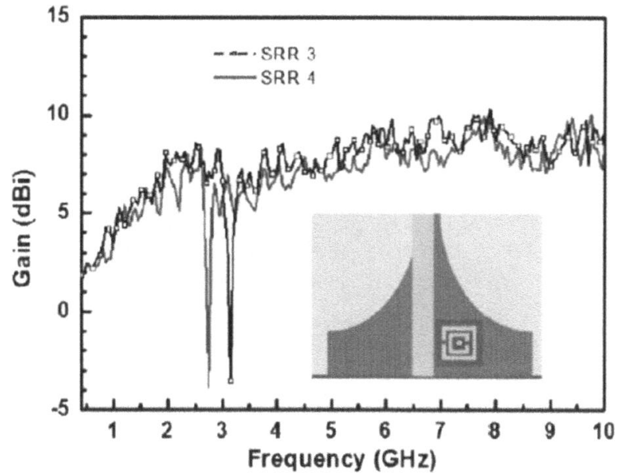

FIGURE 3.55
Measured realized gain as a function of frequency of the various SRR loaded ATSA.

Reproduced by permission of © 2018 The IEEE [61].

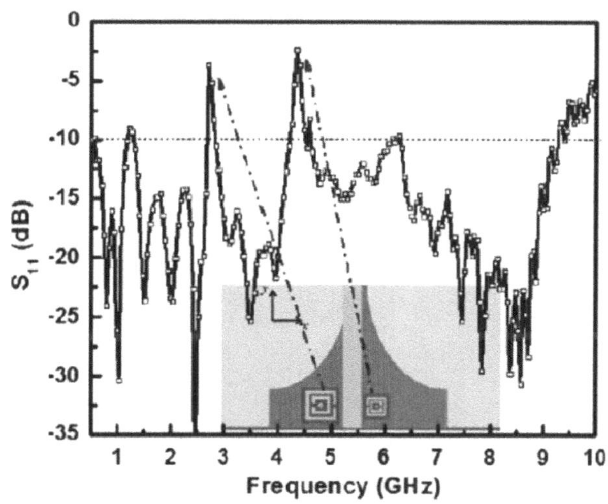

FIGURE 3.56
Measured realized gain as a function of frequency of the various SRR loaded on microstrip line of ATSA.

Reproduced by permission of © 2018 The IEEE [61].

radiating slot mutually cancels the vertical component of the radiated electric field which in turn improves the cross-polar performance of the antenna as proposed in Langley et al. [64]. Furthermore, a spur line is embedded on the feed section of the antenna which yields a frequency notch in the UWB response of the antenna. A quarter-wavelength spur line resonator of length $(a+b)$ and slot width t is loaded at a distance h_t from the feed end of the antenna.

The antenna configuration has two symmetrical ground planes above and below. Due to the modified geometry, the electric field is modified with an additional component from

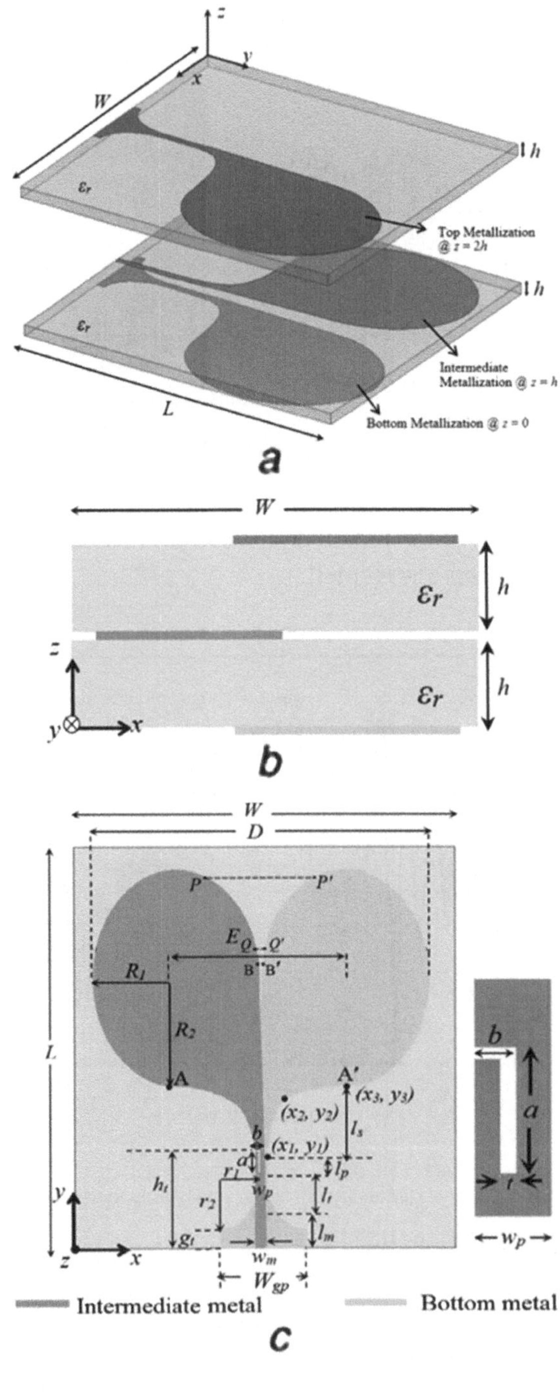

FIGURE 3.57
(a) Exploded schematic showing the layers and the corresponding metallization of the antenna, (b) lateral view of antenna at *x*-*z* plane along the line joining the centers of the elliptical conductors, (c) schematic indicating the corresponding design parameters with spur-line filter.

Reproduced by permission of © 2018 The IET [63].

the signal line to the top ground plane. As the fields are decomposed into co-polar and cross-polar, the cross-polar components, irrespective of the frequencies, cancel out and lead to reduced cross-polar levels as shown in Figure 3.58.

The antenna is fabricated on a RT Duriod (ε_r=2.33 and *tan* δ=0.0012) substrate. Figure 3.59a,b shows the fabricated prototype of the proposed antenna consisting of two ground planes and a signal line. Figure 3.59c shows the spur line on the feed part of the antenna. Figure 3.59d shows a 3-D perspective view of the fabricated antenna just before affixing the top ground plane. Figure 3.59e shows top metallization of the antenna fixed with the index matched screws to avoid an air gap between the layers.

Figure 3.60 shows the simulated and measured reflection coefficient (S_{11}) of BATSA along with a trace of measured S_{11} of ATSA. The ATSA offers wideband impedance characteristics from 830 MHz to 9.8 GHz while the for BATSA, the measured impedance bandwidth is from 920 MHz to 9.7 GHz against 820 MHz to 9.92 GHz for simulation.

Figure 3.61 shows the simulated and measured reflection coefficient of the proposed spur-line loaded BATSA. Due to the loading of the λ/4 spur line in the feeding microstrip line, a distinct frequency notch is observed at 5.50 GHz in simulation, corresponding to the resonance frequency of the spur line. The plot reveals measured S_{11} < −10dB rejection from 5.47–5.96 GHz with notch at 5.60 GHz and the simulation yields a S_{11}<−10dB rejection from 5.28–5.73 GHz with notch at 5.50 GHz. This covers part of the WLAN band (5.15–5.825 GHz).

The measured peak gain as a function of frequency of the proposed antenna with and without the spur line is shown in Figure 3.62. The plot shows a sharp reduction in gain to −7.9 dBi at the notch frequency (5.6 GHz) of the spur-line loaded BATSA. The overall gain of the antenna at other frequencies (0.9–10 GHz) is well above 3 dBi with a peak gain of 10 dBi at around 9 GHZ for both the configurations.

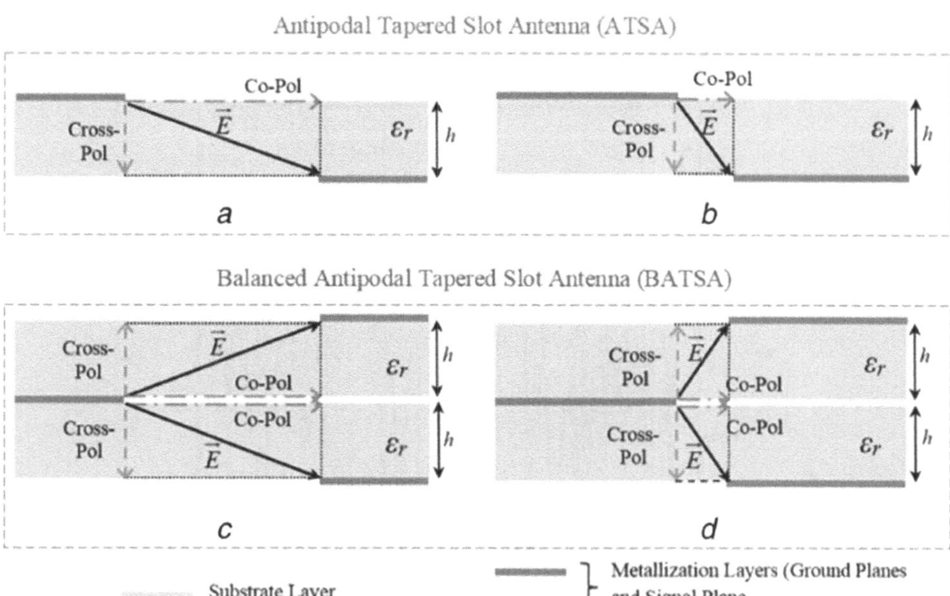

FIGURE 3.58
Comparison of the simulated E-field distribution of the ATSA and BATSA.

Reproduced by permission of © 2018 The IET [63].

FIGURE 3.59
Photograph of the fabricated BATSA.

Reproduced by permission of © 2018 The IET [63].

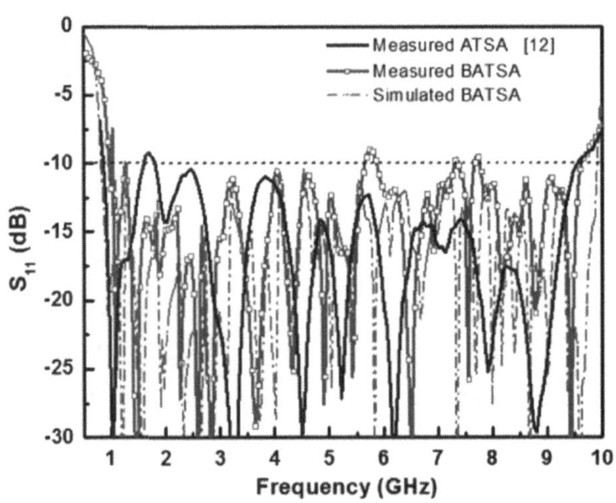

FIGURE 3.60
Measured and simulated S_{11} as a function of frequency of ATSA and BATSA.

Reproduced by permission of © 2018 The IET [63].

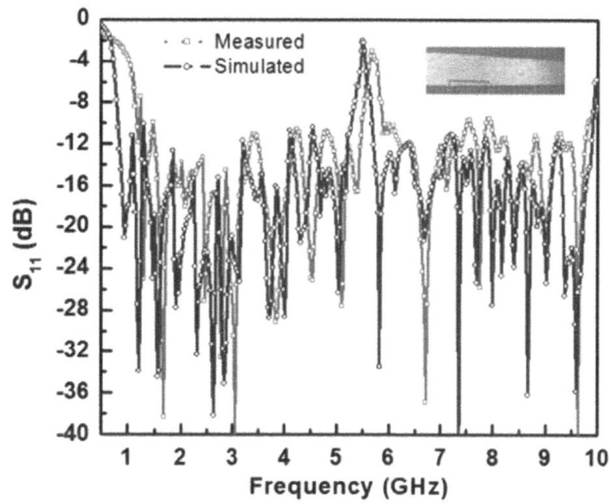

FIGURE 3.61
Measured and simulated S_{11} as a function of frequency of the spur-line loaded frequency-notched BATSA.

Reproduced by permission of © 2018 The IET [63].

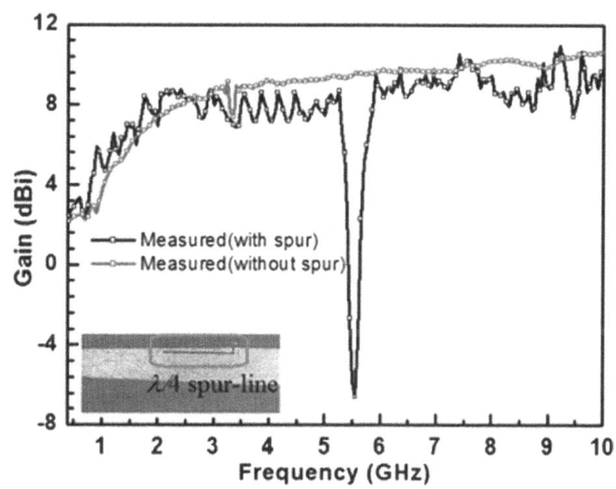

FIGURE 3.62
Comparison of the measured gain as a function of frequency of the conventional and spur-line loaded BATSA.

Reproduced by permission of © 2018 The IET [63].

The simulated and measured normalized E(*x-y*) and H(*y-z*) plane radiation patterns of the ATSA and BATSA is presented in Figure 3.63. Figure 3.63 represents the measured normalized E-plane and H-plane radiation pattern of the ATSA at 4.6 GHz and 9 GHz, respectively. The two frequencies are deliberately chosen as widely separated to examine the cross-polar performance of the antenna at high frequency (9 GHz) compared to that at low frequency (4.6 GHz). The plot reveals that the antenna exhibits directional pattern in both E(*x-y*) plane and H(*y-z*) plane.

The plot also reveals that while at 4.6 GHz cross-polar radiation is significantly lower, at 9 GHz the co- to cross-polar discrimination is only around 10–15 dB. Figure 3.63c and d

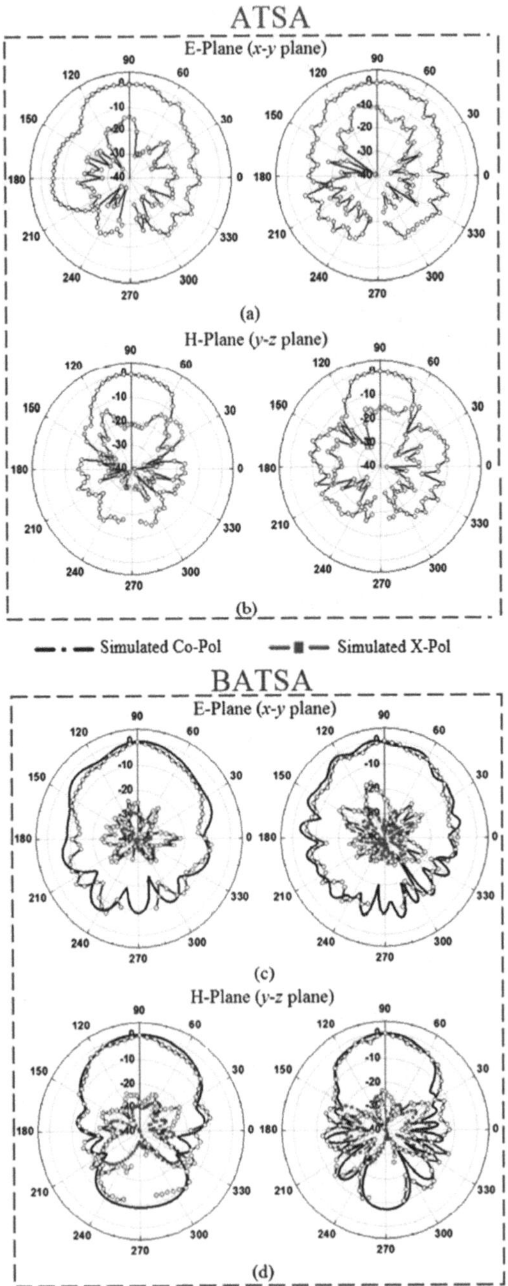

FIGURE 3.63
Comparison of the simulated and measured E-plane and H-plane radiation patterns of ATSA and BATSA.

represent both the simulated and measured normalized E(x-y) plane and H(y-z) plane radiation patterns of the BATSA at same frequencies (4.6 GHz and 9 GHz) as chosen for ATSA. The simulated and measured patterns exhibit a very high correlation. It can be observed from the plots of Figure 3.63c,d that BATSA offers a much improved co- to cross-polar discrimination of around 25 dB, both at lower and higher frequencies. Thus, the proposed antenna offers very low cross-polarization over its entire impedance bandwidth.

3.6.3 Printed Vivaldi Antenna Loaded with SRR on the Microstrip Feedline

Another design of an exponentially tapered slot UWB antenna with frequency notch characteristics is proposed using a pair of SRRs loaded along the microstrip-line feed of the antenna. Two differently shaped SRRs, namely square SRR (S-SRR) and hexagonal SRR (H-SRR), are used to obtain frequency notches [65].

Figure 3.64 shows the schematic of the proposed square/hexagonal SRR loaded exponentially tapered slot antenna (ETSA). The basic radiating element consists of an exponentially tapered slot having initial and terminal widths W_s and W_{eff} excited by a 50 Ω microstrip line having width W_m printed on the opposite side of a dielectric substrate of overall dimension $L \times W \times h$ with L, W, and h indicating length, width and height, respectively, of the overall ETSA. For better coupling and enhanced bandwidth, the microstrip feed is orthogonally oriented with the slot on the other side and both of them are extended one quarter (R_m and R_s) wavelength further and terminated with radial stubs of angular widths. A pair of square/hexagonal split-ring resonators (S-SRR/H-SRR) is printed on both sides of the feeding microstrip line. The strong magnetic resonance phenomenon of the SRRs inhibits signal propagation along the feeding microstrip line around its resonance frequency and contributes to the frequency notch in the UWB spectrum of the antenna.

Figure 3.65 compares the simulated and measured S_{11} versus frequency plots of the S-SRR loaded ETSA. As discussed previously, a magnetically coupled SRR placed along the feeding microstrip line inhibits propagation of signal around its resonance frequency which contributes to a notch in the UWB spectrum of the antenna. As revealed in the S_{11} plot shown in Figure 3.65, the printed S-SRR pairs provide a measured frequency notch at 6.3 GHz against 6.28 GHz for a simulated notch.

The proposed technique of frequency-notched ETSA design is further customized for dual-frequency notched UWB design. This is achieved by invoking two different resonances contributed by two asymmetric S-SRRs (SRR1: $a_{ext}=2.5$ mm, $c=0.3$ mm, $d=0.3$ mm, $g=0.4$ mm; SRR2: $a_{ext}=2.7$ mm, $c=0.4$ mm, $d=0.3$ mm, $g=0.4$ mm) placed on two sides of the feeding microstrip line. The present concept of achieving dual-frequency notched UWB response from an ETSA is very simple and doesn't impose any additional complexity on the basic ETSA design. Figure 3.66 shows the simulated S_{11} versus frequency plot of the proposed asymmetric dual SRR loaded ETSA. As revealed from the plot, the antenna exhibits dual-notched response at 4.27 GHz and 6.1 GHz contributed by two SRRs loaded in the feed region.

The gain versus frequency plot of the basic ETSA is compared with the proposed S-SRR loaded ETSA and is shown in Figure 3.67. The basic ETSA indicates a maximum measured gain of 7.7 dBi and maintains a nearly flat profile over the most of the UWB spectrum. The measured maximum realized gain versus frequency of the S-SRR loaded ETSA exhibits a drastically reduced gain of −4.46 dBi at 6.3 GHz, with acceptable gain in the range of 3.4–6.9 dBi over the rest of the UWB spectrum. The measured H-SRR loaded

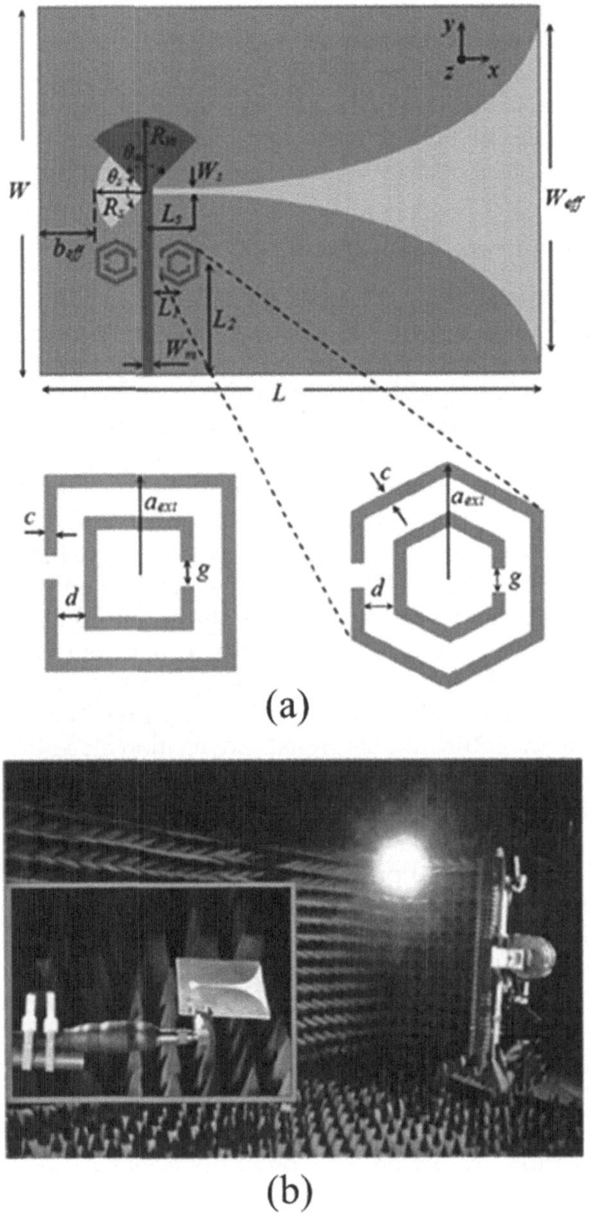

FIGURE 3.64
(a) Schematic of the microstrip-fed ETSA loaded with H-SRR, zoomed view of S-SRR and H-SRR, (b) the fabricated prototype mounted for measurement.

Reproduced by permission of © 2017 The Wiley [65].

ETSA also confirms the presence of a frequency notch at 7.1 GHz, closely corresponding to the H-SRR's resonance frequency, with a sharp fall of gain to –3.9 dBi and a good gain profile ranging between 3.4–7.55 dBi over the rest of the UWB spectrum. The measured E-plane and H-plane radiation patterns are shown in Figure 3.68 for four different frequencies.

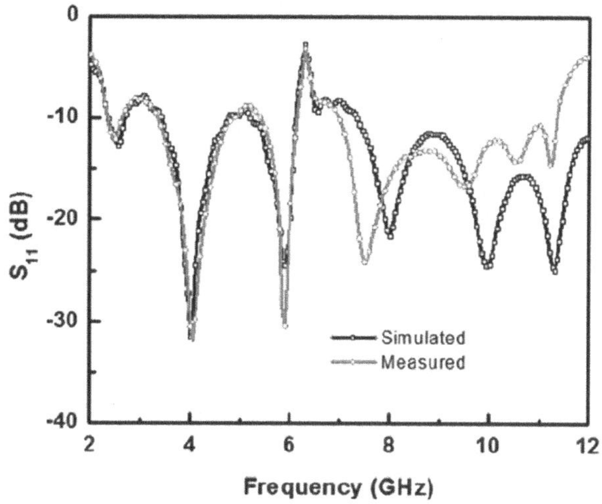

FIGURE 3.65
Measured and simulated S_{11} as a function of frequency of the SRR-loaded ETSA.

Reproduced by permission of © 2017 The Wiley [65].

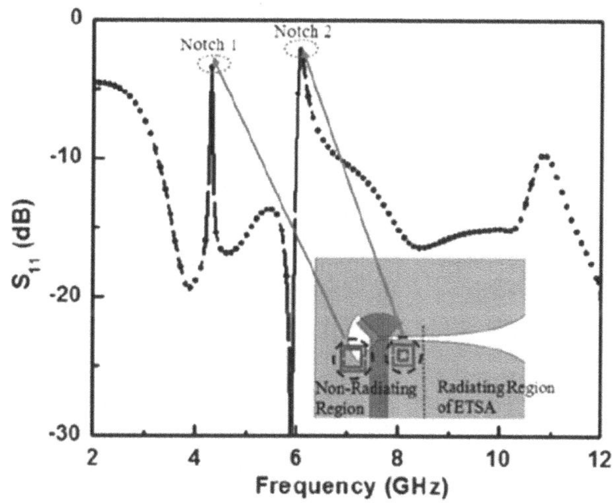

FIGURE 3.66
Simulated S_{11} as a function of frequency of the dual SRR-loaded ETSA.

Reproduced by permission of © 2017 The Wiley [65].

Figure 3.68 shows the E- and H-plane measured normalized radiation pattern of the proposed S-SRR loaded ETSA at 3.8 GHz, 5.8 GHz, 8 GHz, and 9.4 GHz. Measured patterns exhibit a directional nature in both planes and ensure the desired end-fire type radiation with radiation maxima along the center of the slot (Figure 3.69).

Another SRR loaded tapered slot antenna is discussed in Natani et al. [66]. In this design, the slot line is modified by splitting into two separate slot lines, to accommodate the SRRs under them, and then converged to form a part of the tapered slot antenna. Two square-shaped split-ring resonators of different dimensions are placed on the back side of the

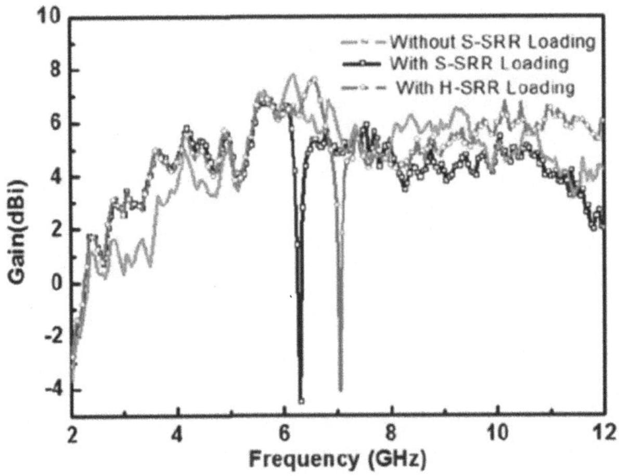

FIGURE 3.67
Measured gain as a function of frequency of the dual SRR-loaded ETSA.

Reproduced by permission of © 2017 The Wiley [65].

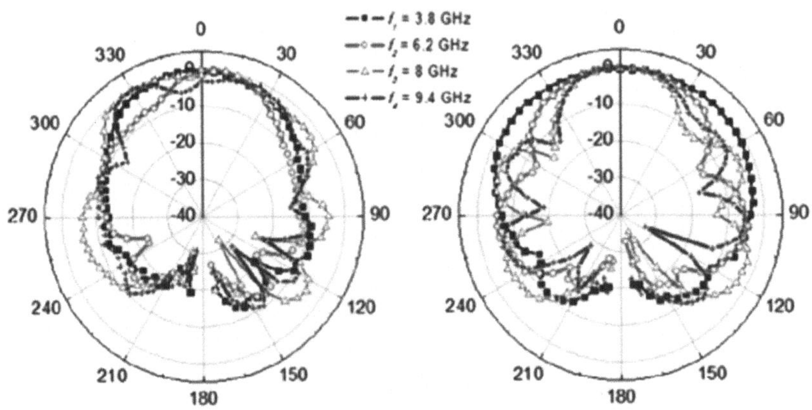

FIGURE 3.68
Measured E-plane and H-plane radiation patterns of the dual SRR-loaded ETSA.

Reproduced by permission of © 2017 The Wiley [65].

modified slot line with their centers coinciding with the center of the slot lines. Dual frequency notch operation is obtained by dividing the slot lines and combining again before it is tapered to obtain UWB performance. This configuration ensures that SRRs are not placed in the radiating region of the antenna and therefore do not interfere with the radiation field of the antenna.

Figure 3.70 compares the simulated S_{11} versus frequency plots of the conventional ETSA of Figure 3.69a and the dual S-SRR loaded modified slotline loaded ETSA of Figure 3.69b.

A CPW-fed dielectric resonator antenna as shown in Figure 3.71 was loaded with a pair of square SRRs on the feedline which resulted in a notch frequency at the SRRs resonance [67].

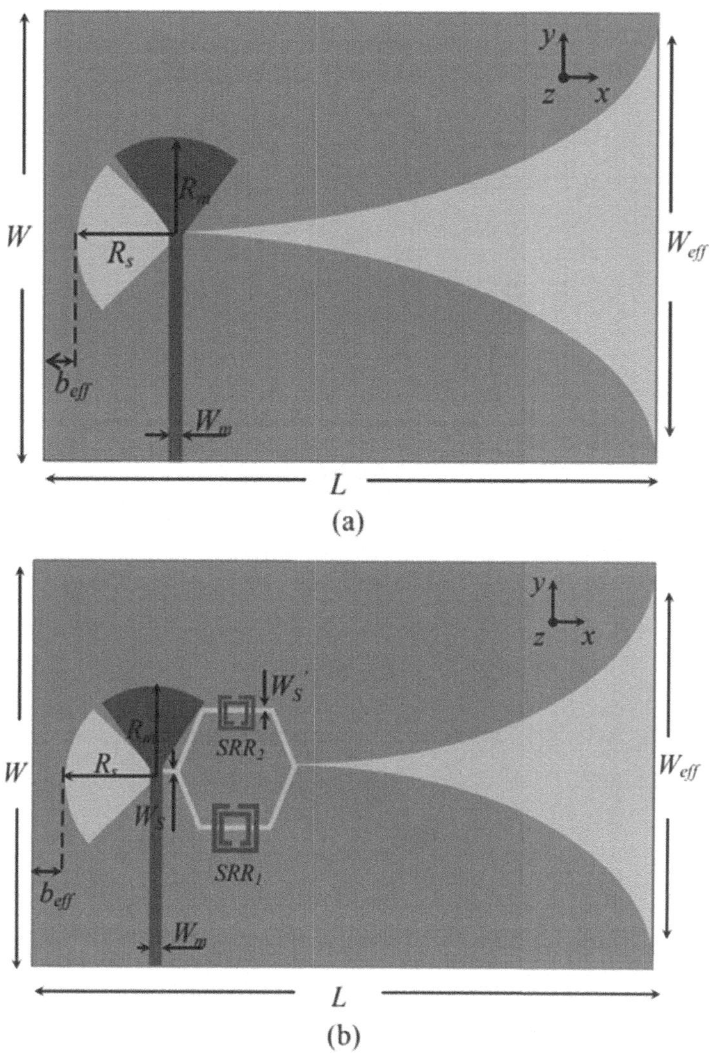

FIGURE 3.69
Schematic of (a) conventional microstrip-fed coplanar tapered slot antenna, (b) proposed new dual SRR-loaded UWB coplanar tapered slot antenna with modified slot line for dual notch application.

Reproduced by permission of © 2016 IEEE [66].

3.7 Comparison of Various Frequency-Notching Techniques

Various techniques and methods have been used to yield frequency notches in the UWB spectrum of printed antennas. Key points highlighting the basic features, advantages and disadvantages of various frequency-notching techniques are as follows:

- Slot-loading on the radiators to design frequency-notched UWB antenna is the most widely used technique. However, in this technique the frequency-notching elements directly perturb the antenna and interact with the radiation field.

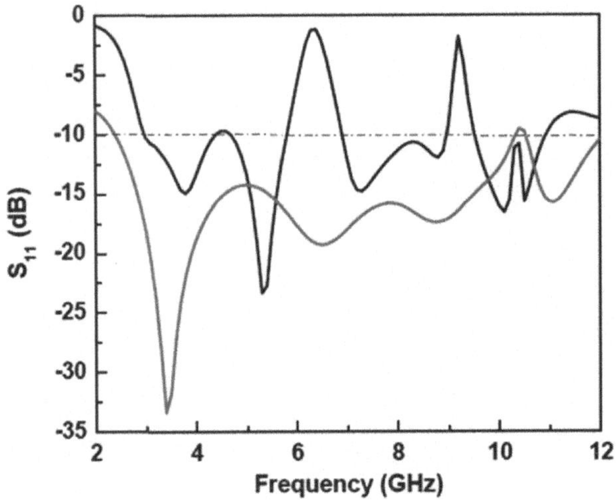

FIGURE 3.70
Simulated S_{11} as a function of frequency of the dual SRR-loaded ETSA.

Reproduced by permission of © 2016 IEEE [66].

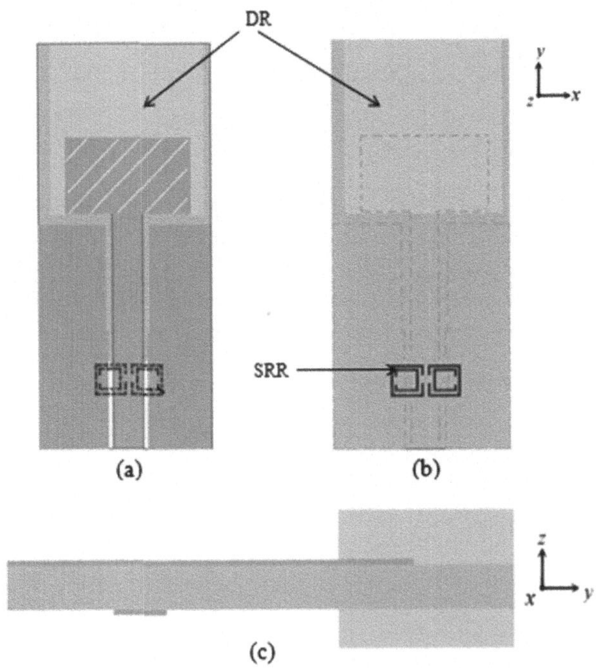

FIGURE 3.71
SRR-loaded CPW-fed frequency-notched DRA antenna: (a) top view, (b) bottom view, and (c) side view.

Reproduced by permission of © 2015 The IEEE [67].

- In the slot-loading frequency-notched antenna design, design of the notching elements highly depends on the antenna design. Moreover, performance of the antenna gain and radiation pattern is compromised.
- In the SRR-loaded techniques, resonant inclusions are strategically placed to exploit their resonance by exciting them using the magnetic field. With SRR-loading notching techniques, coplanar waveguide is the most conventional feed to exploit the SRRs resonance as the SRRs are placed on the slot line. The microstrip feed needs proper aligning of the SRRs close to the microstrip line to exploit the fringing fields for excitation.
- Parasitic resonator-based frequency-notching techniques, such as loading SRRs, are independent of the design of the radiators. Therefore, this technique is quite versatile, robust and hence provides an added degree of freedom to design the radiators and resonators independently.

Appendix A: Calculation of Resonance Frequency of the Square SRR

$$a = \frac{d}{2}; \tag{A.1}$$

$$b = a + c; \tag{A.2}$$

$$k = \frac{a}{b}; \tag{A.3}$$

$$k' = \sqrt{1-k^2}; \tag{A.4}$$

$$k1 = \frac{\sinh\left(\frac{\pi a}{2h}\right)}{\sinh\left(\frac{\pi b}{2h}\right)} = 0.375; \tag{A.5}$$

$$k1' = \sqrt{1-k1^2}; \tag{A.6}$$

$$Kk = \left(\frac{1}{\pi}\ln\left(2\frac{\left(1+\sqrt{k'}\right)}{\left(1-\sqrt{k'}\right)}\right)\right)^{-1} = 0.748 \tag{A.7}$$

$$Kk1 = \left(\frac{1}{\pi}\ln\left(2\frac{\left(1+\sqrt{k1'}\right)}{\left(1-\sqrt{k1'}\right)}\right)\right)^{-1} = 0.674 \tag{A.8}$$

$$\varepsilon_{\text{eff}} = 1 + \frac{\varepsilon_r - 1}{2}(Kk)^{-1}Kk1 = 1.6 \tag{A.9}$$

$$Z_0 = \frac{120\pi}{\sqrt{\varepsilon_{\text{eff}}}}(Kk) = 222.846\ \Omega \tag{A.10}$$

$$C_{\text{pul}} = \frac{\sqrt{\varepsilon_{\text{eff}}}}{v_0 \cdot Z_0} = 1.892 \times 10^{14} \tag{A.11}$$

$$a_{\text{avg}} = a_{\text{ext}} - c - \frac{d}{2}; \tag{A.12}$$

Total equivalent capacitance of the SRR is calculated as follows, Saha and Siddiqui [68–71],

$$C_{eq} = \left(2a_{\text{avg}} - \frac{g}{2}\right) \cdot C_{\text{pul}} + \frac{\varepsilon_0 \cdot c \cdot t}{2 \cdot g} = 9.751 \times 10^{-14} \tag{A.13}$$

$$l = 8 \cdot a_{\text{ext}} - g = 26.5\text{ mm} \tag{A.14}$$

$$L_T = 0.0002 \cdot l \cdot \left(2.303 \cdot \log\left(\frac{4 \cdot l}{c}\right) - 2.853\right) \times 10^{-6}\mu H \tag{A.15}$$

$$f_{\text{SSRR}} = \frac{1}{2\pi}\sqrt{\frac{1}{L_T C_{eq}}} = 4.139 \times 10^9\text{ GHz} \tag{A.16}$$

Equations A.1–A.17 show the computational steps required to compute the resonance frequency $f_{\text{S-SRR}}$ of the square SRR having geometrical parameters $a_{\text{ext}}=3.4$ mm, $c=0.35$ mm, $d=0.6$ mm, and $g_1=g_2=0.7$ mm printed on the dielectric substrate with $\varepsilon_r=2.33$, $tan\delta=$ 0.0012 and height $h=0.787$ mm. The computed value of the resonance frequency obtained is 4.14 GHz. By changing the a_{ext} value of the SRR, one can tune the resonance frequency of the SRR and control the notch frequency.

References

1. M. Chiani and A. Giorgetti, "Coexistence between UWB and narrow-band wireless communication systems", *Proceedings of the IEEE*, vol. 97, no. 2, 231–254, February 2009.
2. J. Kruys and L. Qian, "Sharing RF Spectrum with Commodity Wireless Technologies: Theory and practice", *Signals and Communication Technology*, Springer, 2011.
3. A.F. Molisch, "Ultra-Wideband Communications: An Overview", *The Radio Science Bulletin*, no. 329, 31–42, June 2009.
4. R.A. Scholtz, "Multiple access with time-hopping impulse modulation", *Proceedings of TEEE MILCOM*, 447–450, 1993.
5. M.Z. Win and R.A. Scholtz, "Impulse radio: How it works", *IEEE Communications Letters*, vol. 2, no. 2, 36–38, 1998.
6. M.Z. Win and R.A. Scholtz, "Ultra-wide bandwidth time-hopping spread-spectrum impulse radio for wireless multiple-access communications", *IEEE Transactions on Communications*, vol. 48, no. 4, 679–691, 2000.

7. Y. Zhang, W. Hong, C. Yu, Z.Q. Kuai, Y.D. Don, and J.Y. Zhou, "Planar Ultrawideband antennas with multiple notched bands based on etched slots on the patch and/or split ring resonators on the feed line", *IEEE Transactions on Antennas Propagating*, vol. 56, no. 9, 3063–3068, 2008.
8. P. Wang, G.J. Wen, Y.J. Huang, and Y.H. Sun, "Compact CPW-fed planar monopole antenna with distinct triple bands for WiFi/WiMAX applications", *Electronics Letters*, vol. 48, no. 7, 357–359, 2012.
9. D.T. Nguyen, D.H. Lee, and H.C. Park, "Very compact printed triple band-notched UWB antenna with quarter-wavelength slots", *IEEE Antennas Wireless Propagating Letter*, vol. 11, 411–414, 2012.
10. W.T. Li, X.W. Shi, and Y.Q. Hei, "Novel planar UWB monopole antenna with triple band-notched characteristics", *IEEE Antennas Wireless Propagating Letter*, vol. 8, 1094–1098, 2009.
11. H.W. Liu, C.H. Ku, and C.F. Yang, "Novel CPW-fed planar monopole antenna for WiMAX/WLAN applications", *IEEE Antennas Wireless Propagating Letter*, vol. 9, 240–243, 2010.
12. T. Dissanayake and K.P. Esselle, "Design of slot loaded band-notched UWB antennas", *Digest IEEE Antennas and Propagation Society International Symposium*, Washington DC, 545–548, July 2005.
13. M.-C. Tang, S. Xiao, T. Deng, D. Wang, J. Guan, B. Wang, and G.-D. Ge, "Compact UWB antenna with multiple band-notches for WiMAX and WLAN", *IEEE Transactions on Antennas Propagating*, vol. 59, no. 4, 1372–1376, 2011.
14. Q.-X. Chu and Y.-Y. Yang, "A compact ultrawideband antenna with 3.4/5.5 GHz dual band-notched characteristics", *IEEE Transactions on Antennas Propagating*, vol. 56, no. 12, 3637–3644, 2008.
15. K. Bahadori and Y. Rahmat-Samii, "A miniaturized elliptic-card UWB antenna with WLAN band rejection for wireless communications", *IEEE Transactions on Antennas and Propagation*, vol. 55, no. 11, 3326–3332, 2007.
16. K.S. Ryu and A.A. Kishk, "UWB antenna with single or dual band-notches for lower WLAN band and upper WLAN band", *IEEE Transactions on Antennas and Propagation*, vol. 57, no. 12, 3942–3950, 2009.
17. W.T. Li, Y.Q. Hei, W. Feng, and X.W. Shi, "Planar antenna for 3G/Bluetooth/WiMAX and UWB applications with dual and-notched characteristics", *IEEE Antennas Wireless Propagating Letter*, vol. 11, 61–64, 2012.
18. L. Li, Z.L. Zhou, J.S. Hong, B.Z. Wang, and Y.H. Sun, "Compact dual-band-notched UWB planar monopole antenna with modified SRR", *Electronics Letters*, vol. 47, no. 17, 950–951, 2011.
19. X.J. Liao, H.C. Yang, N. Han, Y. Li, and Y.H. Sun, "UWB antenna with dual narrow band notches for lower and upper WLAN bands", *Electronics Letters*, vol. 46, no. 24, 1593–1594, 2010.
20. L. Luo, Z. Cui, J.P. Xiong, X.M. Zhang, and Y.C. Jiao, "Compact printed ultra-wideband monopole antenna with dual band-notch characteristic", *Electronics Letters*, vol. 44, no. 19, 1106–1107, 2008.
21. C.Y. Hong, C.W. Ling, I.Y. Tarn, and S.J. Chung, "Design of a planar ultrawideband antenna with a new band-notch structure", *IEEE Transactions on Antennas Propagating*, vol. 55, 3391–3397, 2007.
22. L.-N. Zhang, S.S. Zhong, C.Z. Du, and J.H. Chen, "Compact UWB planar monopole antenna with band-notch function", *Microwave and Optical Technology Letters*, vol. 51, no. 8, 1908–1911, 2009.
23. L. Peng and C.L. Ruan, "UWB band-notched monopole antenna design using electromagnetic-bandgap structures", *IEEE Transactions on Microwave Theory Techniques*, vol. 59, no. 4, 1074–1081, 2011.
24. W.C. Liu, "Dual wideband coplanar waveguide-fed notched antennas with asymmetrical grounds for multi-band wireless application", *IET Journals and Magazines*, vol. 1, no. 5, 980–985, 2007.
25. E.A. Daviu, M.C. Fabres, M.F. Bataller, and A.V. Jimenez, "Active UWB antenna with tunable band-notched behaviour", *Electronics Letters*, vol. 43, no. 18, 959–960, 2007.
26. S.A. Schelkunoff and H.T. Friis, *Antennas Theory and Practice*, John Wiley and Sons, New York, 1952.

27. J.B. Pendry, A.J. Holden, D.J. Robbins, and W.J. Stewart, "Magnetism from conductors and enhanced nonlinear phenomena", *IEEE Transactions on Microwave Theory Technology*, vol. 47, 2075–2084, 1999.
28. D.R. Smith, D.C. Vier, T. Koschny, and C.M. Soukoulis, "Electromagnetic parameter retrieval from inhomogeneous metamaterials", *Physical Review Part E*, vol. 71, 036617, 2005.
29. J.D. Baena, J. Bonache, F. Martín, R. Marqués, F. Falcone, T. Lopetegi, M.A.G. Laso, J. García-García, I. Gil, M. Flores, and M. Sorolla, "Equivalent circuit models for split ring resonators and complementary split ring resonators coupled to planar TransmisiónLines", *IEEE Transactions on Microwave Theory Technology*, vol. 53, no. 4, 1451–1461, 2005.
30. R. Marquez, F. Martin, and M. Sorrola, *Metamaterials with negative parameters: Theory, Design and Microwave Applications*, Wiley, New York, 2007.
31. J. Bonache, I. Gil, J. Garcia-Garcia, and F. Martin, "Novel microstrip bandpass filters based on complementary split-ring resonators", *IEEE Transactions on Microwave Theory Techniques*, vol. 54, no. 1, 265–271, 2006.
32. M. Gil, J. Bonache, J. Selga, J. Garcia-Garcia, and F. Martin, "High-pass filters implemented by Composite right/left Handed (CRLH) transmission lines based on Complementary Split Rings Resonators (CSRRs)", *PIERS Online*, vol. 3, no. 3, 251–253, 2007.
33. F. Martin, F. Falcone, J. Bonache, R. Marqués, and M. Sorolla, "Miniaturized coplanar waveguide stop band filters based on multiple tuned split ring resonators", *IEEE Microwave and Wireless Communications Letters*, vol. 13, no. 12, 511–513, 2003.
34. M.M.I. Saadoun and N. Engheta, "A reciprocal phase shifter using a novel pseudochiral or V medium", *Microwave and Optical Technology Letters*, vol. 5, no. 4, 184–188, 1992.
35. M.A. Antoniades and G.V. Eleftheriades, "A broadband series power divider using Zerodegree metamaterial phase shifting lines", *IEEE Microwave Wireless Computing Letters*, vol. 15, 808–810, 2005.
36. K.D. Jang, J.H. King, D.H. Lee, and W.S. Park, "Compact resonant antenna based on composite right/left handed transmission line with a magneto- dielectric material", *Microwave and Optical Technology Letters*, vol. 51, no. 8, 1994–1997, 2009.
37. R.W. Ziolkowski and A. Kipple, "Application of double negative metamaterials to increase the power radiated by electrically small antennas", *IEEE Transactions on Antennas Propagating*, vol. 51, no. 10, 2626–2640, 2003.
38. S. Lim, C. Caloz, and T. Itoh, "Electronically scanned composite right/left handed microstrip leaky-wave antenna", *IEEE Microwave Wireless Computing Letters*, vol. 14, 277–279, 2004.
39. I. Arnedo, J. Illescas, M. Flores, T. Lopetegi, M.A.G. Laso, F. Falcone, J. Bonache, J. García-García, F. Martín, J.A. Marcotegui, R. Marqués, and M. Sorolla, "Forward and backward leaky wave radiation in split ring resonator based metamaterials", *IET Microwaves, Antennas Propagating*, vol. 1, no. 1, 65–68, 2007.
40. C. Caloz and T. Itoh, "Array factor approach of leaky-wave antennas and application to 1D/2D composite right/left-handed (CRLH) structures", *IEEE Microwave and Wireless Components Letters*, vol. 14, no. 6, 274–276, 2004.
41. L. Liu, C. Caloz, and T. Itoh, "Dominant mode leaky-wave antenna with backfire-to-endfire scanning capability", *Electronics Letters*, vol. 38, no. 23, 1414–1416, 2002.
42. A. Sanada, M. Kimura, I. Awai, C. Caloz, and T. Itoh, "A planar zeroth order resonator antenna using a left-handed transmission line", *Proceedings of the European Microwave Conference*, Amsterdam, The Netherlands, 1341–1344, 2004.
43. R.W. Ziolkowski and A.D. Kipple, "Reciprocity between the effects of resonant scattering and enhanced radiated power by electrically small antennas in the presence of nested metamaterial shells", *Physical Review Part E*, vol. 72, no. 3, pt 2, 036602, 2005.
44. R.W. Ziolkowski, "Metamaterials applications to electrically small antennas", *Proceedings of the IEEE Integracija Workshop on Antenna Technology: Small Antennas and Novel Metamaterials*, 7–10, Singapore, March 7–9, 2005.
45. A. Erentok and R.W. Ziolkowski, "Dipole antennas enclosed in double negative (DNG) and single-negative (SNG) nested spheres: Efficient electrically small antennas", *Proceedings of the*

IEEE AP-S Integracija symposium and USNC/URSI National Radio Science Meeting, Washington, DC, July 3–8, 2005.
46. R.W. Ziolkowski and A. Erentok, "A path to an efficient electrically small antenna: A dipole antenna enclosed in a double negative (DNG) or a single- negative (SNG) metamaterial spherical shell", *Proceedings of the Symposium on Antennas and Propagation, ISAP2005,* Seoul, South Korea, Aug. 3–5, 2005.
47. R.W. Ziolkowski, "Design, fabrication, and testing of double negative metamaterials", *IEEE Transactions on Antennas Propagat,* vol. 51, 2572–2581, 2003.
48. F.-L. Zhang, Q. Zhao, Y.-H. Liu, C.-R. Luo, and P. Xiao, "Behaviour of hexagon split ring resonators and left-handed metamaterials", *Chinese Physics Letters,* vol. 21, no. 7, 1330–1332, 2004.
49. K. Chandrasekhar, D. Mishra, D.R. Poddar, and R.K. Mishra, "Design of left-handed metamaterials using hexagonal split ring resonator at S-band frequencies", *Proceedings of the IEEE Applied Electromagnetics Conference,* India, Dec. 19–20, 2007.
50. X.P. Zhao, Q. Zhao, L. Kang, J. Song, and Q.H. Fu, "Defect effect of split ring resonators in Left-Handed Metamaterials", *Physics Letters. Part A,* vol. 346, no. 1–3, 87–91, 2005.
51. C. Zhang, M. Sun, Z. Yuan, and X. Song, "Study on new period structures of left-handed material composed of hexagonal SRRs and metal wires", *Acta Physiologica Sinica,* vol. 58, 1758–1764, 2009.
52. C. Saha, Doctoral Dissertation, University of Calcutta, India, 2012.
53. R. Marqués, F. Mesa, J. Martel, and F. Medina, "Comparative analysis of edge and broadside couple split ring resonators for metamaterial design. Theory and Experiment", *IEEE Transactions on Anterieur Propagating,* vol. 51, no. 10, 2572–2581, 2003.
54. I. Bahl and P. Bhartia, *Microwave Solid State Circuit Design,* chapter 2, John Wiley and Sons, New York, 1998.
55. F.E. Terman, *Radio Engineers' Handbook,* McGraw Hill, New York, 1943.
56. F.W. Grover, *Inductance Calculations: Working Formulas and Tables.* Dover Publications Inc., New York.
57. J.Y. Siddiqui, C. Saha, and Y.M.M. Antar, "Compact SRR loaded UWB circular monopole antenna with frequency notch characteristics", *IEEE Transaction on Antennas and Propagation,* vol. 62, no. 8, 4015–4020, 2014.
58. J.Y. Siddiqui, C. Saha, and Y.M.M. Antar, "Compact dual SRR loaded UWB monopole antenna with dual frequency and wideband notch characteristics", *IEEE Antennas and Wireless Propagating Letter,* vol. 14, 100–103, 2015.
59. L.A. Shaik, C. Saha, J.Y. Siddiqui, and Y.M.M. Antar, "Ultra-wideband monopole antenna for multiband and wideband frequency notch and narrowband applications", *IET Microwaves, Antennas and Propagation,* vol. 10, no. 11, 1204–1211, 2016.
60. M. Barbuto, F. Trotta, F. Bilotti, and A. Toscano, "Horn antennas with integrated notch filters", *IEEE Transactions on Antennas and Propagation,* vol. 63, no. 2, 781–785, 2015.
61. J.Y. Siddiqui, C. Saha, C. Sarkar, L.A. Latheef, and Y.M.M. Antar, "Ultra-wideband antipodal tapered slot antenna with integrated frequency-notch characteristics", *IEEE Transaction on Antennas and Propagation,* vol. 66, no. 3, 1534–1539, 2018.
62. J.Y. Siddiqui, Y.M.M. Antar, A.P. Freundorfer, E.C. Smith, G.A. Morin, and T. Thayaparan, "Design of an ultrawideband antipodal tapered slot antenna using elliptical strip conductors", *IEEE Antennas and Wireless Propagation Letters,* vol. 10, 251–254, 2011.
63. C. Sarkar, C. Saha, L.A. Shaik, J.Y. Siddiqui, and Y.M.M. Antar, "Frequency notched balanced antipodal tapered slot antenna with very low cross-polarised radiation", *IET Microwaves, Antennas and Propagation,* vol. 12, no. 11, 1859–1863, 2018.
64. J.D.S. Langley, P.S. Hall, and P. Newham, "Novel ultrawide-bandwidth Vivaldi antenna with low cross polarization", *Electronic Letters,* vol. 29, no. 23, 2004–2005, 1993.
65. C. Saha, P. Natani, L. Ahmed, Y.M.M. Antar, and J.Y. Siddiqui, "Square / hexagonal split ring resonator loaded exponentially tapered slot ultra wideband (UWB) antenna with frequency notch characteristics", *Microwave and Optical Technology Letters,* vol. 59, no. 6, 1241–1245, 2017.
66. P. Natani, L.A. Shaik, C. Saha, J.Y. Siddiqui, and Y.M.M. Antar, "Dual frequency notched coplanar tapered slot antenna using split ring resonators on modified slotline", *2016 17th International*

Symposium on Antenna Technology and Applied Electromagnetics (ANTEM), Montreal, Canada, 1–3, 2016.
67. A. Kumar, L.A. Shaik, C. Saha, and J.Y. Siddiqui, "CPW fed dielectric resonator antenna with complementary characteristics", *2015 IEEE Applied Electromagnetics Conference (AEMC)*, Guwahati, India, 1–2, 2015.
68. C. Saha and J.Y. Siddiqui, "Versatile CAD formulation for estimation of the resonant frequency and magnetic polarizability of circular split ring resonators", *International Journal of RF and Microwave Computer-Aided Engineering*, vol. 21, no. 4, 432–438, 2011.
69. C. Saha and J.Y. Siddiqui, "A comparative analysis for split ring resonators of different geometrical shapes", *2011 IEEE Applied Electromagnetics Conference (AEMC)*, 1–4, December 2011.
70. C. Saha and J.Y. Siddiqui, "Theoretical model for estimation of resonance frequency of rotational circular split-ring resonators", *Electromagnetics*, vol. 32, no. 6, 345–355, 2012.
71. C. Saha, J.Y. Siddiqui, and Y.M. Antar, "Square split ring resonator backed coplanar waveguide for filter applications", *General Assembly and Scientific Symposium, 2011*, Istanbul, Turkey, 1–4, August 2011.

4

UWB Antennas for Multifunctional Operations

4.1 Introduction

A major technological challenge associated with wideband and UWB antenna design is the efficient rejection of interference from various narrowband services having an overlapped spectrum with the UWB band. As discussed in detail in Chapter 3, this interference rejection is performed by introducing an integrated notch-filter section inside the UWB antenna. Design of such frequency-notched UWB antennas appeared as one of the most intriguing research topics for antenna engineers over the last decade. Unprecedented expansion of the wireless market and industry, mainly driven by demands for high-speed high data rate wireless services, acts as continuous motivation for EM and wireless engineers to explore next-generation technology and upgrade the UWB system in a more efficient manner. One of the key approaches attempted over the last few years to enhance the potential of UWB systems further is the efficient utilization of the UWB spectrum in an opportunistic manner. This concept of efficient utilization of the UWB spectrum has evolved due to the observation by Giupponi and Ibars [1] that many frequency bands in the spectrum are largely unoccupied most of the time, and some frequency bands are partially occupied. Cognitive radio (CR) technology, which scans for availability of the unused spectrum from the primary user and allocates it to the secondary user, is an excellent solution toward efficient utilization of the UWB spectrum. Such spectrum search and then allocation of the idle spectrum in a dynamic manner to the secondary user requires compatible antenna systems. The antenna design for such applications is quite complicated as it requires multiple antennas to be housed in a common platform. The antenna(s) here should be capable of wideband "scanning" to check the idle spectrum (this is known as "white space") and simultaneously communicating over a limited bandwidth in a reconfigurable fashion. Thus, for cognitive radio applications, the antenna modules need to be multifunctional.

This chapter deals with the design and development of antennas capable of performing multiple antenna functionalities. As cognitive radio (CR) in software-defined radio (SDR) technology is the major application of these antennas, it is important to be familiar with the basics of CR and SDR technology, especially in the area of antenna requirement. Keeping this in mind, a section focuses exclusively on antenna requirement for SDR and CR technologies. Two different scenarios of CR technologies, spectrum interweave and spectrum underlay [2], are briefly discussed in Section 4.2. Various arts and techniques of designing multifunctional antennas are discussed in detail in Section 4.3. Advantages of multifunctional antenna (MFA) design with a single radiating element over MFA realized with multiple radiators is discussed in Section 4.5.

4.2 Multifunctional Antennas: Concepts and Evolution

Multifunctional antenna in a broad sense is a special type of reconfigurable antenna which, as the name suggests, can perform multiple antenna functionalities, such as UWB operations, frequency-notched UWB operations, and reconfigurable narrowband operation. However, in contrast to conventional reconfigurable antenna, where the aim is to achieve reconfiguration with respect to particular antenna characteristics such as radiation frequency, pattern, polarization, etc., multifunctional antenna focuses more on its reconfigurations based on functions and applications. Sometimes, the MFA should have complementary antenna response [3]. For example, a frequency-notched UWB antenna and a narrowband antenna which radiates at the notch frequency of the notched UWB antenna [3], are two complementary antennas. In addition to yielding multiple cases of input matching in various configurations, an MFA also has to comply with other antenna parameters such as pattern, peak gain, polarization, etc. in different modes of operation.

A multifunctional antenna, similar to a regular reconfigurable antenna, achieves the multiple antenna functionalities with various tuning techniques such as electrical tuning, mechanical tuning, etc. A detailed discussion on various techniques of antenna tuning in a conventional reconfigurable antenna is included separately in Chapter 5. Figure 4.1 shows the basic block diagram of a multifunctional antenna. It comprises of two main elements:

1. a wideband or UWB antenna
2. a tuning block for activating various modes/functionalities.

Though wideband or UWB antenna can be chosen from various families of popular UWB candidates, circular disc monopole or its variants is mostly preferred due to its omnidirectional radiation pattern and consistent gain. The tuning block of the MFA is its most important part as this makes the UWB antenna operate in various "configurations" which includes all the desired functionalities such as frequency-notched functions, narrowband function as well as default UWB function. The tuning block provides such multifunctionalities by altering the current distribution on the antenna which in turn contributes to the desired impedance profiles. Two major techniques used to achieve the multifunctional responses are:

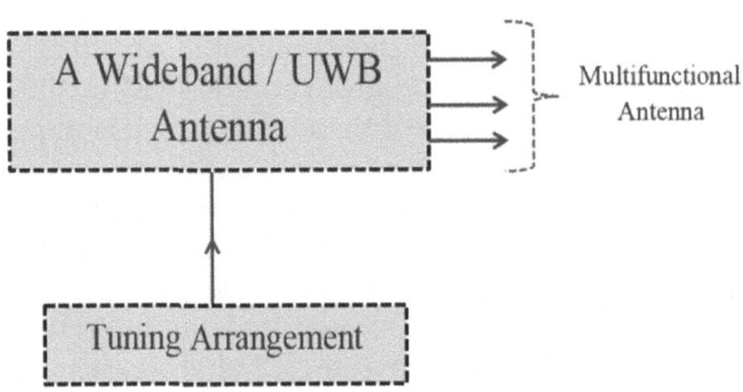

FIGURE 4.1
A conceptual block diagram of multifunctional antenna.

4.2.1 Employing Multiple Radiating Elements on a Common Substrate

This technique normally uses a dedicated radiator to provide the UWB response and multiple narrowband radiators to cover the narrowband functionalities. Figure 4.2 shows a schematic diagram explaining this technique. Here, antenna 1 is a microstrip-fed circular monopole antenna providing UWB response. Antenna 2 provides multiple narrowband response with a mechanical arrangement. As indicated in the schematic diagram, antenna 1, and antenna 2 are housed in a common substrate. As a portion of the substrate is rotatable, different radiators marked as 1, 2, ...*n* come in contact with the feedline and excite various narrowband resonances. The narrowband radiators are designed to resonate at different frequencies within the UWB band based on the applications and command from the sensing antenna.

Apart from the desired multifunctional response from such an antenna system, the designer also needs to take care to ensure a good amount of isolation between the UWB antenna and the narrowband radiators, and minimize the impact of (*n*-1) unused radiators in proximity when not in use. Moreover, positioning of the motor, its size, weight, and impact on the overall performance of the antennas are to be decided based on the overall system specifications.

4.2.2 Employing a Single Radiating Element

Multiple antenna functionalities can be derived from a single radiating element by an improvised design of the feedline. Different combinatorial loading of the split-ring resonator (SRR), along a planar transmission line and switches on the feedline and gaps of the SRR, invokes various filtering responses, such as a notched response, narrow band pass response and all-pass response. When such a feedline excites a UWB antenna, based on the status of the switches, the same antenna can provide multiple antenna responses. Figure 4.3 shows the schematic diagram of various combinations of the SRR and switches loaded on a feedline of a monopole antenna. Figure 4.3a shows the structure of a popularly known CPW-fed

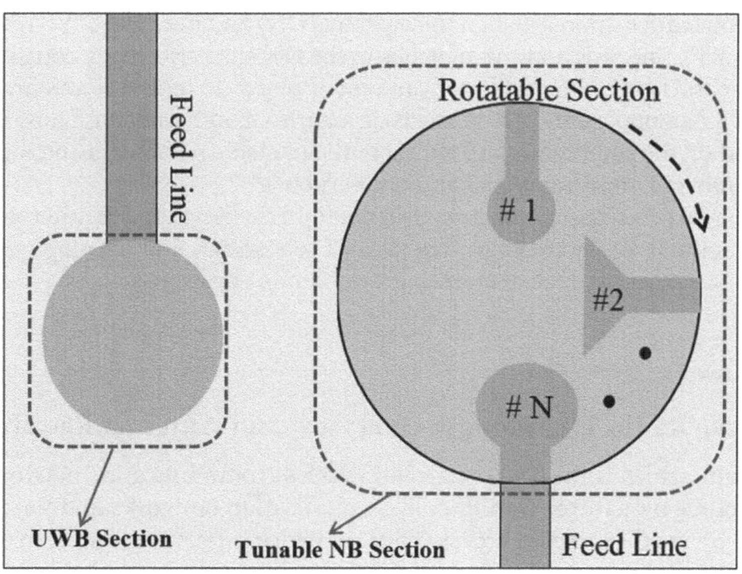

FIGURE 4.2
A conceptual block diagram of multifunctional antenna with multiple radiating elements on a common substrate.

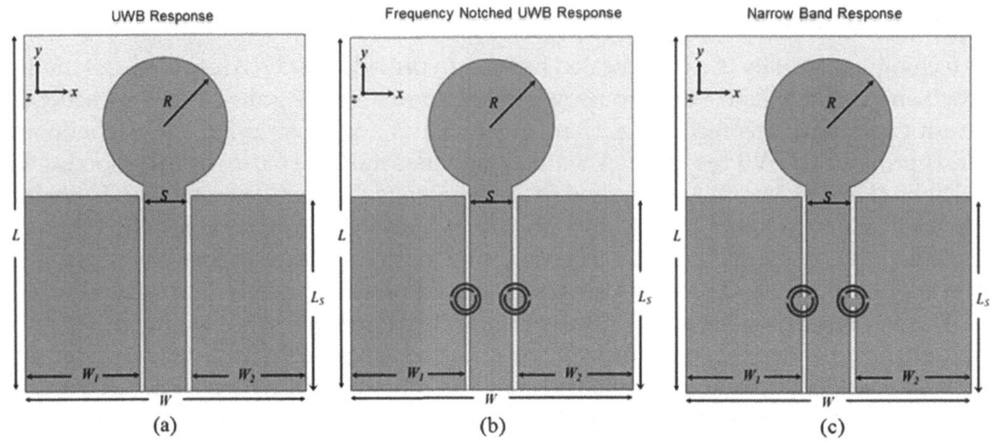

FIGURE 4.3
A conceptual block diagram of multifunctional antenna with single radiating elements. The multifunctionality is due to the improvised feed-line of the antenna.

circular disc monopole antenna which covers the entire UWB band and radiates an omnidirectional beam. As shown in Figure 4.3b, and discussed in detail in Chapter 3, when the antenna is loaded with a pair of split-ring resonators, the antenna acts as frequency-notched UWB antenna. The propagating EM signal along the feedline excites the resonance in the SRR which in turn inhibits signal propagation at its resonance frequency and contributes to the frequency notch in the UWB band. For strong excitation of the SRR resonance, it is placed in such a way that the center of the SRR placed opposite to the feedline is located at the center of the slot gaps. As discussed in Chapter 3, the notch frequency can be controlled by selecting the geometrical shape and size of the SRR [4]. When a pair of switches are placed in a precise location between the central line and ground planes of the CPW feedline, based on the status (ON/OFF) of the switches, the notched UWB antenna can be made to toggle between narrowband response (when the switch is ON) and notched-UWB response (when the switch is OFF). The SRR and the switches in the ON state effectively constitute a narrow bandpass filter and hence allow a filtered amount of energy to reach the antenna. Due to this, the same UWB monopole antenna now acts as a narrowband antenna. Thus, a single UWB radiator, based on the combinatorial loading of the parasitic resonators and switches on the feedline, can provide multifunctional antenna response.

Multifunctional antennas, as discussed in detail in the next section, are the most crucial requirement at the front end of cognitive radio. Their design and development are quite a new concept and requires special attention by antenna engineers.

4.3 Cognitive Radio Technology: Overview and Antenna Requirement

Radio spectrum, which is the backbone of all wireless communication network services, is a limited resource by nature. Traditional communication networks and wireless services are managed by fixed spectrum access (FSA), in which a portion of available spectrum is assigned to one/multiple users in a predefined manner. In FSA schemes, as the spectrum is completely owned by the assigned users, they enjoy the spectrums, while other users are not entitled to use it, even if the assigned users are idle. A statistical report by a task

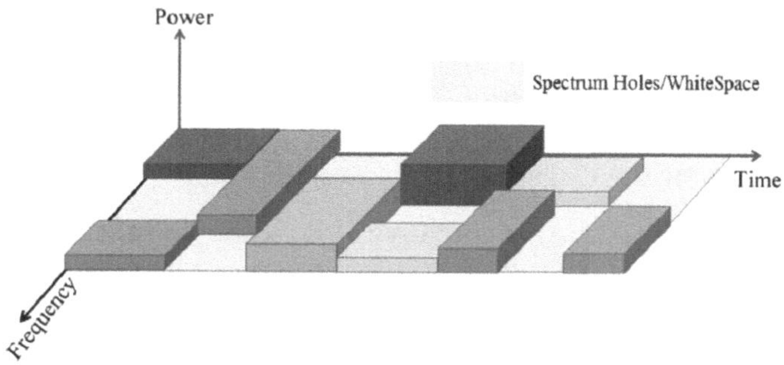

FIGURE 4.4
Spectrum holes/white space in a combined time-frequency diagram.

force, formed by the FCC to monitor usage of spectrum, reveals severe underutilization of many frequency bands with fixed spectrum allocation techniques. In some geographical locations, sometimes the allocated spectrum to the licensed users remains idle for more than 90% of the time. Figure 4.4 shows the typical plot of power against combined time and frequency scale associated with a radio frequency (RF) environment. As indicated in the plots, there are various unused spectrums which may change over time. Opportunistic and intelligent utilization of these unused spectrums of the licensed users, known as spectrum holes/white space, to the secondary users can significantly help in mitigating the challenges related to wireless traffic management.

4.3.1 History and Motivation

The term "cognitive" radio was first introduced by J. Mitola in 1999. As explained by Mitola in an article [5] and later on, in his Ph.D. dissertation [6], cognitive radio was touted as the most efficient technology to achieve much improved traffic capacity in a wireless channel by exploiting the idle spectrum [7]. As indicated before, CR technology achieves this by allowing the unlicensed users to access the frequency bands dedicated to the licensed users. Mitola's new concept of cognitive radio and the report by the task force formed by the FCC in 2002, as mentioned previously, are considered the two most important landmarks in the development of the new CR-based wireless technology. To explore feasibility, challenges and new directions, a workshop on cognitive radio technology [8] was organized in Washington, DC in May 2003. This was followed by a conference focusing exclusively on cognitive radio in March 2004 [9]. It was soon understood that practical realization of the theoretical concepts of CR technology is a gigantic task and demands collaborative research efforts from various research wings such as signal processing, computer networks, communication theory, RF circuits, and antenna engineering.

4.3.2 What is a Cognitive System?

Before we introduce the technical details and antenna requirements for cognitive radio technology, let us understand the meaning of 'cognition'. According to the Oxford Dictionary, the word "cognition" implies:

- The mental action or process of acquiring knowledge and understanding through thought, experience, and the senses
- A perception, sensation, idea, or intuition resulting from the process of cognition.

Thus, a cognitive system corresponds to an intelligent system which learns from the previous experience in an operating environment and adapts itself according to the system requirement to provide maximum throughput.

4.3.3 Brief Overview of the Cognitive Radio System

Dynamic spectrum sharing (DSA) of cognitive radio technology makes it one of the most promising technologies for future wireless communications. In line with the literal meaning of "cognition", as defined by Haykin [1], a cognitive radio is:

> "[...] an intelligent wireless communication system that is aware of its surrounding environment (i.e., outside world), and uses the methodology of understanding-by-building to learn from the environment and adapt its internal states to statistical variations in the incoming RF stimuli by making corresponding changes in certain operating parameters (e.g., transmit-power, carrier frequency, and modulation strategy) in real-time, with two primary objectives in mind:
>
> - highly reliable communications whenever and wherever needed;
> - efficient utilization of the radio spectrum."

A cognitive radio network, unlike a conventional radio network working on the principle of fixed spectrum access (FSA), supports two different kinds of users:

- *Primary user:* Primary users own the spectrum and therefore have the highest priority to use a particular piece of spectrum/frequency band allotted to them. Primary users have no constraint on the radiated power level. However, the spectrum is not exclusively granted to the primary users; when idle, upon request, it should be released to the other users.
- *Secondary user:* In CR technology, the secondary users are allowed to use particular pieces / a band of spectrum on a dynamic basis when the primary user of that particular band is not using it. Alternatively, secondary users can get access to the primary user's spectrum on a sharing basis with the primary user, provided that the primary user's interest is protected. This can be ensured by forcing the secondary users to operate within a specific interference margin allowed by the primary users. In CR technology, secondary users are referred as the "cognitive user" as they must adjust in dynamic manner to transmit/receive in real time based on the availability of the idle spectrum.

Thus, for a cognitive radio device to be useful a in real-time scenario, it must have "cognition" capability and sufficient built-in intelligence to evolve using previous observation and adapt to the changing RF and wireless environment. The "cognition" property of a CR system is achieved through four main functional blocks: spectrum sensing, spectrum management, spectrum sharing, and spectrum mobility. It uses four cognitive cycles in achieving this: Observe⟶Decide⟶Act⟶Learn as shown in Figure 4.5.

From an RF and antenna engineering standpoint, the first step "Observe" implies scanning and sensing the RF environment, popularly known as "spectrum sensing". This action requires compatible antennas. The second step "Decide" indicates the intelligence of the CR system. Based on the previous experience, the cognitive radio has to come to a course of action, which is normally executed through various algorithms such as neural network, particle swarm optimization (PSO), genetic algorithm, etc. "Act" indicates the response of the cognitive radio based on the result obtained from scanning the RF

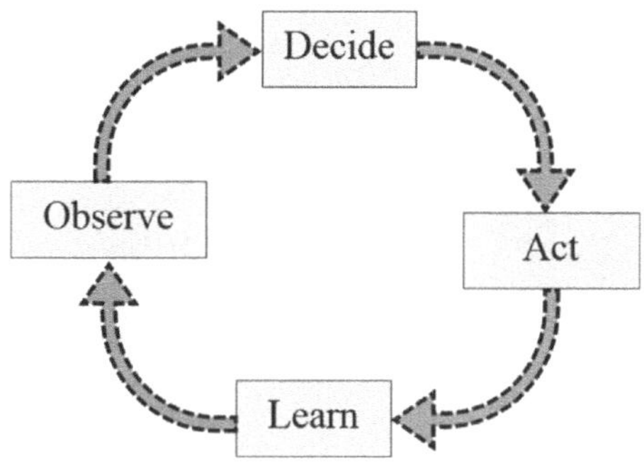

FIGURE 4.5
Cognitive cycles in a cognitive radio environment.

environment. This step demands "reconfiguration" block in the CR as it has to work on a changed operating condition such as switching from one frequency to another, directing the beam in the proper direction, etc. The fourth step "Learn" implies special capability of a CR-based system to be able to learn from past observation and desires and anticipate, predict, and correct present standards of communication, mode of operation and other associate RF parameters accordingly.

4.3.4 Transmission Techniques in CR System

Cognitive radio systems employ three different strategies/models to support transmission from secondary/cognitive users and at the same time ensuring priority of the primary users. These are:

- *Overlay model:* Overlay model, also known as interference model, supports concurrent transmission of primary and secondary users. In this model, secondary users use part of their transmitted power for relaying the data of primary users and the remaining power is used for transmitting their own data.
- *Underlay model:* This model is also known as interference avoidance model. Like the overlay model, this also supports concurrent transmission of primary and secondary users over the ultrawideband (UWB) spectrum. However, secondary users' power level is strictly maintained below a threshold level, decided by the primary user. This enforcement of spectral mask to the secondary users ensures the signal power from the secondary users remains below the noise floor of the primary user and thereby primary users are fully protected. Low power level of the secondary signals restricts the underlay model from being applicable for low data rate and short-range applications [10].
- *Interweave model:* In this model, the cognitive radio scans the entire allocated spectrum to the primary user and searches for the unused spectrum, customarily known as "white space" or spectrum hole. Based on the available white space at any instant in time, the cognitive radio communicates through this band in a dynamic fashion. The main advantage of the interweave model is that, here, the secondary users' power level is not restricted. However, the main challenge in this model is to sense very fast and accordingly design a highly agile system to tune the frequency of the secondary user. Figure 4.6 shows the power spectral density (PSD) as a function of the frequency for the underlay and interweave models.

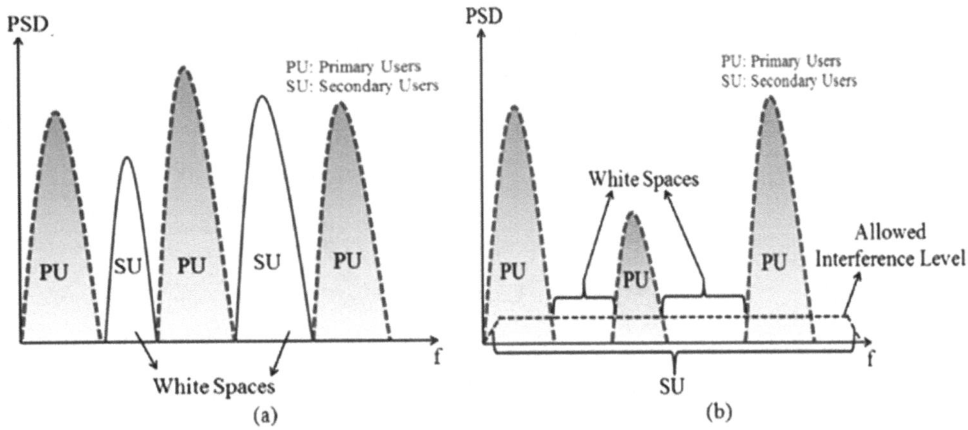

FIGURE 4.6
Spectrum holes/white space in a combined time-frequency diagram [2].

Reproduced by permission of © 2014 The IEEE [2].

As revealed from Figure 4.6, in the interweave model secondary users can only use the spectrum holes of the primary users without any restriction in the power level; while in the underlay model, secondary users operate over the same band as that of primary users but within the allowed interference level.

4.3.5 Software-Defined Radio to Cognitive Radio: Evolution

RF and antenna engineers consider cognitive radio (CR) technology as an evolution and improvement on software-defined radio (SDR). Though the field of SDR and CR is quite versatile and multi-disciplinary, we should have a brief idea on the evolution from conventional fixed spectrum radio operating in FSA mode to modern cognitive radio, working on the principle of dynamic spectrum access. The evolution of modern cognitive radio (CR) can be described by the following stages:

- Hardwire driven radio (HDR): This is the conventional radio for which the frequency, modulation techniques, and various related RF parameters are solely determined by hardware and cannot be changed without a change in hardware.
- Digital radio (DR): A digital radio performs signal processing and transmission in full or part using digital techniques. However, it is not programmable and intelligent like a "cognitive" radio.
- Software-defined radio (SDR): This has evolved due to the necessity to implement radios that can be controlled by programming to reconfigure so that it can be operated over a wide frequency range, various modulation schemes, etc. In this way, SDR can support multiple standard wireless radios.
- Cognitive radio (CR): Cognitive radio (CR) is an improvement on SDR due to the ability to learn from the observed environment and autonomously reconfigure its hardwire, including antennas and RF circuits, to support the optimal mode of communication. CR can be defined as: "An SDR that is aware of its environment, internal state, and location and can autonomously adjust its operations to achieve a number of designated objectives" [11, 12].

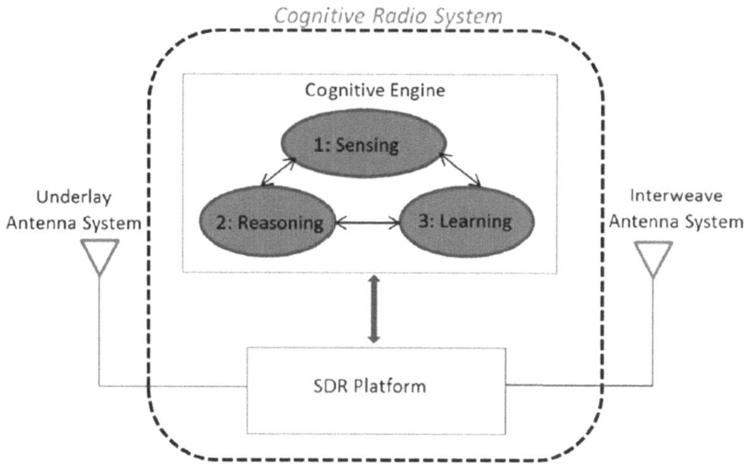

FIGURE 4.7
A CR platform evolved from an SDR platform.

Reproduced by permission of © 2014 The IEEE [2].

Figure 4.7 shows the block diagram of a CR system. It comprises the SDR platform and the cognitive engine which controls the antenna system and specifies the operating or notch frequency of the communicating antenna along with power level of the primary and secondary users.

4.3.6 Antenna Requirements for CR Models

Different models of cognitive radio technology require different types of antennas to be housed within the cognitive radio wireless device. Figure 4.8 shows a basic block diagram of the antenna module in a CR system. In general, a CR system comprises a sensing UWB antenna and a reconfigurable antenna.

As shown in Figure 4.8, CR technology requires two sets of complementary antennas: a UWB antenna and a reconfigurable antenna for efficient spectrum utilization. The UWB antenna, called the "sensing antenna", continuously scans over the ultrawideband to search the idle spectrum not currently in use by the primary user and allocates it to the secondary user which immediately starts communicating at this frequency, using the reconfigurable antenna, also known as the "communicating antenna". Thus, efficient utilization of the idle spectrum dedicated to the primary user and distributing it dynamically to the secondary user is the key in CR technology to achieve higher spectrum utilization.

To utilize the available idle spectrum in a dynamic manner, the communicating antenna in the CR antenna system needs to be reconfigurable in nature so that its operating frequency can be synchronously tuned with the command obtained from the sensing antenna. As discussed previously, the CR system operates in three modes, i.e. overlay, underlay, and interweave. In the interweave mode of operation, as shown in Figure 4.6a, the device searches for white spaces, and decides which white space (frequency range) to allocate to secondary users. The secondary users can transmit without any power constraint in this case. The spectrum-sensing techniques can be wideband or narrowband by requiring a wideband or narrowband sensing antenna. As the secondary user transmits in the available white spaces, the communicating antenna should be a reconfigurable

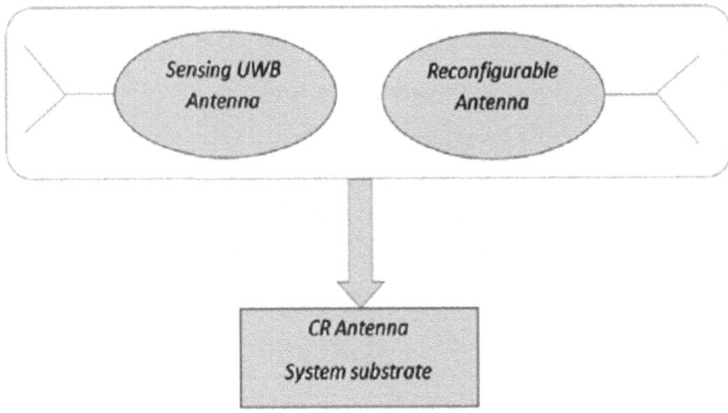

FIGURE 4.8
General antenna modules in a cognitive radio system.

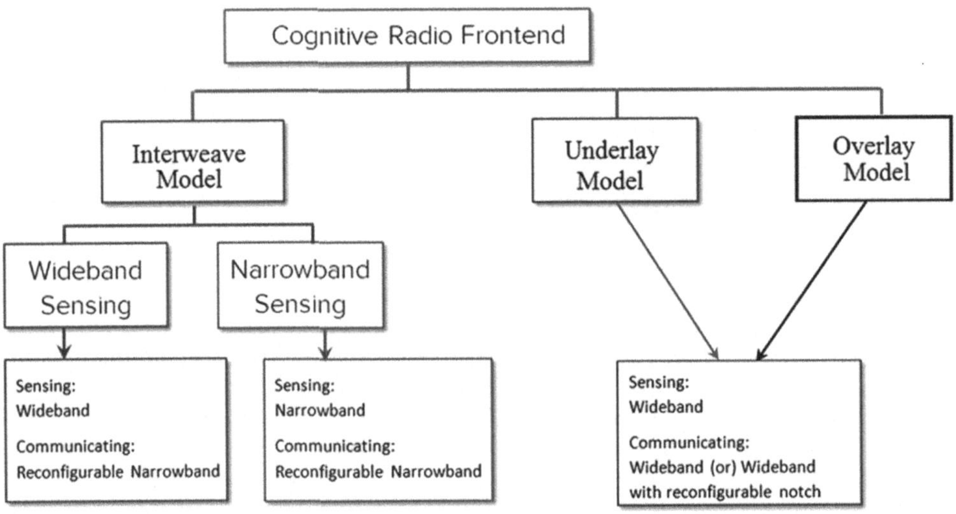

FIGURE 4.9
Antenna requirement for various model of CR systems.

narrowband antenna. In the underlay mode of operation as shown in Figure 4.6b, the device searches for white spaces. Overlay mode of operation acts almost in an identical way. In these cases, there is a constraint on the transmitted power by the secondary users, since both primary and secondary users can occupy the same frequency bands. However, if under certain circumstances primary users cannot tolerate any level of secondary users' interference, the secondary users are forced to transmit over the whole frequency band except for the restricted channels occupied by the primary users. The spectrum-sensing techniques are wideband thereby leading to wideband-sensing antenna. As the secondary user can transmit over the entire spectrum with specified interference levels or the entire spectrum excluding certain portions, the communicating antenna should be a wideband or wideband with notched narrowband functionality. Figure 4.9 summarizes the different antenna required for various models of CR system.

4.4 Multifunctional Antenna Design Techniques

We have already touched on multifunctional antennas in Section 4.2. As briefly mentioned, multifunctional antennas are a special class of reconfigurable antennas which can perform multiple antenna functionalities and yield various antenna responses, such as UWB response, frequency-notched UWB response, and narrowband response. These attractive features of multifunctional response are almost in line with features of the antenna module of a cognitive radio. Thus, due to the diversified antenna response, multifunctional antennas have great potential for cognitive radio (CR) applications. Research and development of multifunctional antennas are driven by requirements of the wireless industry and cater to the software-defined radio and cognitive radio. While antennas for software-defined radio mainly need to be band-switchable, design of CR antennas are more complicated as they must have multiple antenna response. Switched band antennas can alter their operating frequency based on various techniques and are a good candidate for software-defined radio. Such antenna, if designed with a switching speed of greater than or equal to twice the maximum of signal bandwidth, can receive/transmit more than one signal simultaneously [13, 14].

In this section, we will focus on the design techniques of multifunctional antennas. Since such antennas need to have UWB, narrowband, and/or notched UWB response, traditionally, they are designed using multiple radiating elements on a common substrate. Alternatively, with improvised design of the radiator/feedline and combinatorial loading of the switches and parasitic elements, a single radiator can yield multiple antenna response. We will discuss both the techniques of MFA design. Moreover, we will discuss the applicability of the various designs in the cognitive radio environment.

4.4.1 MFA Using Multiple Radiating Elements

As briefly mentioned in Section 4.2 and indicated in Figure 4.2, in this case multiple radiating elements are placed on a common substrate. Different antennas can yield the various antenna responses needed. For interweave modeled CR applications, one of the radiators is designed as a UWB antenna for spectrum sensing while other radiator(s) act as the "communicating antenna" with frequency reconfigurable features.

Figure 4.10 shows a multifunctional antenna comprising multiple radiators [15]. A slotted UWB monopole as shown in Figure 4.10b, evolved from 4.10a, acts as the "sensing antenna". The slotted UWB monopole antenna is obtained by removing a circular portion from the basic UWB radiator. This circular region is accommodated with two radiating elements placed on a rotatable circular substrate. Rotation of the circular substrate sequentially brings the two radiating elements in contact with the feeding line and thereby excites two different narrowband configurations. Figure 4.11 shows the fabricated prototypes of the antenna in which the central rotatable portion acts as the communicating antenna. Figures 4.12 and 4.13 show the simulated and measured return loss of "sensing" antenna and reconfigurable "communicating" antenna, respectively.

For antenna systems comprising multiple radiating elements on a common substrate, ensuring minimal coupling is also important. As shown in the plot of Figure 4.14, the CR antenna system provides a coupling better than –10 dB for the entire UWB band.

In Figure 4.15 [16], a similar concept of rotating the radiating elements to achieve reconfiguration of the communicating antenna accommodated with the UWB sensing antenna is shown. Here, a microstrip-fed elliptical monopole provides UWB response which

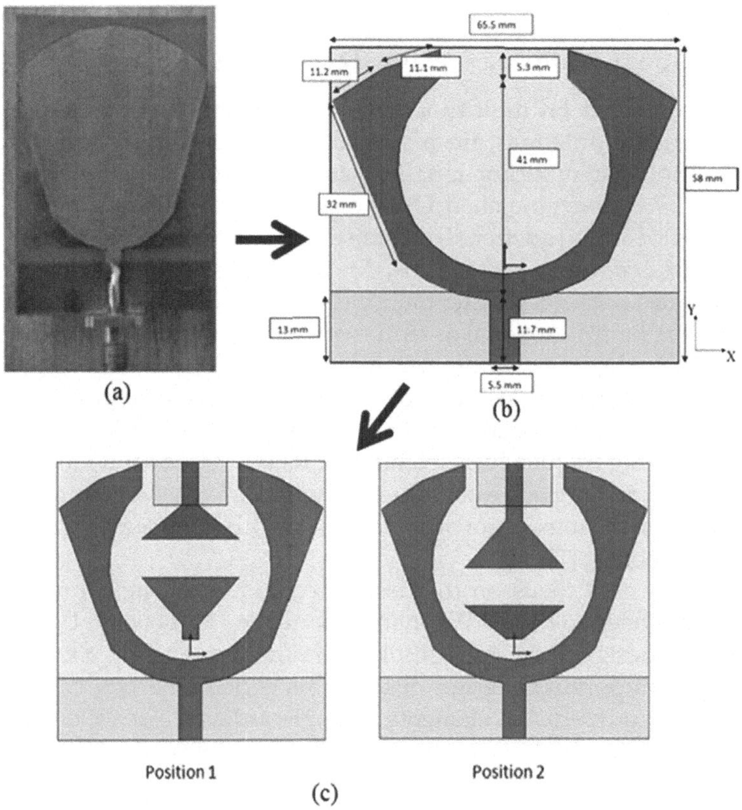

FIGURE 4.10
Sensing and reconfigurable communicating antenna for CR application.

Reproduced by permission of © 2009 The IEEE [15].

performs the spectrum sensing. A stepper motor augments rotational arrangements of various shaped radiators printed on a circular portion of substrate which is placed inside the circular hole of the main substrate, providing the reconfigurable "communicating" function of the multifunctional antenna. As shown in Figure 4.16, the stepper motor is suitably programmed to actuate a different amount of angular rotation in various steps so that different radiating elements come into contact with the feedline of the communicating antenna. Figure 4.17 shows the top and bottom surface of the fabricated prototype of the antenna along with the motor embedded on the substrate. The sensing antenna provides good impedance matching for the entire UWB band while the reconfigurable communicating antenna covers five narrow bands for five different phases of the stepper motors.

A reconfigurable slot resonator incorporated inside the elliptically shaped microstrip-fed monopole antenna [17] serves as a good communicating antenna. Here, a varactor diode loaded inside the slot yields the desired tunability of the communicating antenna while the elliptical antenna covering the entire UWB band acts as the scanning antenna.

Another useful technique in designing the communicating antenna stage of the CR antenna system is to employ reconfigurable filtennas. A tunable bandpass or band stop section loaded in the feedline of the communicating antenna along with a UWB antenna

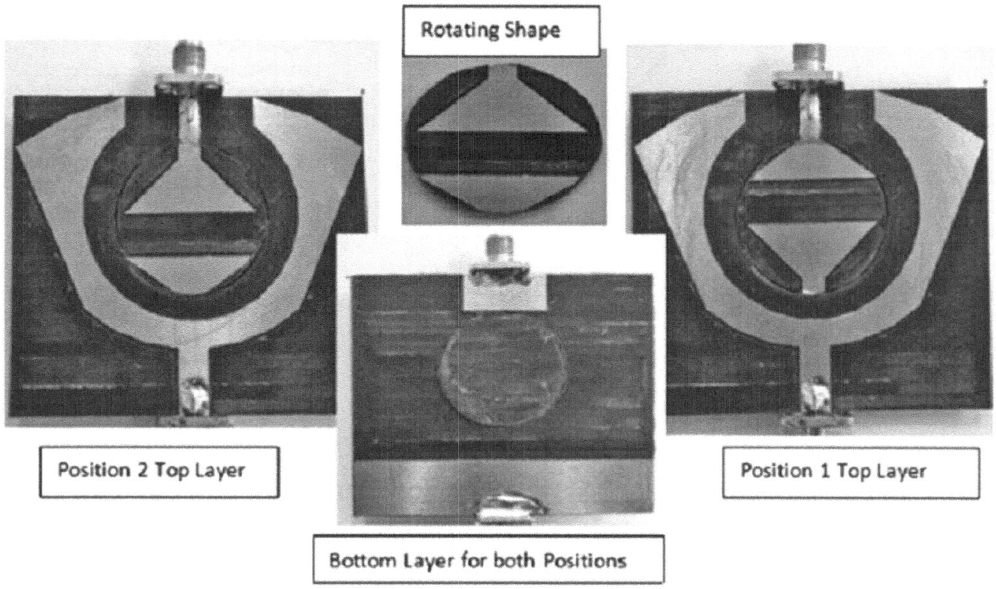

FIGURE 4.11
Fabricated prototype of the sensing and reconfigurable communicating antenna for CR application.

Reproduced by permission of © 2009 The IEEE [15].

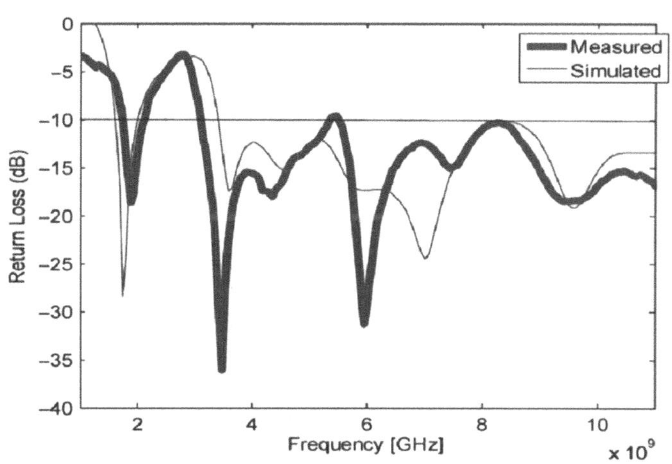

FIGURE 4.12
Return loss of the sensing antenna.

Reproduced by permission of © 2009 The IEEE [15].

together forms an excellent CR antenna for spectrum underlay and spectrum interweave models [2]. Figure 4.18 shows the feedline of the communicating antenna. The feed section consists of three microstrip sections in series in which the central section is gap coupled and loaded with a T-shaped slot. Pin diodes (*S1* and *S2*) loaded inside the slot change the effective dimension of the slot and thereby tune the passband response of the filter. Figure 4.19 shows the *S*-parameters of the bandpass filter. This filter section loaded microstrip

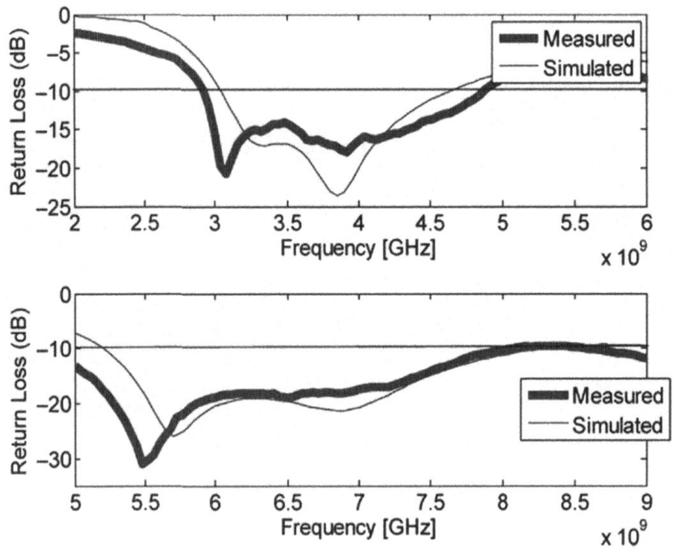

FIGURE 4.13
Return loss of the "communicating" antenna for two different positions.

Reproduced by permission of © 2009 The IEEE [15].

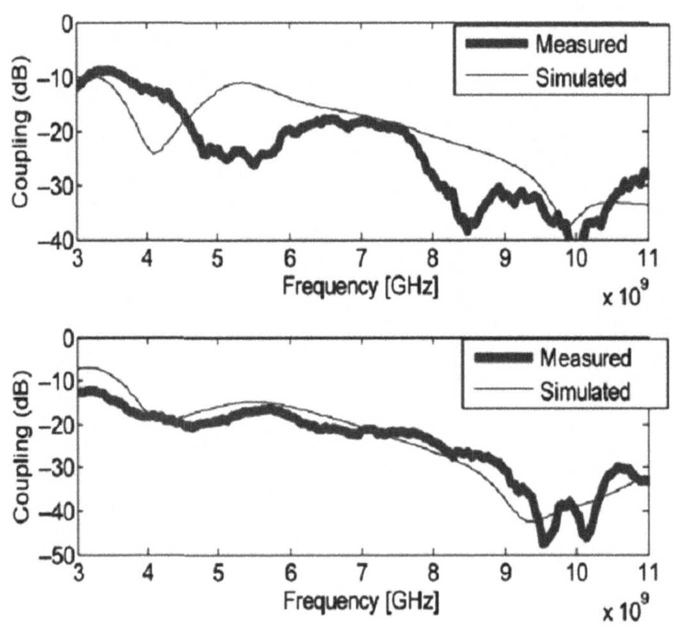

FIGURE 4.14
Coupling between the "sensing" and "communicating" antenna for two different positions.

Reproduced by permission of © 2009 The IEEE [15].

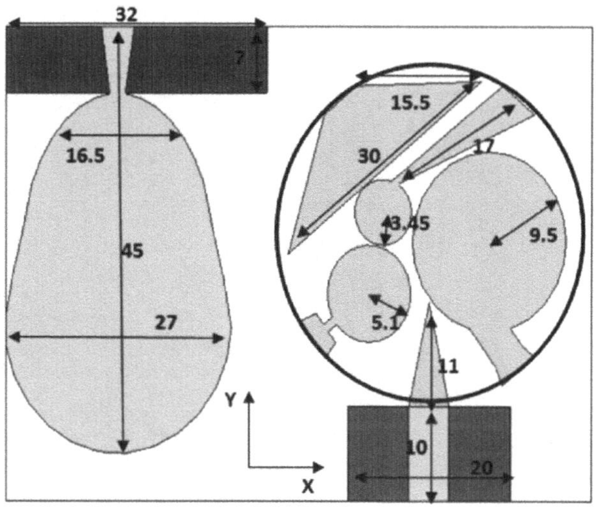

FIGURE 4.15
Multifunctional antenna structure with multiple radiating elements.

Reproduced by permission of © 2011 The IEEE [16].

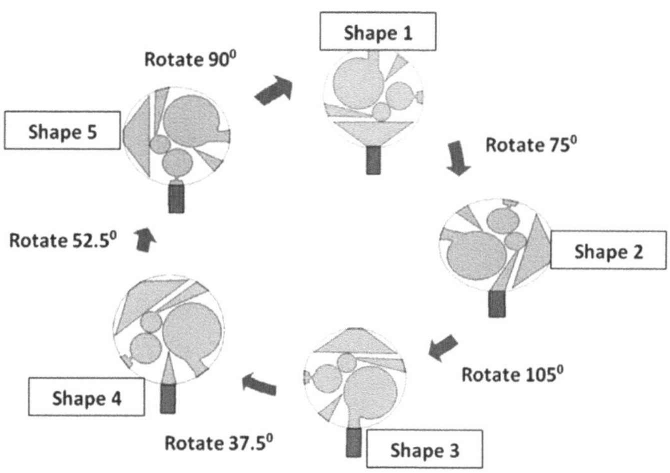

FIGURE 4.16
Reconfiguration process of the communicating antenna [16].

Reproduced by permission of © 2011 The IEEE [16].

acts as the feedline of the communicating antenna which in turn contributes to the desired reconfiguration in the communicating antenna.

4.4.2 MFA Using Single Radiating Elements

The design concept of multifunctional antenna using single radiating elements was briefly discussed in Section 4.2 of this chapter. Since, in this case, the same antenna contributes to multiple antenna functionalities (UWB, notched UWB, and/or narrowband response),

FIGURE 4.17
Fabricated antenna prototype with rotational arrangement [16].

Reproduced by permission of © 2011 The IEEE [16].

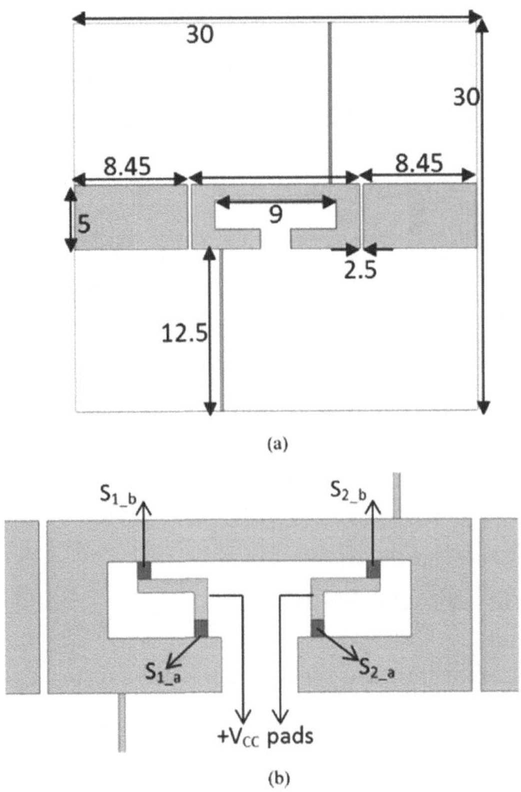

FIGURE 4.18
A tunable band-pass section working as the feed of a communicating antenna of a CR system.

Reproduced by permission of © 2014 The IEEE [2].

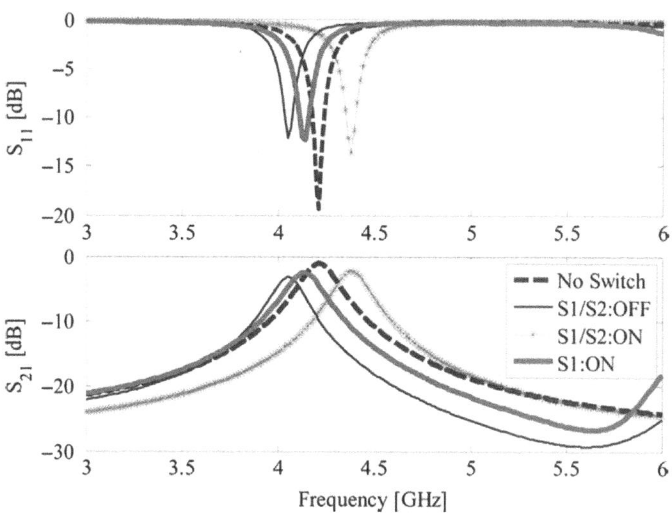

FIGURE 4.19
Fabricated antenna prototype with rotational arrangement[n].

Reproduced by permission of © 2014 The IEEE [2].

this technique employs a UWB radiator as the basic element. When the feed section of this basic UWB radiator is loaded with band-notch and/or band-pass filter sections, the same antenna contributes to multiple antenna response. Mechanical or electrical switching of the feed section to toggle around all-pass, band-notched, and band-pass response with sufficient speed is key in the practical realization of such multiple antenna responses. In addition, sufficient care should be taken to ensure minimal impact of the switching and tuning elements on the radiator. Key advantages of this technique of MFA design are:

- minimum antenna footprint as one radiator is enough to achieve multiple antenna response
- no isolation-enhancing elements are needed like the MFA design with multiple radiators.

Figure 4.20 shows a multifunctional printed UWB antenna [3]. The basic UWB radiator, which is a CPW-fed printed monopole antenna, is loaded with split-ring resonators and switches to derive three different antenna responses. As discussed in Chapter 3, the UWB monopole when loaded with pair of SRRs with precise alignment on the feedline, so that the center of the SRRs and the center of the CPW slot lines coincide, the UWB monopole acts as frequency-notched UWB antenna. Here, the strong magnetic coupling of the propagating electromagnetic signal along the CPW excites SRR resonance which in turn inhibits the signal propagation at the resonance frequency. As shown in Figure 4.21, when this SRR-loaded antenna is coupled with a pair of copper strips, which effectively act like closed switches between the CPW signal line and the ground planes, the same UWB monopole antenna is converted into a narrowband antenna. SRRs and switches at RF frequency together constitute a narrow band pass filter around the SRR's resonance frequency which allows filtered energy to reach the UWB antenna. Thus, the same antenna now acts as a narrowband antenna. When the switches are open/OFF, the antenna acts as a frequency-notched UWB antenna.

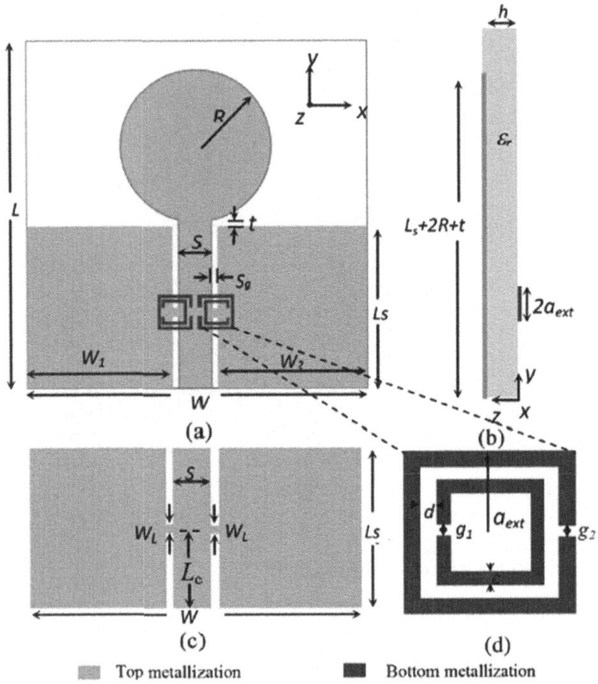

FIGURE 4.20
Schematic of a printed circular monopole fed by SRR coupled and shunt strip shorted CPW (configuration A): (a) top view with SRR on bottom metallization, (b) side view with SRR separated by h from the CPW-fed Monopole, (c) schematic of a CPW transmission line shorted with shunt strips of width W_L (d) Schematic of a unit cell square SRR.

Reproduced by permission of © 2015 The IEEE [3].

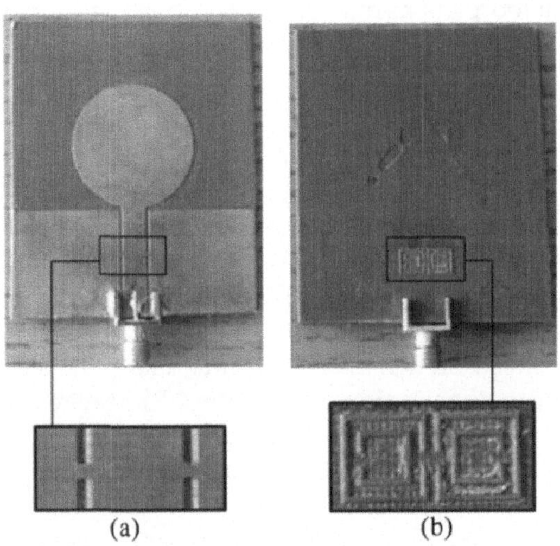

FIGURE 4.21
Fabricated prototype of a printed circular monopole fed by SRR coupled and shunt strip shorted CPW (configuration A): (a) top view, (b) bottom view.

Reproduced by permission of © 2015 The IEEE [3].

TABLE 4.1

Illustration of Different Combinative Loading of SRRs, CRRs and Shunt Strips on the CPW Feed of the Printed Circular Monopole Antenna and their Different Characteristic Response

Antenna Topologies	Schematics of the Loaded/ Unloaded CPW Feed (Top, Bottom)	SRR/CRR	Shunt Strip	Antenna Response
Configuration A	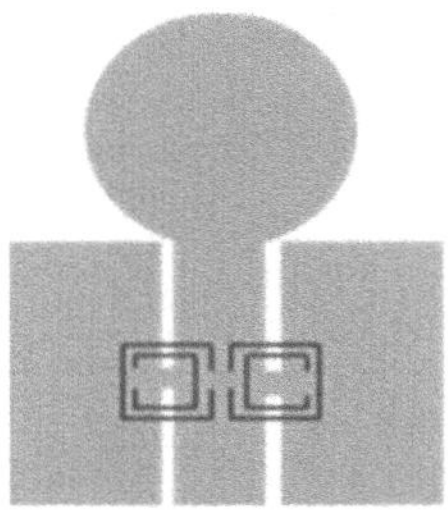	SRR (g_1=g_2≠0)	YES (W_L≠0)	Narrowband
Configuration B	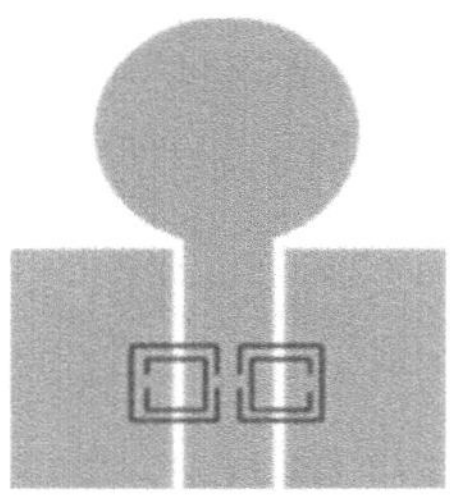	SRR (g_1=g_2≠0)	NO (W_L=0)	Frequency- Notched UWB
Configuration C	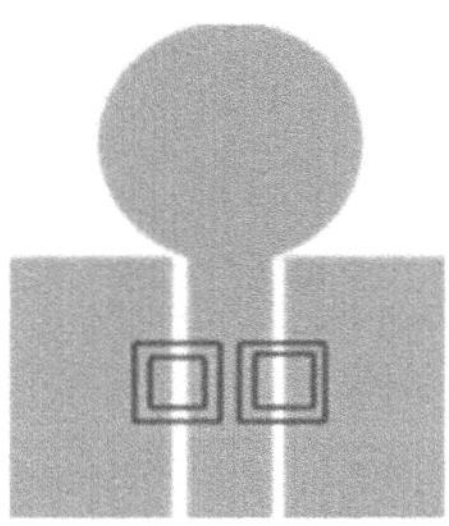	CRR (g_1=g_2≠0)	NO (W_L=0)	UWB

Source: Reproduced by permission of © 2015 The IEEE [3].

The SRR-loaded CPW antenna, which provides a frequency-notched UWB response, can again be converted into a UWB antenna by closing the gaps of the SRR. Closing the SRR gaps converts it into a closed-ring resonator and thereby annihilates its resonance behavior and as a result, the antenna again works as a UWB antenna. Table 4.1 summarizes the performance of the antenna and clearly shows three different antenna functionalities corresponding to three different configurations named as configuration A, B, and C, respectively.

Figure 4.22 shows the measured S_{11} of the antenna for frequency-notched UWB and narrowband mode. As shown in the figure, the same UWB antenna exhibits notched UWB and narrowband antenna responses, yielding a complementary impedance profile. Another interesting feature of this antenna is that the notch/narrow band frequency of the antenna can be smoothly controlled by changing the geometrical parameters of the SRR. As discussed in Section 3.4 of Chapter 3, change in the SRR's geometrical parameters changes its resonance frequency which in turn controls the notch/narrow band frequency

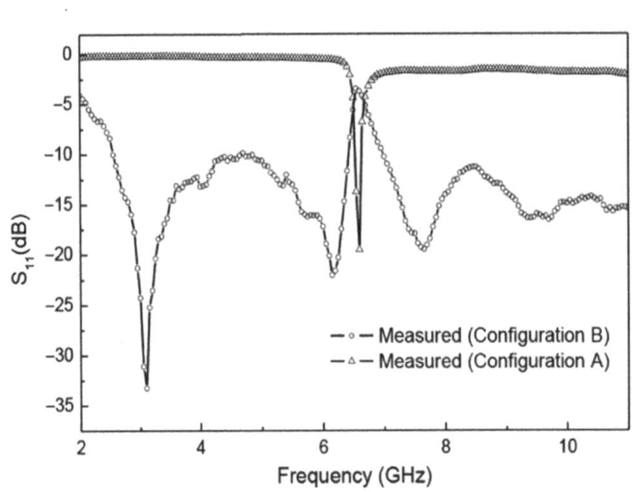

FIGURE 4.22
Measured *S11* of the CPW-fed circular monopole loaded with SRRs only (configuration B) and with SRRs and shunt strips (configuration A).

Reproduced by permission of © 2015 The IEEE [3].

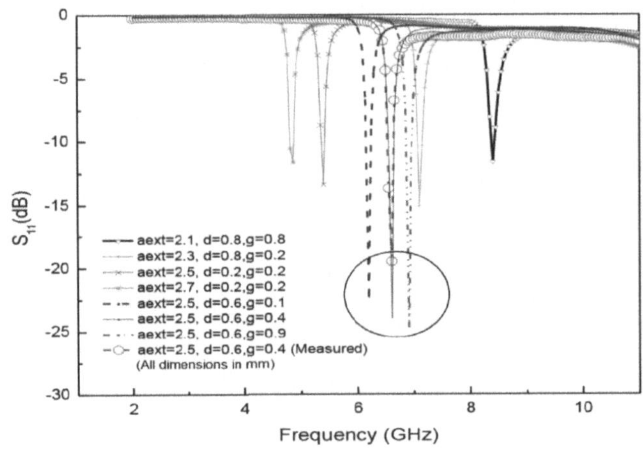

FIGURE 4.23
Simulated S_{11} of the CPW-fed circular monopole loaded with SRRs and shunt strips (configuration A) for variable SRR parameters and compared with measured prototype of configuration A.

Reproduced by permission of © 2015 The IEEE [3].

of the antenna with the switch ON/OFF position. Figure 4.23 shows the S_{11} response of the antenna in the narrow band mode for various combinations of SRR dimensions. As shown from the plot in Figure 4.23, the antenna exhibits a narrowband tuning in the range of 4.4 GHz to 8.4 GHz. Practically, such frequency tuning is possible with same SRR by loading a varactor diode across it and changing the bias voltage of the varactor diode. Figure 4.24 plots the simulated and measured S_{11} of antenna configuration C (SRR is closed, switches between CPW signal and ground planes are open) along with the measured response of the basic monopole. The response confirms that the antenna configuration C provides

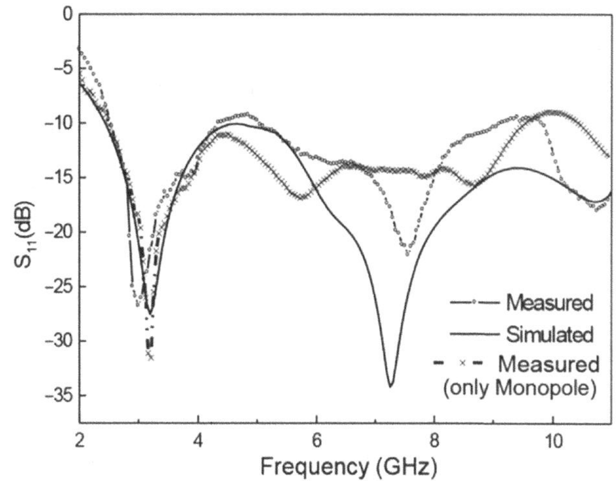

FIGURE 4.24
Measured and simulated S_{11} of the CPW-fed circular monopole loaded with closed-ring resonators (shorted split gaps, $g_1=g_2=0$) and without shunt strips (configuration C) compared with CPW-fed monopole only without SSRs and shunt wires.

Reproduced by permission of © 2015 The IEEE [3].

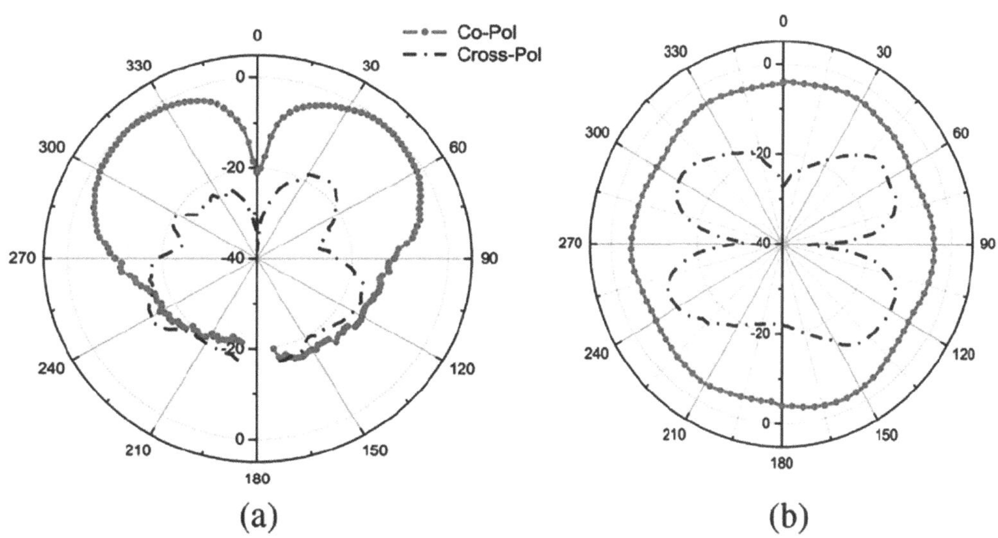

FIGURE 4.25
Measured co-polarized and cross-polarized radiation patterns for: (a) *x-y* plane (E-plane), and (b) *x-z* plane (H-plane) of the narrowband antenna (configuration A) resonating at 6.59 GHz.

Reproduced by permission of © 2015 The IEEE [3].

the UWB response. Thus, this technique is extremely effective in deriving three different antenna responses from the same UWB antenna. Figure 4.25 plots the measured co- and cross-polarized radiation pattern of antenna configuration A for E- and H-planes. Since the narrowband response is derived by exploiting the filtering action in the feed region of the basic UWB monopole, the antenna in all three configurations provides a consistent

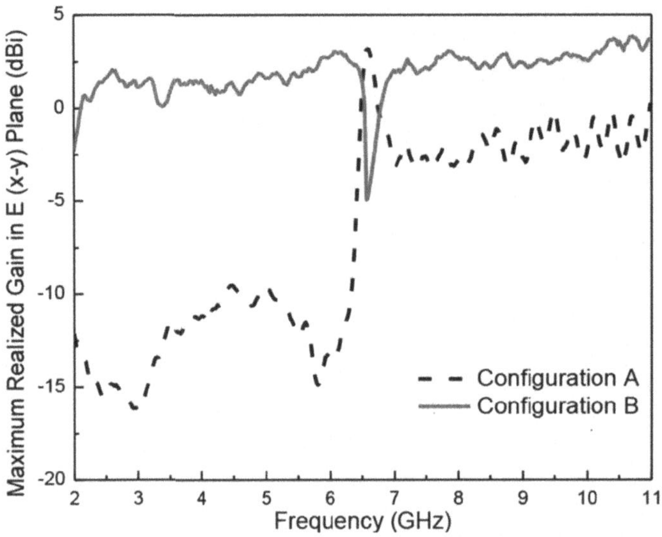

FIGURE 4.26
Measured maximum realized calibrated gain characteristic in the E- (*x*-*y*) plane as a function of frequency of the CPW-fed circular monopole loaded with SRRs and shunt strips (configuration A) and CPW-fed circular monopole loaded with SRRs only (configuration B).

radiation pattern, i.e. directional radiation in the E-plane with axial null and an omnidirectional pattern in the H-plane. Figure 4.26 plots the measured maximum realized gain of the antenna for configurations A and B, respectively. As observed from the plot, in configuration B the antenna yields high gain for the entire UWB band except for the notch, whereas in configuration A, it exhibits an exactly complementary gain profile.

The complementary nature of antenna configurations A and B can be explained with more physical insight by observing the magnitude of the Poynting vector along one of the slots of the feeding CPW line of the antenna over the entire UWB band as a function of the distance from the feed position for both the antennas, as illustrated in Figure 4.27. As shown in Figure 4.27a, the diamagnetic behavior of the SRRs at its resonance frequency prohibits propagation at that frequency, whereas the rest of the spectrum is transmitted along the slot, giving rise to a notched UWB response. However, with the positioning of the shunt wire, which acts as an electrical wall on the transmission line, the filtered energy over a narrow band is propagated along the line and coupled to the antenna ensuring a narrowband operation of the antenna. The rest of the energy is reflected back to the input port as depicted in Figure 4.27b, with the dashed line *AA* indicating the central position of the shunt wire at a distance L_c =13.25 mm from the input port.

The dual complementary antenna shown by Siddiqui et al. [3] exhibiting three different antenna functionalities is extended to multiple narrowband and/or multi-notched UWB antenna using multiple SRR pairs and switches along the feedline of the MPA [18]. Figure 4.28 shows the fabricated prototype of the antenna realized with three different SRR pairs with changing rotational angle of the inner ring and copper strips acting as the switch between the CPW signal and ground lines. Table 4.2 shows the dimensions of the SRR and the angular orientations. As discussed in Chapter 3, varying dimensions and angular orientations change the resonance frequency of the SRR. A pair of SRRs and switch, as

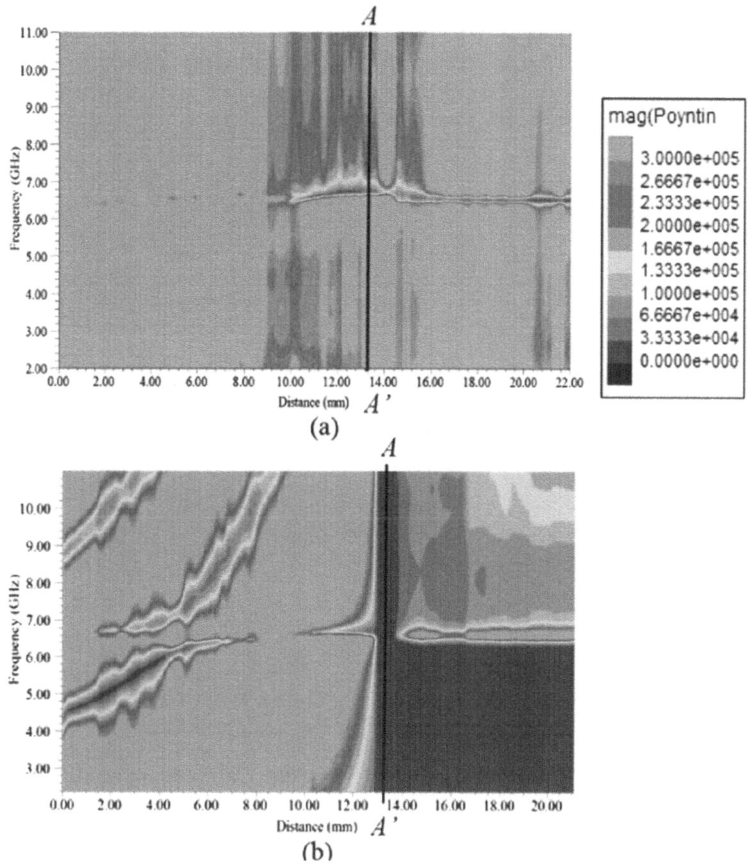

FIGURE 4.27
Simulated contour plots of the magnitude of the Poynting vectors of the propagating electromagnetic energy through the longitudinal dimension of one of the slots as a function of frequency: (a) antenna configuration B, and (b) antenna configuration A. (All plots have uniform field intensity).

Reproduced by permission of © 2015 The IEEE [3].

discussed in Siddiqui et al. [3], constitute a narrow band pass filter which when loaded along the CPW feedline of the UWB monopole, the UWB monopole acts like a narrowband antenna.

To get a more comprehensive insight into the operation of the antenna, we can represent the SRR and switch loaded UWB monopole antenna with an equivalent circuit. Due to symmetry in the structure along the axis of the monopole, we can apply the magnetic wall concept and draw the equivalent circuit of the structure. The equivalent circuit diagram of the antenna with the switch in the OFF and ON positions is shown in Figure 4.29a,b. The split-ring resonator, magnetically coupled with the CPW line, is represented by a tank circuit made up of L_{SRR} and C_{SRR} and their coupling is represented by coupling coefficient k. R_0 (= 50 Ω) and Z_A represent the feed port and antenna input impedance, respectively. L and C are associated with the inductance and capacitance of the CPW line. L_P in the circuit in Figure 4.29b represents the switch inductance in the ON position. By computing the circuit parameters and using a circuit simulator, the notched-UWB and narrowband response of the same antenna for various statuses of the switches can be confirmed [19].

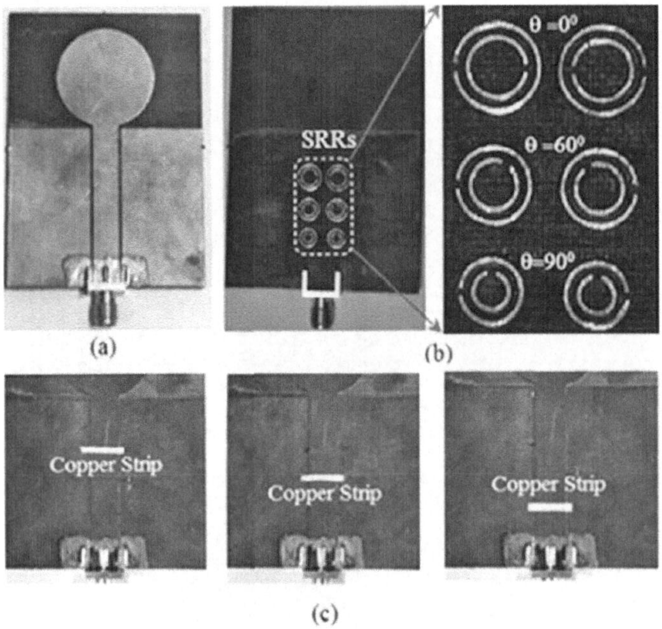

FIGURE 4.28
Fabricated prototype of the rotational SRR-loaded multifunctional monopole antenna with SRR and copper switch loading [18].

Reproduced by permission of © 2016 The IET [18].

TABLE 4.2
Dimensions and Orientation of the SRR Pairs Loaded Along the Feedline for Multifunctional Antennas [18]

SRR Parameters			
Parameter	**SRR_1**	**SRR_2**	**SRR_3**
R_{ext}	2.4 mm	2.8 mm	3.2 mm
c	0.5 mm	0.5 mm	0.5 mm
d	0.4 mm	0.4 mm	0.4 mm
g	0.4 mm	0.4 mm	0.4 mm
Theta (θ)	90°	60°	0

Using this concept, as indicated in the block diagram in Figure 4.30, three different narrowband responses (indicated as configurations C, D, and E) are derived from the same monopole antenna with different positions of the shunt strip connected between the signal and ground plane.

Figure 4.31 plots the surface current distribution and contoured plot of the Poynting vector along the feedline of the antenna. As shown in the plot, due to the band-pass filtering section of various SRR and switch loaded sections, the same antenna is excited with three different narrowband signals which in turn contributes to the multiple functionalities in the antenna.

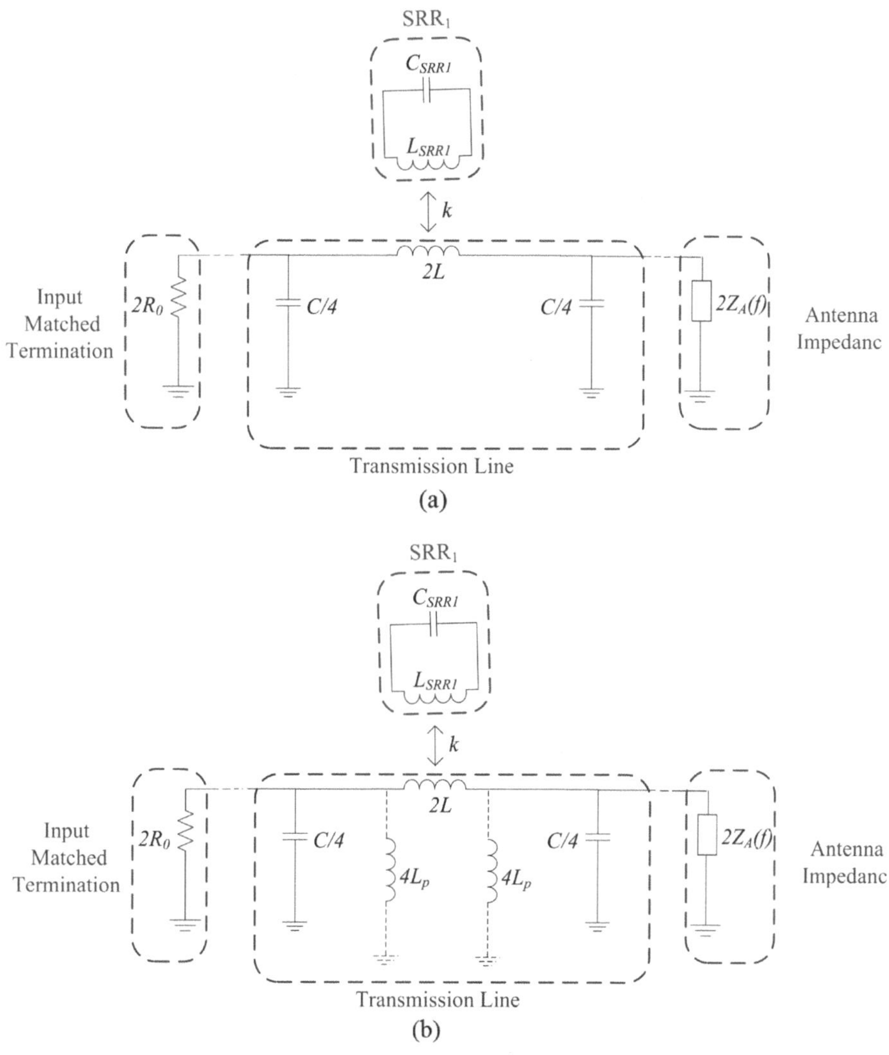

FIGURE 4.29
Lumped element equivalent model of the multifunctional SRR and switch-loaded CPW-fed monopole antenna. Due to the symmetry, the magnetic wall concept has been used and the circuit corresponds to one half of the structure. Shunt inductance *Lp*, representing the inductance of the copper strip between signal line and ground plane connected by dotted lines, is present in narrowband configuration: (a) switch open, (b) switch closed.

Figure 4.32 shows a multifunctional antenna in which the SRRs are loaded in two different layers, beneath the substrate and on the superstrate [19]. Loading of two differently sized SRR pairs and switches along the feedline of the annular ring-shaped monopole antenna yields seven different antenna responses. Figure 4.33 shows the cross-sectional view of various antenna topologies. Representing the presence and absence of an element as "1" or "0", the multifunctional antenna response is summarized in Table 4.3

The main advantage of the SRR and switch loaded multifunctional antenna design is its applicability to various antennas. Using this concept, multifunctional dielectric resonator antennas [20] can be designed. Figure 4.34 shows a multifunctional dielectric resonator

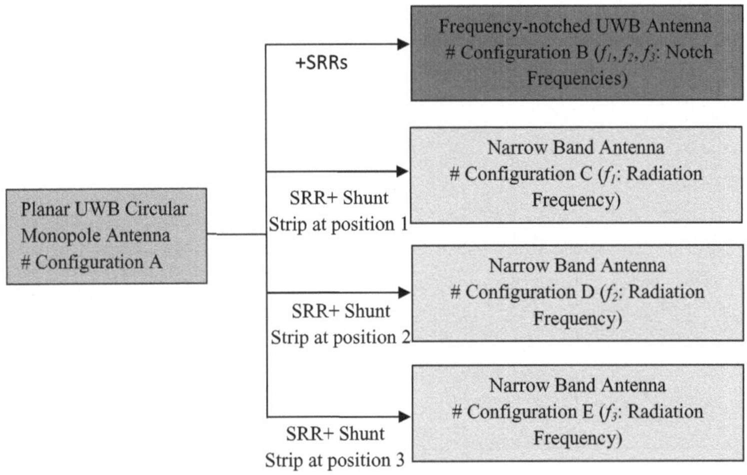

FIGURE 4.30
Block diagram of the proposed antenna configurations invoking various antenna performances using different combinative loading of SRRs and shunt strip on the feed section of the printed circular monopole antenna.

Reproduced by permission of © 2016 The IET [18].

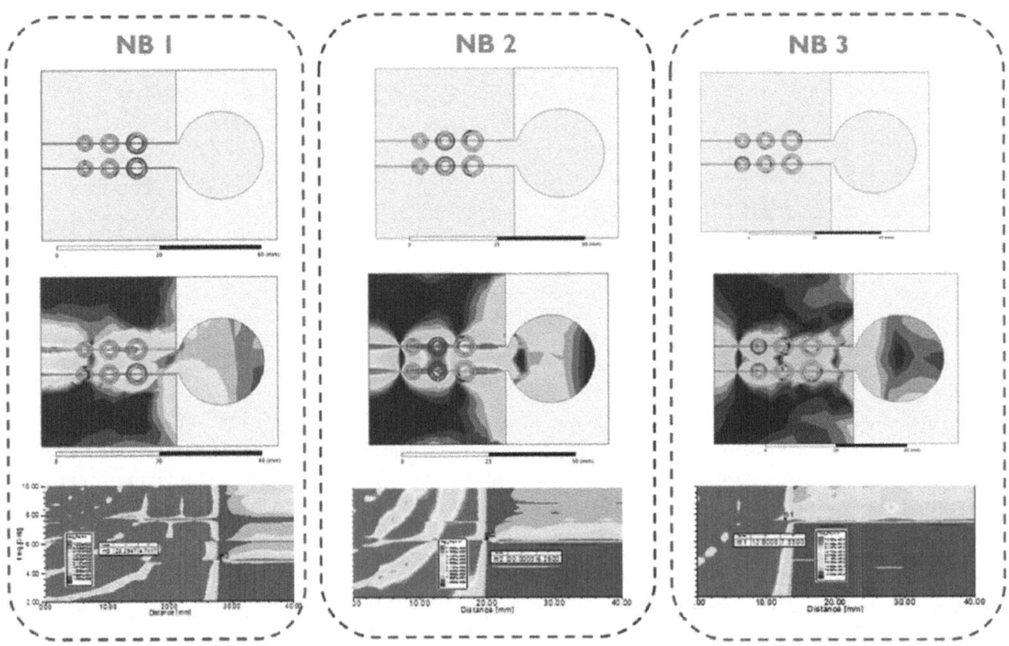

FIGURE 4.31
Excitation of various "SRR+switch" pairs for three different NB response of the antenna along with surface current and contoured pointing vector plot.

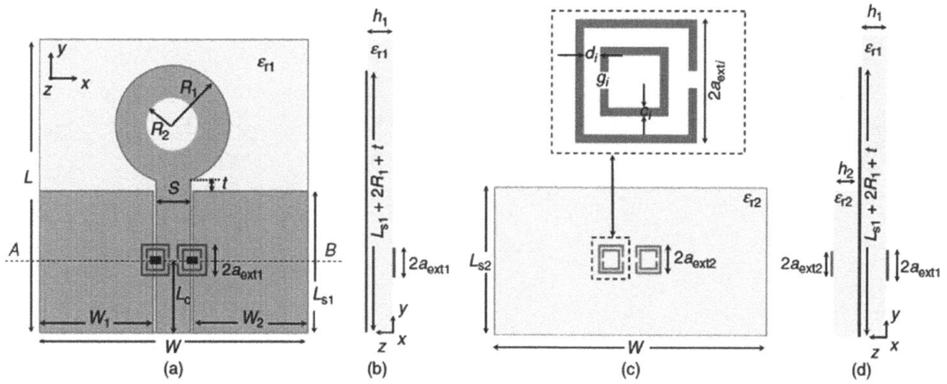

FIGURE 4.32
(a) Geometrical layout of a CPW-fed ARMA loaded with SRR, (b) cross-sectional view of the configuration with printed SRRs and CPW-fed monopole antenna on either sides of the substrate, (c) superstrate with SRR loading, (d) cross-sectional view showing the loading of superstrate on antenna. Zoomed version of unit cell of SRR is indicated in the dotted box.

Reproduced by permission of © 2018 The IEEE [19].

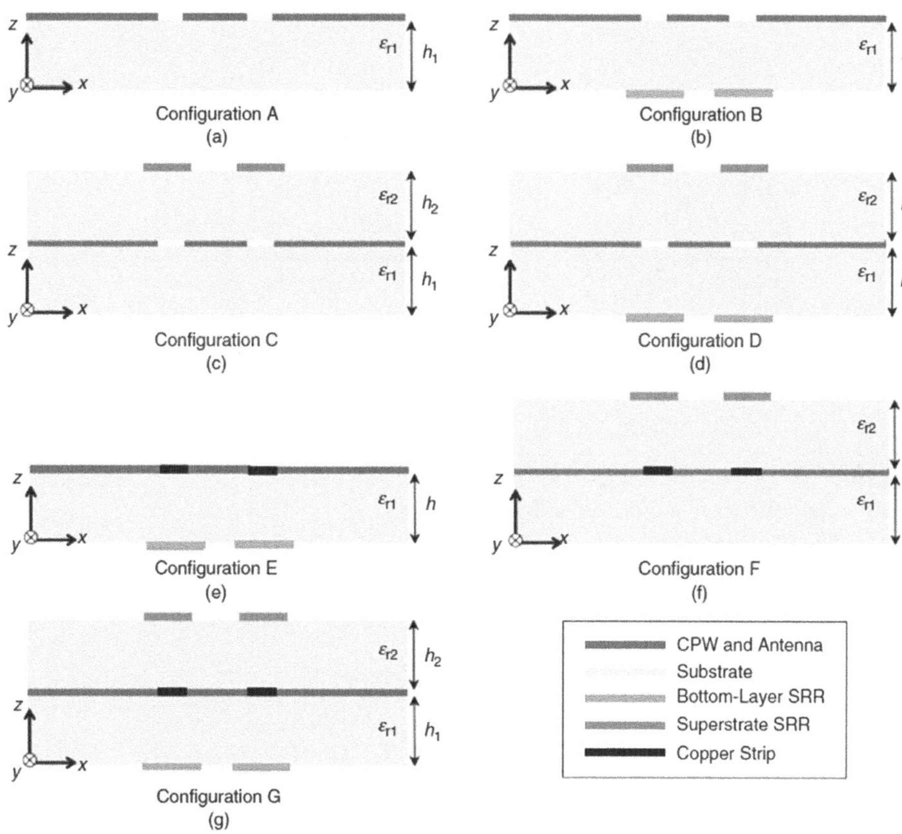

FIGURE 4.33
Evolution of multiple configuration of operation from a single antenna illustrated through a cross-sectional view of all cases.

Reproduced by permission of © 2018 The IEEE [19].

TABLE 4.3

Functional Boolean Table of the Square SRRs Loaded CPW-fed Annular Ring Monopole Antenna Demonstrating Seven Antenna Configurations

Configuration	Shunt Strip	Top SRR	Bottom SRR	Functionality
A	0	0	0	UWB
B	0	0	1	Notched UWB (f_1)
C	0	1	0	Notched UWB (f_2)
D	0	1	1	Notched UWB ($f_1 f_2$)
E	1	0	1	Narrow Band (f_1)
F	1	1	0	Narrow Band (f_2)
G	1	1	1	Dual Narrow Band (f_1, f_2)

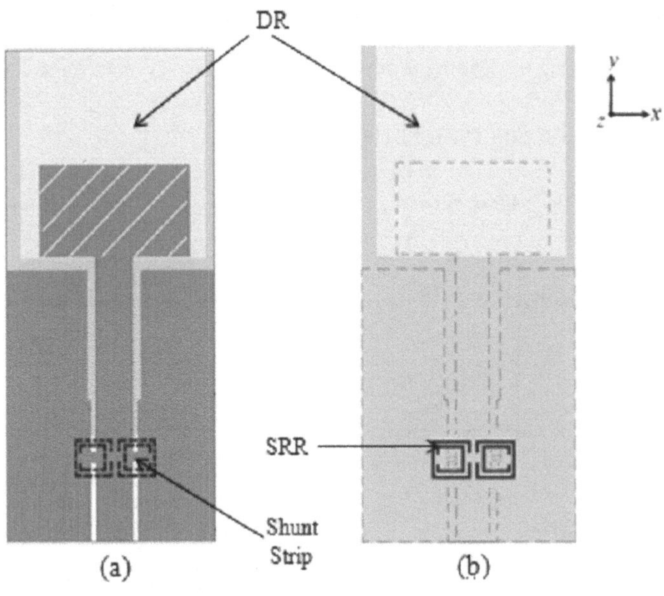

FIGURE 4.34
A multifunctional dielectric resonator antenna with SRR and shunt strip loading.

Reproduced by permission of © 2015 The IEEE [20].

antenna. As indicated in the S_{11} response in the plot in Figure 4.35, the same antenna contributes to three different antenna responses for various combinations of the SRR and switch loading.

4.5 Advantages and Applications of Multifunctional Antennas

Multifunctional antennas, as discussed in previous sections, are a special class of reconfigurable antenna which can be operated for various antenna responses required in a modern wireless radio. The main advantages of MFAs are

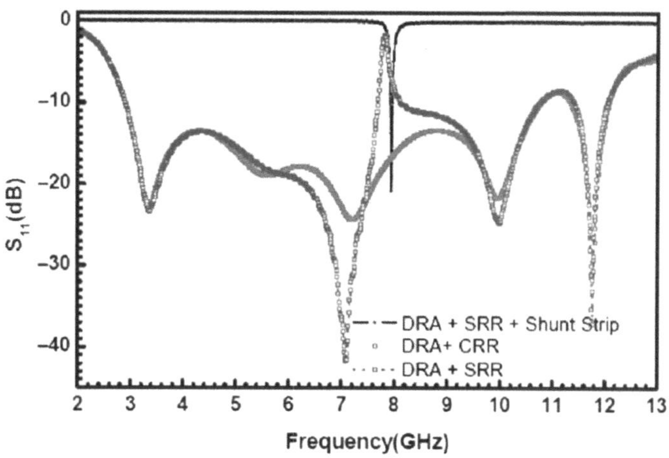

FIGURE 4.35
S_{11} response of the multifunctional dielectric resonator antenna with SRR and shunt strip loading.

Reproduced by permission of © 2015 The IEEE [20].

- As the same antenna can be used for multiple antenna responses, MFAs help in realizing a compact receiver.
- MFAs act like effective clustered antennas without requiring any extra design steps to minimize inter-port coupling.

MFAs are increasingly being used in modern wireless radios, such as cognitive radio (CR) and multi-standard radios.

References

1. S. Haykin, "Cognitive radio: Brain-empowered wireless communications", *IEEE Journal on Selected Areas in Communications*, vol. 23, no. 2, 2005, 201–220.
2. Y. Tawk, J. Costantine, and C.G. Christodoulou, "Reconfigurable filtennas and MIMO in cognitive radio applications",. *IEEE Transactions on Antennas and Propagation*, vol. 62, no. 3, 2014, 1074–1083.
3. J.Y. Siddiqui, C. Saha, and Y.M.M. Antar, "A novel ultrawideband (UWB) printed antenna with a dual complementary characteristic", *IEEE Antennas and Wireless Propagation Letters*, vol. 14, 2015, 974–977.
4. C. Saha and J.Y. Siddiqui, "A comparative analysis for split ring resonators of different geometrical shapes", *Applied Electromagnetics Conference (AEMC), Kolkata, India 2011 IEEE*, 1–4, IEEE, 2011.
5. J. Mitola and G.Q. Maguire, "Cognitive radio: Making software radios more personal", *Personal Communications, IEEE*, vol. 6, no. 4, 1999, 13–18.
6. J. Mitola, "Cognitive radio: An integrated agent architecture for software defined radio", Ph.D. Dissertation, Royal Institute of Technology (KTH), Stockholm, Sweden, 2000.
7. J. Mitola, *Cognitive Radio Architecture*, John Wiley, New York, NY, 2006.
8. US Federal Communications Commission (FCC), Cognitive radio workshop, May 19, 2003 [Online]. Available: http://www.fcc.gov/searchtools.html.

9. *Proceedings of the Conference Cognition Radios*, Las Vegas, NV, March 15–16, 2004.
10. Y.-C. Liang, K.-C. Chen, G. Ye Li, and P. Mähönen, "Cognitive radio networking and communications: An overview", *IEEE Transactions on Vehicular Technology*, vol. 60, no. 7, 2011, 3386–3407.
11. L. Giupponi and C. Ibars, "Cooperative cognitive systems, cognitive radio systems", *InTech*, Wei Wang (Ed.) 2009, ISBN: 978-953-307-021-6.
12. H. Arslan, *Cognitive Radio, Software Defined Radio, and Adaptive Wireless Systems*, Springer, New York, NY, 2007.
13. M. Manteghi, "A switch-band antenna for software defined radio applications", *IEEE Antennas Wireless Propagating Letter*, vol.8, 2009, 3–5.
14. K.R. Boyle, P.G. Steeneken, Z. Liu, Y. Sun, A. Simin, T. Huang, E. Spits, O. Kuijken, T. Roedle, and F. van Straten, "Reconfigurable antennas for SDR and cognitive radio", *Proceedings of the 2nd European Conference Antennas Propagating*, Edinburgh, UK, Nov. 2007, 1–6.
15. Y. Tawk and C.G. Christodoulou, "A new reconfigurable antenna design for cognitive radio", *IEEE Antennas and Wireless Propagation Letters*, vol. 8, 2009, 1378–1381.
16. Y. Tawk, J. Costantine, K. Avery, and C.G. Christodoulou, "Implementation of a cognitive radio front-end using rotatable controlled reconfigurable antennas", *IEEE Transactions on Antennas and Propagation*, vol. 59, no. 5, 2011, 1773–1778.
17. E. Erfani, J. Nourinia, C. Ghobadi, M. Niroo-Jazi, and T.A. Denidni. "Design and implementation of an integrated UWB/reconfigurable-slot antenna for cognitive radio applications", *IEEE Antennas and Wireless Propagation Letters*, vol. 11, 2012, 77–80.
18. L.A. Shaik, C. Saha, J.Y. Siddiqui, and Y.M.M. Antar, "Ultra-wideband monopole antenna for multiband and wideband frequency notch and narrowband applications", *IET Microwaves, Antennas and Propagation*, vol. 10, no. 11, 2016, 1204–1211.
19. L.A. Shaik, C. Saha, Y.M.M. Antar, and Jawad Y. Siddiqui, "An antenna advance for cognitive radio", *IEEE Antennas and Propagation Magazine*, vol. 1045, no. 9243/18, 2018.
20. A. Kumar, L.A. Shaik, C. Saha, and J.Y. Siddiqui, "CPW fed dielectric resonator antenna with complementary characteristics", *Applied Electromagnetics Conference (AEMC), 2015 IEEE*, 1–2, IEEE, Guwahati, India, 2015.

5

Reconfigurable UWB Antennas Design

5.1 Introduction

In Chapter 4, we were introduced to a new class of antennas, known as multifunctional antennas, along with their design techniques and recent applications. As briefly mentioned in Section 4.1, multifunctional antennas are a special class of reconfigurable antenna capable of performing the task of multiple antennas and thereby simplifying the antenna module of modern wireless radio aimed to perform over multiple bands/frequencies. One of the major applications of UWB-based multifunctional antenna is in the area of the cognitive radio (CR) environment of software-defined radio (SDR). CR is basically an intelligent radio, capable of efficiently utilizing the available spectrum using dynamic spectrum allocation policy. While the multifunctional antenna is the newest member in the UWB antenna family, reconfigurable wideband/UWB antenna has been relatively well studied over the last decade. Reconfigurable UWB antennas, due to various inherent advantages such as efficient power and spectrum utilization, are being used in a vast array of applications, both for the terrestrial and space environment [1]. Some of the recent applications of reconfigurable antennas include software-defined Radio (SDR), cognitive radio (CR), multiservice/multiband/multi-standard radio (MSR), and multiple-input-multiple-output (MIMO) technology. Moreover, modern UWB-technology-based sensing and imaging systems prefer reconfigurable systems which also demand compatible wide and UWB reconfigurable antennas. The advancement of RF and optical electronics, especially in the area of compact, high-speed electronics/electro-mechanical/optical switches and ease of integration, has served as a catalyst in the fast growth of research into reconfigurable antennas.

This chapter deals with the arts and techniques of design and realization of reconfigurable UWB antennas with a clear focus on recent and upcoming technological requirements. After an detailed introduction on the classical approach and techniques of reconfigurable antennas, this chapter focuses exclusively on various kinds of reconfigurable UWB antennas. UWB antennas with tunable/reconfigurable notch frequencies and reconfigurable antennas with multiple bands are discussed in detail. A section dedicated to a recent technique of designing dual reconfigurable printed antenna, which can work as a tunable notched UWB antenna and reconfigurable narrowband antenna, is included in the chapter. In addition, filter integrated UWB reconfigurable antenna, popularly known as reconfigurable filtenna, is discussed in Section 5.6. Along with the fundamental concepts and physical insight on the design principles, contemporary contributions by various researchers are also highlighted.

5.2 Introduction to Reconfigurable Antennas

The term "reconfigurable" indicates the capability of an entity or a system to manifest itself in various shapes, configurations, or modes based on the changes in the environment in which the system works. The same concept is applicable to reconfigurable antennas which are capable of changing some of their impedance (S_{11}) or radiation characteristics (pattern, polarization, etc.) so that they can be operated for wireless applications. In a reconfigurable antenna, reconfiguration is achieved by various kinds of switching and/or tuning elements or by physical alternation of the radiating aperture or by using smart materials such as ferrites and liquid crystals [1].

Though the concept and realization of reconfigurable antennas dates back to several years ago [2, 3], it is the introduction of UWB technology in 2002 [4] and its immediate diversified wireless applications which renewed the research interest in reconfigurable antennas.

5.2.1 Types and Classification

A reconfigurable antenna can achieve the desired configurability with various kinds of radiator geometry or apertures [5]. Based on the types of radiator used, a reconfigurable antenna can be classifies as follows:

- Reconfigurable aperture antennas [6, 7]
- Reconfigurable traveling wave antennas [8, 9]
- Reconfigurable microstrip antennas [10]
- Reconfigurable slot antennas [11, 12]
- Reconfigurable monopole/dipole antennas [13, 14]
- Reconfigurable arrays [15, 16]

The design concept and practical realization of various reconfigurable antennas evolve based on the specific applications and functionalities they demand. Based on the functionalities and reconfiguration features, as shown in Figure 5.1, reconfigurable antenna can be broadly classified into four categories:

1. Frequency reconfigurable antenna
2. Pattern reconfigurable antenna
3. Polarization reconfigurable antenna
4. Combined reconfigurable antenna

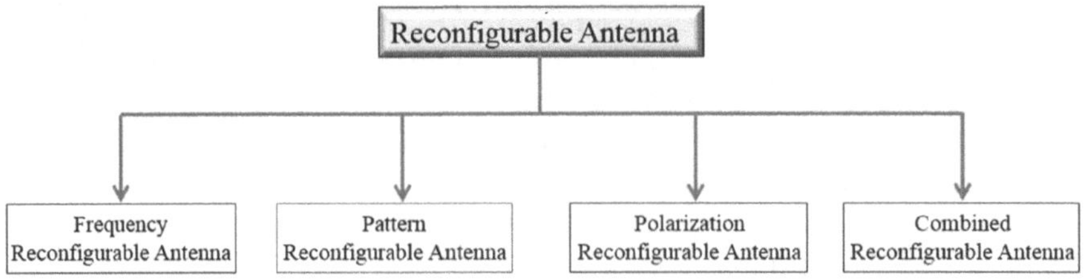

FIGURE 5.1
Classification of reconfigurable antennas based on functionalities.

Irrespective of the type of reconfigurable antenna, the basic principle of achieving all types of reconfiguration is the redistribution of the antenna surface currents or fields on the radiating aperture, which in turn impacts the antenna's impedance and radiation properties. In the case of frequency reconfigurable antenna, the operating frequency is modified by changing the electrical length or the surface current distribution on the antenna. In the case of pattern reconfigurable antenna, radiation properties such as the shape of the pattern, directivity, and direction of maximum radiation are altered by changing the radiating edge or feed network of the radiator. In the case of polarization reconfigurable antenna, the polarization of the antenna is controlled by intelligently altering the feed network or the radiator geometry. In the last category, the antenna is configured to attain multiple reconfigurability by a combination of various basic reconfiguration techniques.

5.2.1.1 Design Steps and Challenges

Based on the desired type of reconfigurable feature and various other restrictions imposed by the overall system designer, designing a reconfigurable antenna in general involves many steps and sometime iterations. The steps involved are

- Selecting a suitable basic radiator from available families of antenna
- Designing the basic radiator based on system specifications, which includes input matching, pattern, gain, and ensuring an acceptable radiation efficiency
- Choosing a reconfigurable mechanism and techniques
- Selecting suitable devices/components for various kinds of switching/tuning along with bias and its assembling issues on the radiator
- Ensuring faithful reconfigurable performance. Due to the coexistence of tunable components, their housing, and biasing, etc., reconfigurable antenna design is much more complicated than a conventional antenna design and often involves multiple iterations of the steps just mentioned, before reaching the final optimized design. Another important criterion when designing a reconfigurable antenna with particular reconfiguration functionality is to ensure minimal impact on other characteristics of the antenna. For example, when designing a frequency reconfigurable antenna, the designer should ensure that the radiation pattern of the antenna including maximum realized gain is acceptable over the entire frequency tuning range.

Advantages of reconfigurable antennas

Despite various challenges and cost involved in mitigating these challenges, reconfigurable antennas offer various advantages as summarized below [1].

Flexible utilization of resources:

- Minimizes cost
- Minimizes space requirement
- Allows easier integration
- Provides good isolation between different wireless standards

Lower front-end complexity:

- No need for front-end filtering
- Good out-of-band rejection

Multifunctional capabilities:

- Ability to change functionality as per the requirement
- Provides wideband or narrowband operation

Suitable candidate for recent wireless standards and protocols, including:

- Software-defined radio (SDR)
- Cognitive Radio
- MIMO and multi-standard radio

5.2.1.2 UWB Reconfigurable Antennas

The surge in UWB technology, in particular since the dedicated spectrum allocation by the FCC in 2002 and the inception of a plethora of wireless technologies to cater to the growing needs of high-speed high data rate communications, has redefined the challenge of designing reconfigurable antennas. Due to the availability of a wider frequency spectrum, reconfigurable antennas with a wide or ultra-wide band tuning range is desired for many applications. As discussed in Chapter 3, avoiding interference from various narrowband services having overlapped frequency spectrum with the UWB band demands an inbuilt filtering section in the UWB antenna. Such antennas, known as frequency-notched UWB antennas, can be benefited immensely by the concept of reconfigurability. Incorporating a tuning technique to control the notch frequency of this antenna, enables it to reject desired narrow band/bands from the UWB spectrum dynamically, based on the applications. Such UWB antennas are known as UWB antennas with reconfigurable notch characteristics. Integration of the switching and tuning elements in the notch filter section of the antenna results in such UWB reconfigurable antenna. UWB antennas with reconfigurable/static notch frequency and a narrowband reconfigurable antenna integrated in common platform are popular candidates for the spectrum underlay scenario of cognitive radio applications. Such antennas are traditionally designed by incorporating separate radiators for UWB-sensing to detect the white space and narrowband communicating antenna. A single radiator capable of performing both functionalities is an attractive alternative [17] and is known as a dual reconfigurable antenna. The recent concept of integrating a filter section inside the UWB antenna converts the antenna into a UWB filtenna. A filtenna in which the filtering section is tunable is another important class of reconfigurable UWB antenna. Due to the specific focus of this book on antennas for modern UWB technology, we will restrict our discussion on reconfigurable UWB antennas. However, as most of the design concept/techniques evolved from traditional reconfigurable antennas, whenever required, we will recall the associated basic concepts.

5.3 Various Techniques of Antenna Reconfiguration

Reconfigurable antenna, from a simplistic approach, can be considered as a conventional antenna in which various switching and tuning devices/elements are accommodated to achieve a desired reconfigurable functionality. Though the switching and/or tuning elements along with the basic antenna together constitute the reconfigurable antenna, their design is not independent, as their performance is significantly changed in composite state compared to their standalone behaviors. Therefore, thorough understanding of various reconfiguration techniques, including undesired impacts due to the proximity of the antenna, and including these in the final design process is crucial for reconfigurable antenna design.

Reconfiguration techniques in a reconfigurable antenna, as indicated in Figure 5.2, can be broadly classified into following four categories:

1. Electrical reconfiguration
2. Optical reconfiguration
3. Physical reconfiguration
4. Material change based reconfiguration

5.3.1 Electrical Reconfiguration

In electrically reconfigurable antenna, reconfiguration is achieved by employing switches and tuning elements on the antenna radiating aperture and/or feedline. Based on the status of the switch (ON/OFF) connected between various sections of an antenna, current distribution on the antenna is altered, which in turn yields the reconfigurable functionality. The popular choice of electrical switches is PIN diodes, varactor diodes, and RF-MEMS. The selection of a particular category switch is normally based on type of the applications, nature of the reconfiguration, as well as factors such as cost, ease of design, switching speed, ON and OFF resistance of the switches, etc. Table 5.1 presents the salient features of PIN diodes, RF-MEMS, and varactor diodes along with their application scenarios in reconfigurable antenna design.

PIN diodes offer very low loss, good dynamic reconfigurablity, and high switching speed. Varactor diodes are utilized where continuous reactive tuning is required; they,

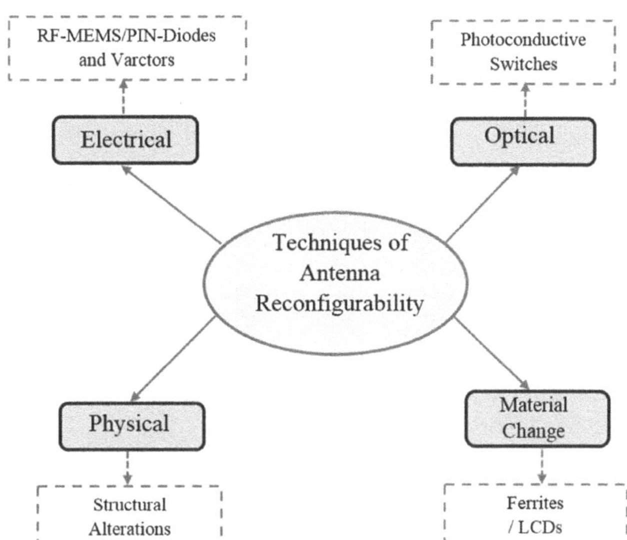

FIGURE 5.2
Various techniques of achieving antenna reconfigurability.

TABLE 5.1

Features of Different Kinds of Switches/Tuning Elements used in Electrically Reconfigurable Antennas

PIN Diodes	RF-MEMS	Varactors
• In most of the applications, it acts as an ON/OFF switch • Switch actuation is controlled by DC voltage like any other semiconductor device • Offers fast switching, with switching speed of 1–100 ns • Good dynamic reconfiguration ability • Requires a constant DC current for biasing • Good candidate for fast • switching	• In most of the applications, it acts as an ON/OFF switch • Switch actuation is controlled by the mechanical movement of the switch • Switching is relatively slower having a speed of 1–200 μs • Provides very good isolation • Operating voltage is relatively high • Very low power consumption • Good candidate for low power and high-isolation applications	• In most of the applications, it acts as a tuning element by providing a variable capacitance against a variation of applied DC voltage • Offers a vast tuning range • Suffers from poor linearity. To mitigate this, ferroelectric variable capacitors are used • Good candidate for fast switching along with tunable response

however, suffer from poor linearity which can be mitigated by using ferroelectric variable capacitors instead, which, in turn, suffer from limited tuning range and higher loss than varactors. MEMS devices offer advantages including very low loss, wide bandwidth, and low power consumption, but suffer from high operating voltage, low reliability, and higher cost than semiconductor devices.

Many recent designs use RF-MEMS [18–20], PIN diodes [21–23], and varactors [24, 25] for achieving reconfigurable antenna functionality. While properly biased PIN diodes and RF-MEMS-based switches provide different antenna aperture based on the ON/OFF mechanism and hence the desired antenna reconfiguration, varactor-based reconfiguration utilizes variable capacitance of the varactor due to the change in the bias voltage providing a vast tuning range. Frequency tunable antenna based on a CPW stub loaded MEMS capacitor is demonstrated in Erdil et al. [18], where the electrical length of the stub is controlled by a MEMS capacitor. In Huff and Bernhard, and Won Jung et al. [19, 20], the MEMS switches are embedded in the arms of square spiral antenna, thereby enabling the pattern reconfigurable antenna depending on the actuation of the switches. A single-fed resonant slot loaded with a series of PIN diode switches is utilized by Peroulis et al. [21] to change effective electrical length, resulting in reconfigurability. By incorporating three PIN diode switches along the meandered tuning stub loaded on top of the half circular patch located within the circular slot, multi-frequency operation is realized in Han et al. [22]. Two PIN diodes are incorporated in the V-shaped feed of annular slot antenna to implement three different polarizations, including one linear polarization and dual orthogonal circular polarizations. Dual-band reconfigurable slot antennas are realized in Behdad and Sarabandi [24] by loading a slot antenna with a varactor at a certain location along the slot. In White and Rebeiz [25], varactors are utilized in the slot antenna to attain single- and dual-polarized slot-ring antennas.

5.3.2 Optical Reconfiguration

Optical reconfigurable antenna utilize photoconductive switches to achieve the desired reconfigurable mechanism. The optical switches can be activated by various means: non-integrated optical fiber [26], integrated optical fiber [27], and integrated optical fiber [28]. In this technique, an optical beam emitted from a source, such as a laser diode, controls the status (ON or OFF) of a photoconductive switch placed in an antenna. Emitted photons of suitable frequency interact with the valance band electrons and excite them into conduction band which in turn takes the optical switch from the non-conducting (OFF) state to the conducting (ON) state. Thus, the optical beam and the switch placed on the antenna is capable of controlling the current and field yielding the desired reconfiguration.

To guide the light onto the switch, an optical fiber fixture is mounted on the ground plane just beneath the switch and a hole is drilled on the substrate [26]. Upon illumination of the switch through the hole, the switch is turned ON and provides the reconfigurable antenna operation. Figure 5.3 shows the placement of a pair of optical switches along with a pair of holes through which the laser beam is guided [26].

5.3.3 Physical Reconfiguration

This is another effective technique to achieve antenna reconfigurations in which some part of the antenna metalizations is not fixed, i.e. physically movable. The physical movement of the various parts of the antenna metalization provides a structural modification which in turn reconfigures the current and/or radiating edges in the current. The most attractive feature of this technique is that it does not require any switching components, associated bias lines, or integration of optical fibers/laser diode on the antenna. However, in this technique, motors, actuators, and various other components and accessories are required for physical movement of the various parts.

Physical reconfiguration techniques are quite effective for various types of antennas. Some important examples are

- In a parabolic reflector antenna, a rotating reflector can be used to cater multiple sources placed at the vertex.

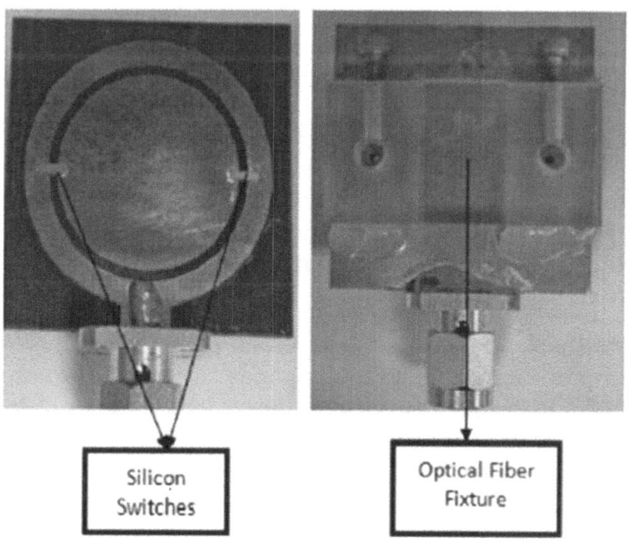

FIGURE 5.3
Top and bottom layers of optical switch integrated reconfigurable antenna [26].

- The response of the antenna is altered by changing the position and spacing of the ground plane with respect to the antenna [28, 29].
- Microstrip patch antenna can be pattern [30] and polarization reconfigured [31].
- A UWB antenna can be reconfigured into a dual reconfigurable antenna by physical movement of split-ring resonators printed on a superstrate layer [17].

This technique doesn't requirement DC bias to control the switches as is the case for electrical/optical based techniques. However, driving the actuator/motor involves power consumption. This technique offers a slower tuning compared to switch-based configurations due to actuation and motor control mechanism. However, the achieved tuning speed is more than sufficient for most of the applications.

5.3.4 Material Change Reconfiguration

This is another alternative technique of achieving antenna reconfiguration which is based on changing the constitutive parameters (permittivity, permeability, and conductivity) of the antenna substrate. The change in the material is affected by change in permittivity and permeability of smart/tunable materials such as liquid crystals, ferrites, BST (Barium Strontium Titanate) employing electric/and or magnetic field. For example, the dielectric constant of liquid crystals is a function of applied voltage. Similarly, for ferrite materials permittivity/permeability can be modulated by applying a proper static electric/magnetic field. Similar tunable material properties in graphene plasmonics [32] and organic semiconductor polymer [33] are also an attractive choice when designing reconfigurable antennas. Some important examples of reconfigurable antenna design employing this technique are

- Design of reflectarray with switchable radiation pattern [34]
- Microstrip antenna for frequency sweep applications [35]
- Design of leaky-wave reconfigurable antenna using graphene plasmonics [32]
- Beam steering applications [36]

5.4 Design of Reconfigurable UWB Antennas

In the previous sections of the chapter, we had a general introduction to reconfigurable antennas, discussed their classifications as well as various techniques in achieving the antenna reconfiguration mechanism. While general concepts of antenna reconfiguration have matured and extensive research has been done on various kinds of functional reconfiguration using all these techniques, the focus has mostly been on narrow and/or multiband applications, with lesser attention given to wideband or UWB reconfigurable antennas. Rapid growth of UWB technology over last decade and recent concepts and trends in designing intelligent and fast wireless radio demands different types of UWB-based reconfigurable antennas for various applications. In this section, we will focus exclusively on the design of reconfigurable UWB antennas for these modern applications.

5.4.1 UWB Antennas with Reconfigurable Notch Characteristics

Spectrum overlapping of various narrowband services, such as WLAN (5.15–5.85 GHz), WiMAX (3.3–3.7 GHz), X-band satellite down link (7.25–7.75 GHz) with the FCC-licensed ultrawideband spectrum of 3.1–10.6 GHz, can cause severe interference between these services, if due care is not given. As discussed in detail in Chapter 3, interference from these narrowband services can be mitigated by introducing an inbuilt filtering mechanism (without an external filter section) in the UWB antenna. Such UWB antennas with inbuilt capability to reject interference from narrowband services are known as frequency-notched UWB antenna. Various mechanisms such as introducing resonant slots/stubs on the radiator and/or ground plane, and parasitic elements on the feedline, etc., are widely used to design such frequency-notched UWB antenna. Though most of these techniques are quite efficient in rejecting a particular band or bands, some of the designs [37–39] are radiator independent, and the notch frequency is easily scalable, notch frequencies are static. In other words, such frequency-notched UWB antennas can be used to reject interference from a predefined narrow band only. Such frequency-notched UWB antennas with static notch frequency are not very useful in rejecting a notch frequency dynamically. On the other hand, an inbuilt tuning mechanism to control the notch frequency in such antenna makes them more useful as the same antenna can be used to reject the interference of various narrowband services as demanded for a specific application and circumstances. Incorporating various tuning elements, as discussed in Section 5.3, in the filtering section of the notched UWB antenna, would qualify them for such applications.

Figure 5.4 shows a MEMS switch actuated reconfigurable band-notched antenna [40]. The radiator consists of a CPW line fed elliptical monopole. As shown in Figure 5.4a, the MEMS switch integrated in the U-slot of the antenna effectively controls the presence/absence of the slot. Figure 5.5 shows the geometry and a photograph of the fabricated MEMS switch. When the switch is open or up, the slot acts as a $\lambda/2$ resonator and contributes to a notch frequency around 5 GHz. On the other hand, with the switch closed or down, the resonance of the U-slot is destroyed and the antenna provides UWB response. In the second configuration shown in Figure 5.4b, MEMS switches are integrated between the elliptical radiators and the $\lambda/4$ stubs. The status of the switch connects or disconnects the stubs with the radiator which contributes to band-notched UWB and UWB response, respectively. Figure 5.6a,b shows the measured S_{11} (dB) plot of both the antennas in various states of the MEMS switch. This antenna, with integration of both the U-slot and L-stub and controlling their electrical status using the MEMS switch, can be operated as a dual-notched/UWB/single-notched antenna with choice of the notch frequencies as per the requirement.

Frequency-notched UWB antennas can be designed by introducing various parasitic resonators along the feedline to the UWB antenna. Such resonators inhibit signal propagation at its resonance frequency corresponding to its electrical length and thereby contribute to a frequency-notched UWB response. Such resonators, when loaded with varactor diodes in a suitable manner, contribute to excellent tunability of the notch frequency. Figure 5.7 shows a varactor tuned tunable frequency-notched UWB antenna [41]. The radiator comprises a circular monopole and is fed by a microstrip line. A pair of open loop resonators (OLR) loaded along the feeding line is utilized to achieve frequency-notched UWB response. The OLR is further loaded with varactor diodes to tune the notch frequency. Change of the applied DC voltage in the varactor changes its capacitance which in turn changes the electrical length of the resonators contributing to change of its resonance frequency. A low pass filter section comprising of two inductors (L_1 and L_2) and a capacitor C

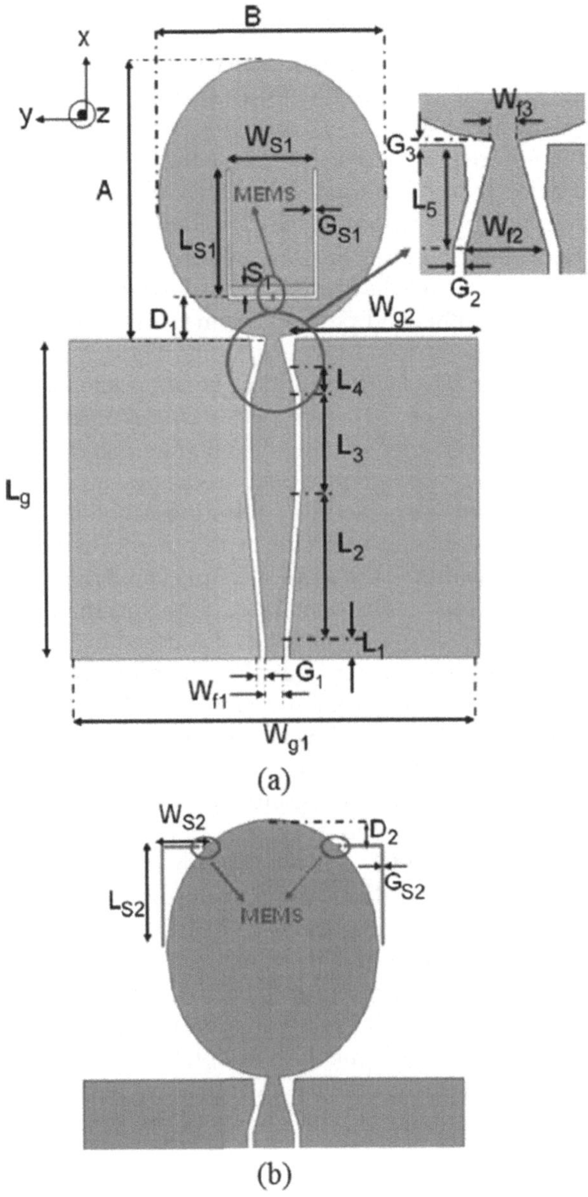

FIGURE 5.4
Photograph of the MEMS switch integrated reconfigurable UWB antenna. MEMS switch is integated (a) in the U-shaped slot, (b) between the $\lambda/4$ stub and the elliptical radiator.

is used to provide required DC to RF isolation. Figure 5.8 shows the fabricated prototype of the antenna with DC bias lines and connectors added.

Figure 5.9 shows the simulated and measured VSWR plot of the OLR loaded antenna with varying bias voltage to the embedded varactor diodes along with the reference antenna. As revealed from the plots, the center frequency of the notch can be continuously tuned from 5.1 to 5.9 GHz by changing the bias from 0–30 V.

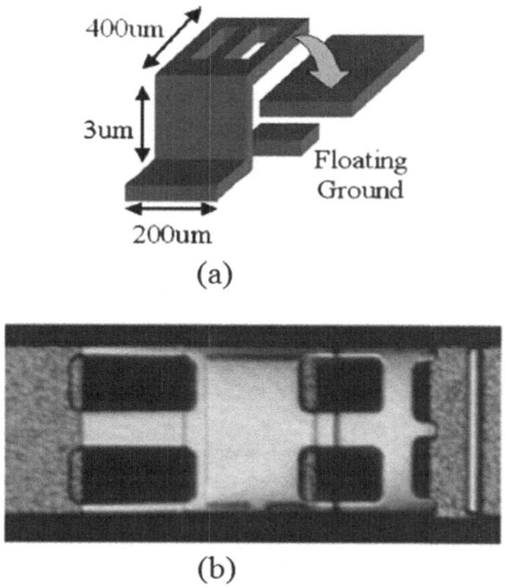

FIGURE 5.5
(a) Geometry of the MEMS switch, (b) fabricated MEMS switch used in the antenna.

Reproduced by permission of © 2009 The IEEE [40].

As discussed in Section 5.3, employing optical switches on the antenna or feed region can be effectively used to alter the current/field on the aperture which in turn can contribute to reconfigurable performance. Figure 5.10 shows a laser-controlled photoconductive switch-embedded CPW-fed UWB antenna with reconfigurable notch frequency [42]. Notch frequency in the antenna is achieved by an embedded symmetric U-slot on the monopole. The electrical length of the monopole is controlled by changing the status of the optical switches embedded on the U-slot using laser beams. As indicated in Figure 5.10b, five photoconductive silicon chips are strategically placed in the U-slot. Actuation of various photoconductive chips using a laser beam (808 nm wavelength), guided through optical fibers, can contribute to various combinations of electrical length which in turn provides a tunable notch frequency. Figure 5.11 shows the simulated and measured VSWR of the antenna for five different cases of switch combinations. As revealed in the plots, this antenna is capable of providing tunable notches at WLAN-2.4 GHz, WiMAX, WLAN 5 GHz, and ITU 8 GHz for various combinations when switched in the ON state.

Figure 5.12 shows an optically controlled reconfigurable band-notched UWB antenna [43]. The basic radiator consists of CPW-fed planar aperture, from which a pair of complementary split-ring resonators (CSRR) are etched from the inner metal patch. The two CSRR are designed to have different resonance frequency so that the basic radiator acts as dual-notched UWB antenna. Two optically controlled microwave switches (OCMS) embedded in the outer slots of the CSRRs, whose status is controlled by semiconductor lasers of 980 nm wavelength, provide four different antenna functionalities as follows: full UWB response (when both the OCMS are ON); notched UWB response at f_1 where f_1 is the resonance frequency of CSRR1 (when OCM 1 is ON and OCM2 is OFF); notched UWB response at f_2 where f_2 is the resonance frequency of CSRR2 (when OCM 2 is ON and OCM1 is OFF); and dual-notched UWB response (when both the OCMS are OFF). Figure 5.13 shows the

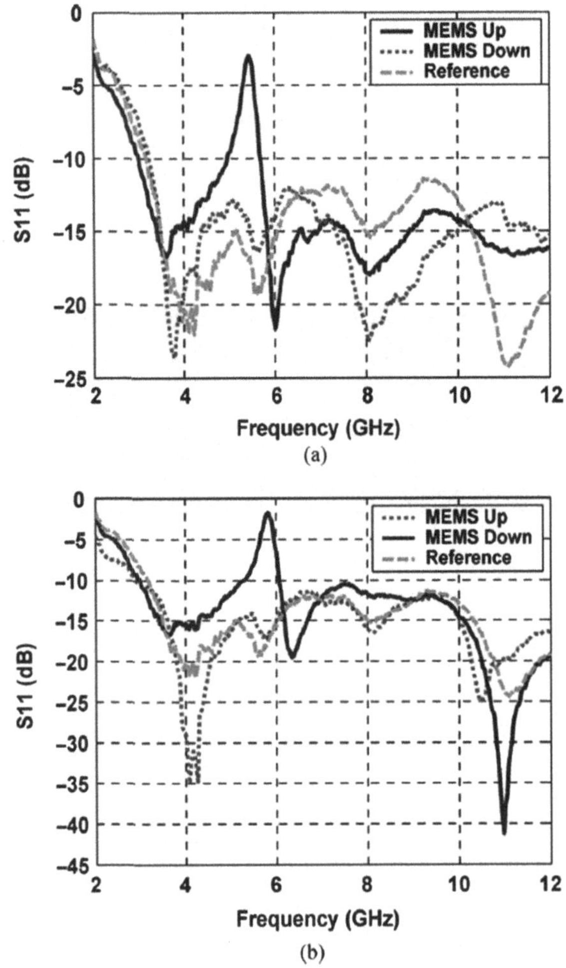

FIGURE 5.6
Return loss of the antenna with MEMS: (a) reconfigurable slot, (b) MEMS reconfigurable stubs.

Reproduced by permission of © 2009 The IEEE [40].

measured VSWR of the antenna for these cases and clearly indicates reconfigurable features of the UWB notched antenna.

5.4.2 Reconfigurable Antennas with Multiple Bands

Antennas with reconfigurable characteristics over a very wide band/UWB spectrum can serve multiple wireless services and naturally act as an automatic choice in various modern RF front-end designs. For example, modern mobile internet devices, such as smartphones, might need to access multiple wireless spectrum, such as WLAN, WiMAX, Bluetooth, GPS, etc. Antennas for such applications should be capable of dynamically adjust their resonance frequency to the desired multiple bands with a discrete or continuous tuning. In addition, the radiation performance of the antenna, such as pattern, gain, polarization, etc. should conform to the desired level, based on the specifications and applications.

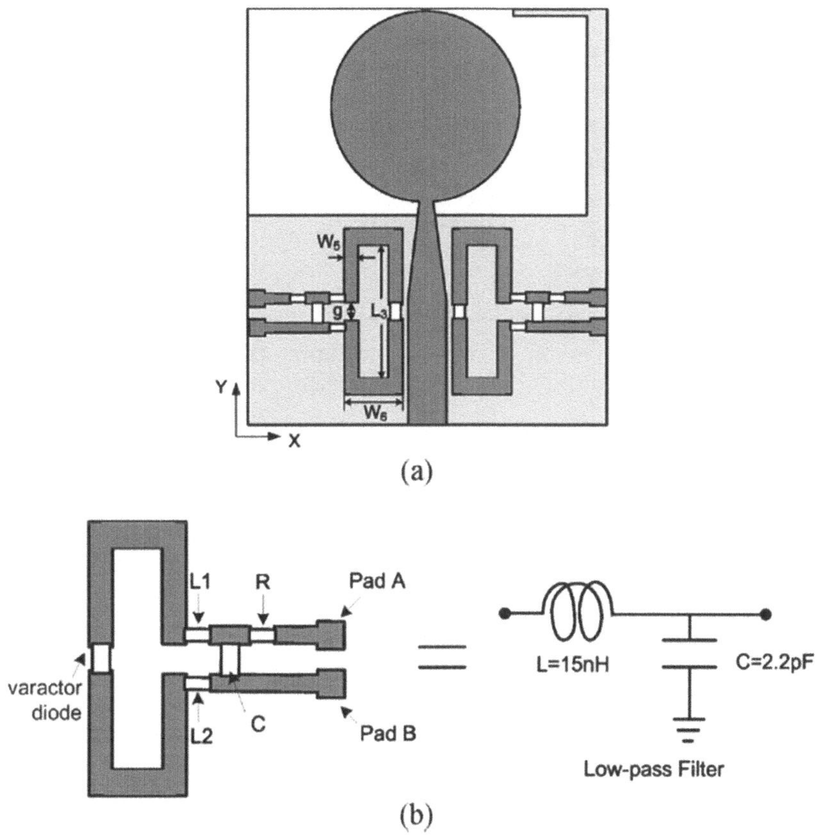

FIGURE 5.7
Open loop resonator loaded varactor tuned band-notched UWB antenna with fabricated prototype of the proposed antenna: (a) schematic, (b) DC bias arrangement and equivalent circuit.

Reproduced by permission of © 2012 The IEEE [41].

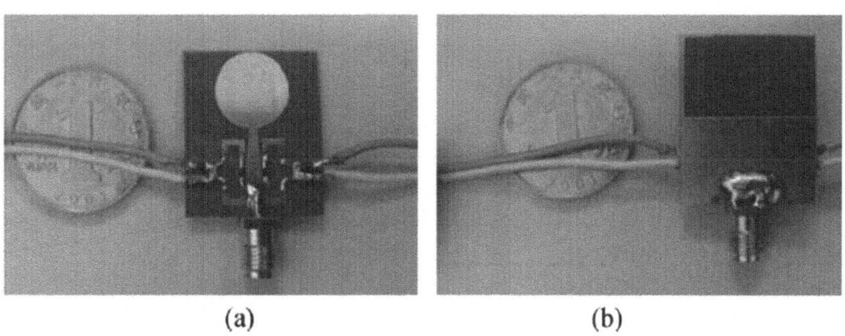

FIGURE 5.8
Fabricated prototype of the proposed antenna: (a) top view, (b) bottom view.

Reproduced by permission of © 2012 The IEEE [41].

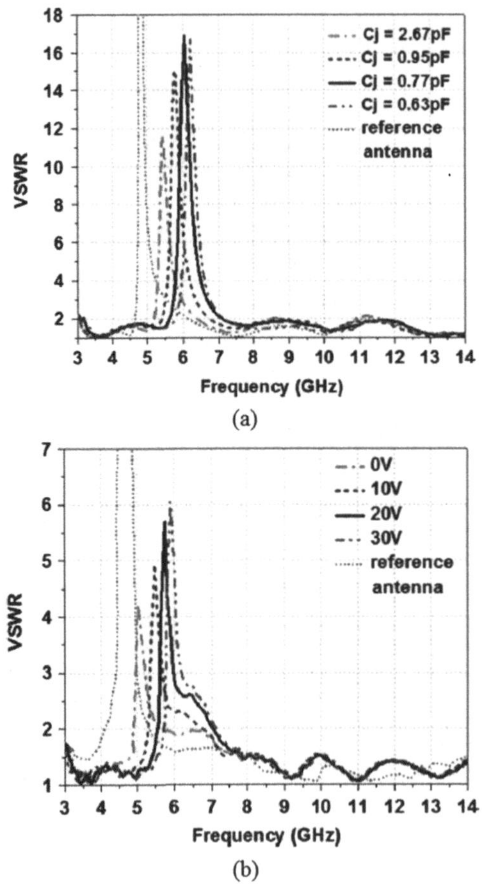

FIGURE 5.9
(a) Simulated, (b) measured VSWR of the varactor tuned OLR loaded frequency-notched UWB antenna.

Moreover, miniaturized design and ease of integration with other electronic module/sections are necessary features too. While multiband antenna appears to be a potential solution for such applications, frequency reconfigurable multiband operation is a much better choice as it provides noise reduction at other bands while not in use. This eliminates the requirement of an additional filter section in front-end circuits. In this section, we will discuss various designs of wideband/multiple band reconfigurable antenna based on contemporary arts and techniques.

Multiband reconfigurable antenna with electronic tuning arrangement can be sequentially used for tuning to various combinations of GSM, DCS, PCS, UTMS, Bluetooth, and wireless local area network [44]. Using a switched feed mechanism and/or switched ground mechanism [44], the antenna current is altered which contributes to multiple modes corresponding to multiple resonance frequency. A PIN and varactor diode actuated Quasi-Yagi folded dipole antenna demonstrates continuous or discrete tuning over a wideband [45]. Figure 5.14 shows the block diagram of an FPGA controlled PIN switch actuated frequency reconfigurable antenna [46]. Four patch sections are connected/disconnected

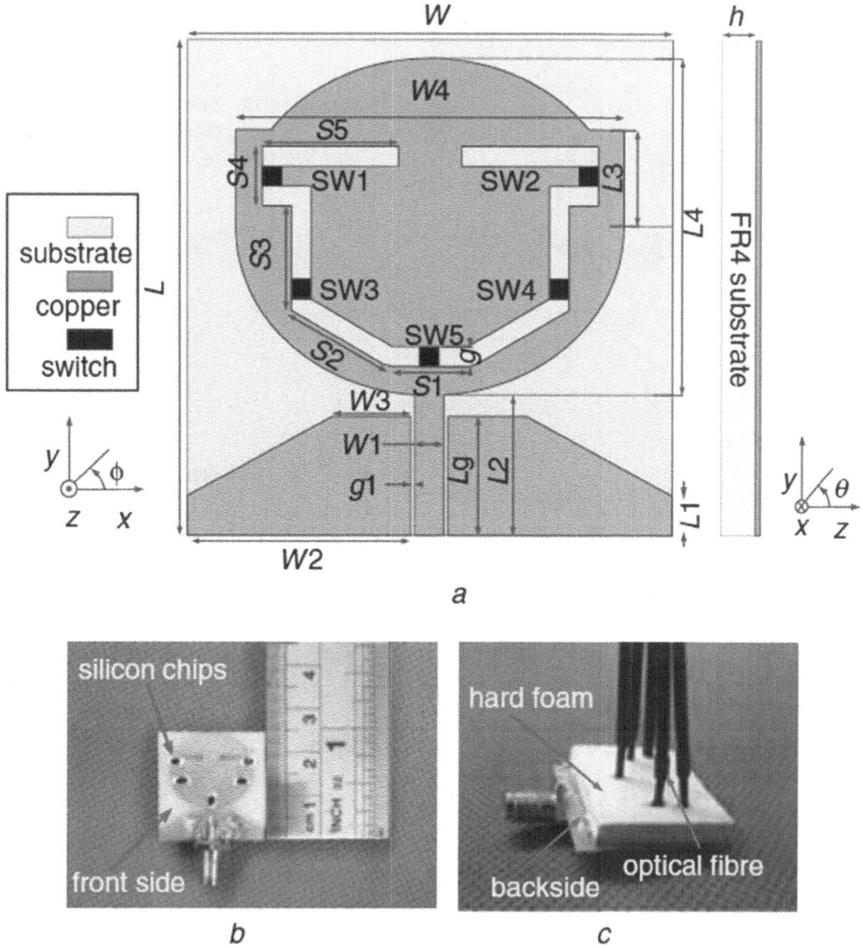

FIGURE 5.10
(a) Antenna configuration, (b) fabricated image, (c) photograph of the back-side optical feeding arrangements of reconfigurable band-notched UWB antenna.

Reproduced by permission of © 2014 The IEEE [42].

using PIN switches to an intermediate metallic section to achieve the desired frequency reconfiguration.

As indicated in the block diagram of Figure 5.14, joint test access group (JTAG) ports of the field programmable gate array (FPGA) kit enable various bit-streams and provide a reconfiguration feature of various components. Figures 5.15 and 5.16 show the schematic and fabricated prototype of the proposed antenna. Figures 5.17 and 5.18 show the measured and simulated S_{11} response of the antenna for two different bit-streams: 0010 and 0110. The bits-stream controls the states of the PIN diodes connected between the intermediate section and four patches and provides the antenna reconfigurations.

Such PIN actuation mechanism can be applied to bow tie antenna [47] to design reconfigurable multiband antenna with a wide tuning range. The combination of PIN actuation and varactor tuning together provides a very wide tuning range of 3.04–5.89 GHz with stable pattern and gain.

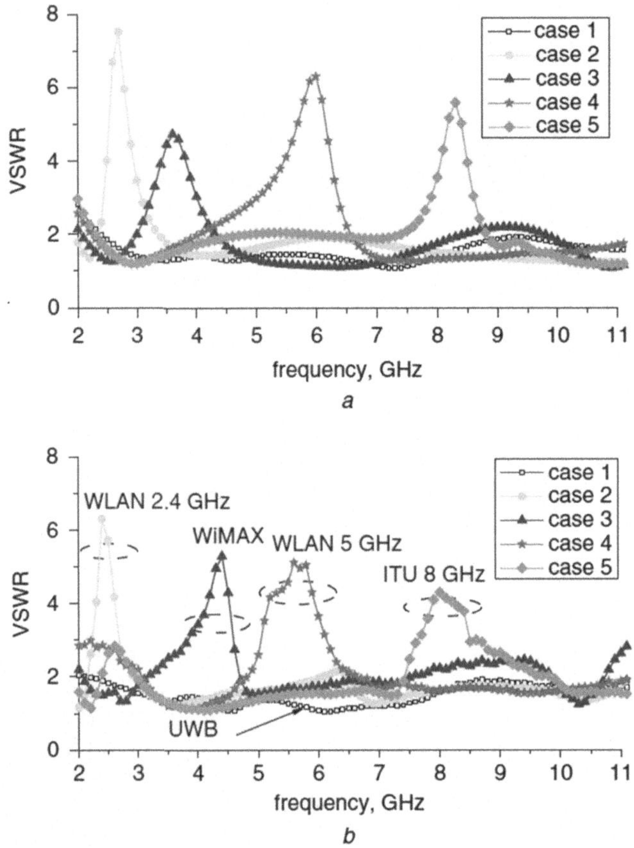

FIGURE 5.11
(a) Simulated and (b) measured VSWR of the varactor tuned OLR loaded frequency-notched UWB antenna.

Reproduced by permission of © 2014 The IEEE [42].

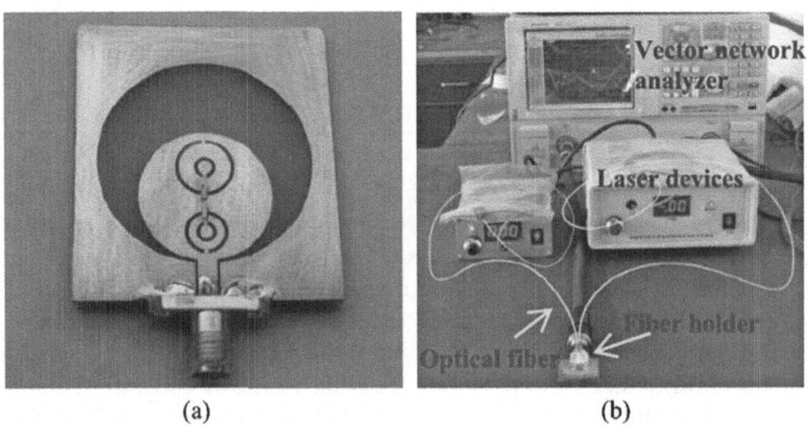

FIGURE 5.12
(a) Prototype of the optically controlled band-notched UWB antenna, (b) experimental setup.

Reproduced by permission of © 2014 The IEEE [43].

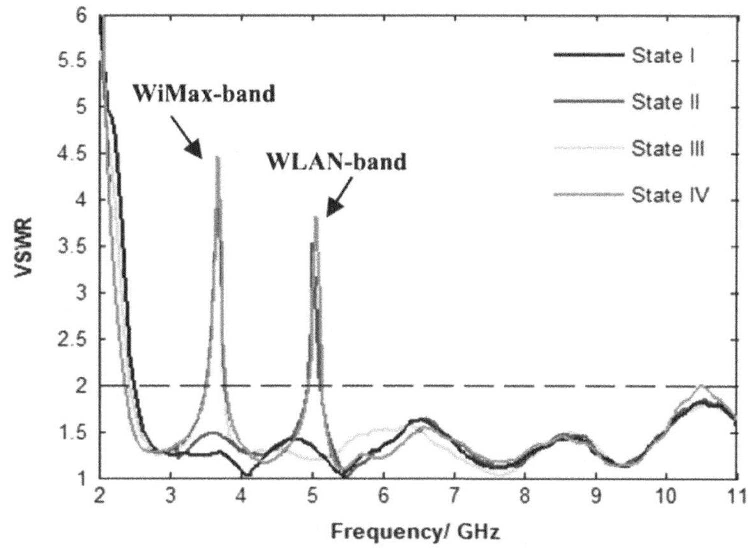

FIGURE 5.13
Measured VSWR of the proposed optically tunable notched UWB antenna [43].

Reproduced by permission of © 2014 The IEEE [43].

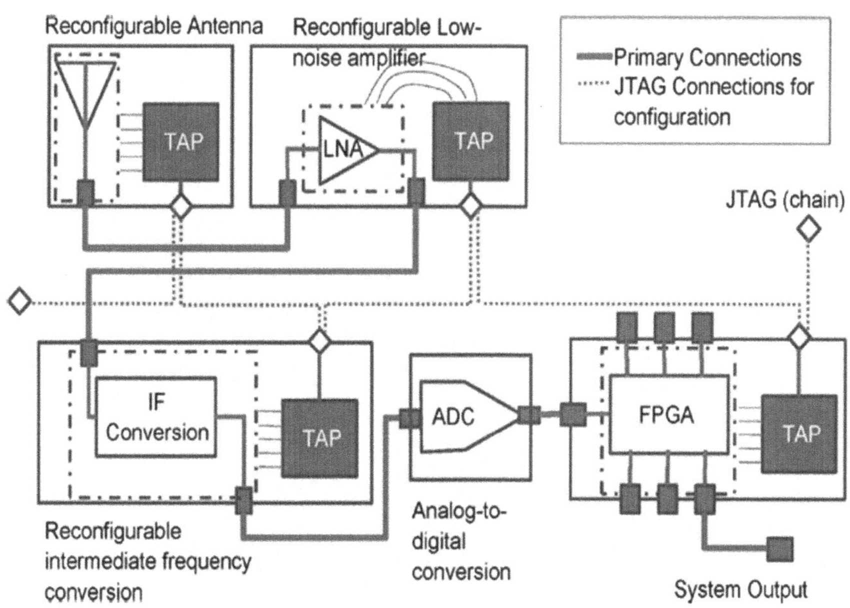

FIGURE 5.14
Block diagram of FPGA controlled switch reconfigured antenna [46].

Reproduced by permission of © 2010 The IEEE [46].

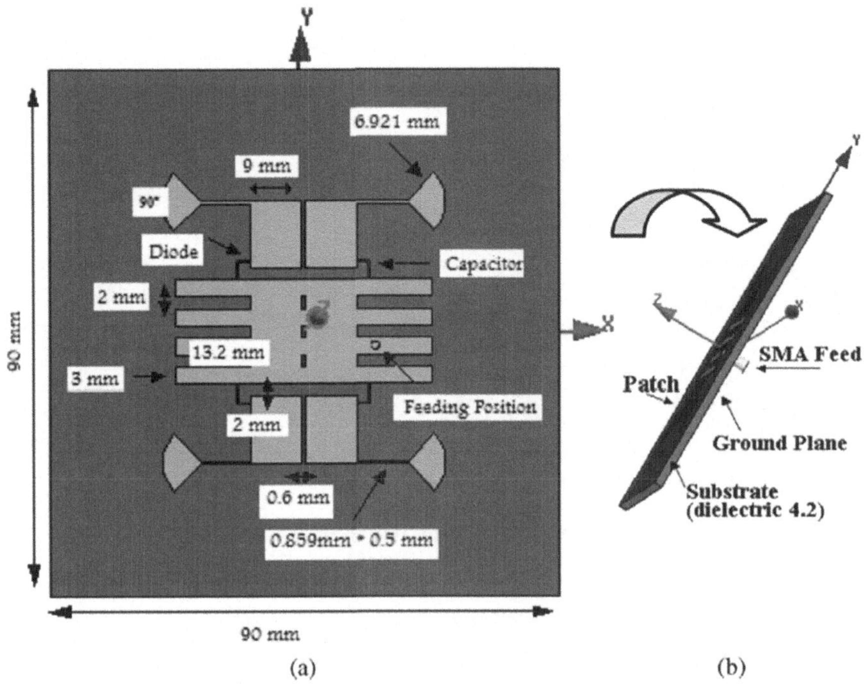

FIGURE 5.15
(a) Top view, (b) side view of the antenna [46].

Reproduced by permission of © 2010 The IEEE [46].

FIGURE 5.16
Fabricated prototype of FPGA controlled reconfigurable antenna [46].

Reproduced by permission of © 2010 The IEEE [46].

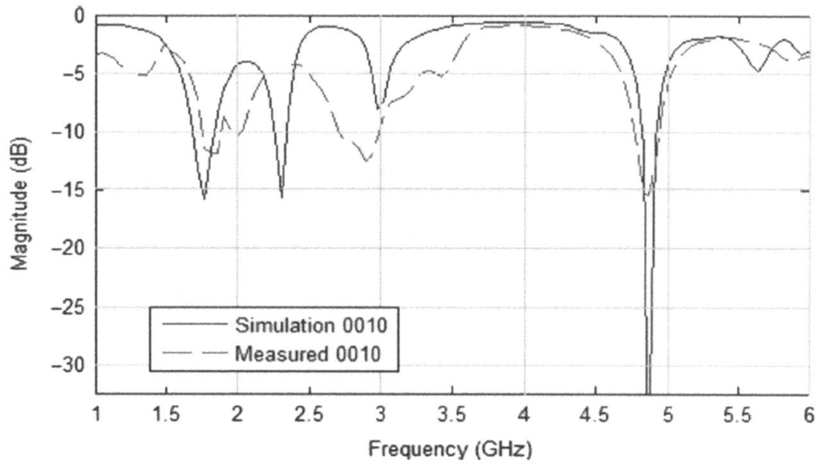

FIGURE 5.17
Measured and simulated S_{11} of FPGA controlled antenna in 0010 state [46].

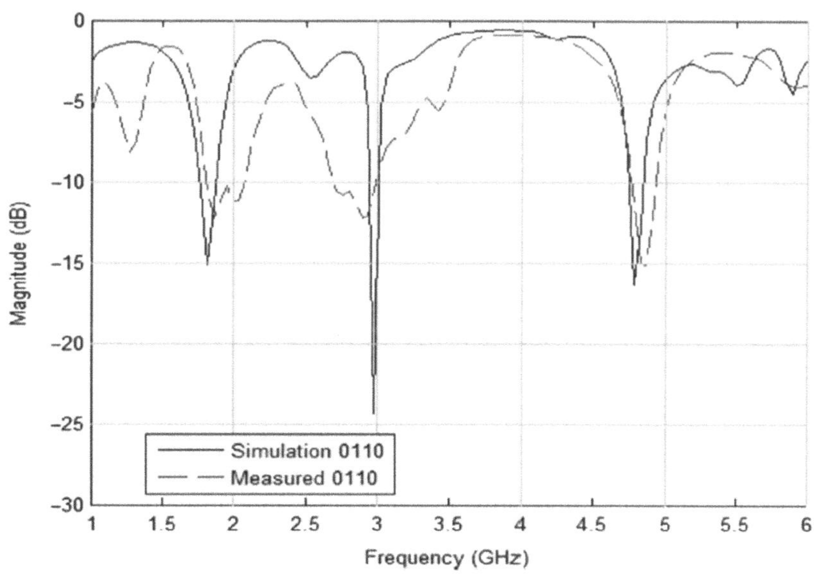

FIGURE 5.18
Measured and simulated S_{11} of FPGA controlled antenna in 0110 state [46].

Another useful technique of designing multiband/wideband tunable antenna is to integrate a tunable band-pass filter in the feeding network of the antenna. The main advantage of this technique is that it doesn't directly impact the radiation aperture of the antenna, and the overall antenna is very compact as the reconfigurable filter section is integrated in the feed network. Figure 5.19 shows a tunable monopole antenna integrated with band-pass filter sections [48]. The feed network of the antenna is intelligently designed to actuate

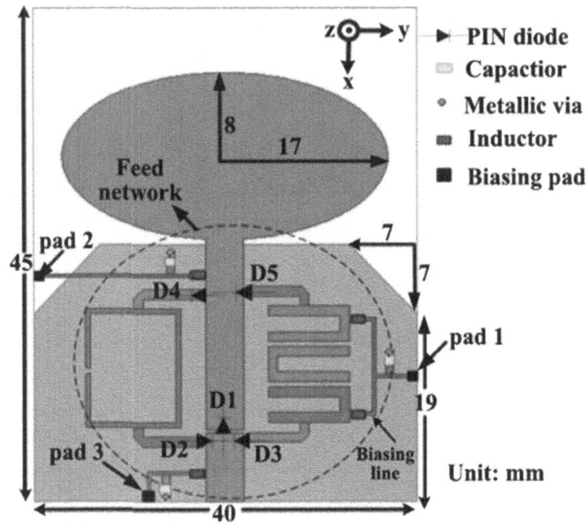

FIGURE 5.19
Bandpass filter integrated reconfigurable monopole antenna [48].

different RF paths which in turn yields three different antenna responses, namely, UWB band, 2.4 GHz WLAN band, and 5.8 GHz WLAN narrowband. The integrated feed network of the antenna consists of a 50 Ω microstrip line, an open bandpass filter, and a three-pole bandpass section. Varying impedance characteristics of the three different RF paths contribute to three different antenna responses. Selection of various RF paths are actuated through DC controlled PIN diodes. Figure 5.20a–c shows the simulated and measured S_{11} response of the antenna. When diode D1 is ON and other diodes are off, the 50 Ω microstrip line effectively feeds the monopole, which as shown in Figure 5.20a, yields UWB response. On the other hand, when diodes D2 and D4 are ON and other diodes are OFF, the RF path containing open loop BPF is actuated which contributes to 2.4 GHz WLAN response of the antenna. Similarly, with D3 and D5 ON, the hair-pin loaded RF path is actuated which provides 5.8 GHz WLAN response. The antenna, for all three states, exhibits good maximum realized gain, around 2 dBi and an acceptable efficiency.

Figure 5.21 shows a multi-UWB and multi-dual-band frequency reconfigurable annular sector slot antenna [49]. As shown in the schematic diagram, the basic radiator consists of an annular sector slot antenna (ASSA) fed by a 50 Ω microstrip line. Six PIN diodes are loaded in radial fashion to connect/disconnect three stubs placed at different locations of the ASSA. Figure 5.22 shows the fabricated prototype of the PIN diode loaded antenna along with DC bias lines, DC blockers, and chokes. The antenna is capable of supporting three UWB states in the range of 3.48–5.5 GHz, 5.18–7.35 GHz, and 6.06–8.05 GHz, along with 18 independent tunable dual-band response (Figure 5.23).

5.4.3 Dual Reconfigurable Printed Antenna

As briefly introduced in Section 5.1, a single antenna capable of providing tunable band-notched UWB response and tunable narrowband response is known as a "dual reconfigurable antenna". Such antennas are excellent candidates for the spectrum underlay scenario of CR applications as the notched UWB antenna functions can sense the unutilized "white"

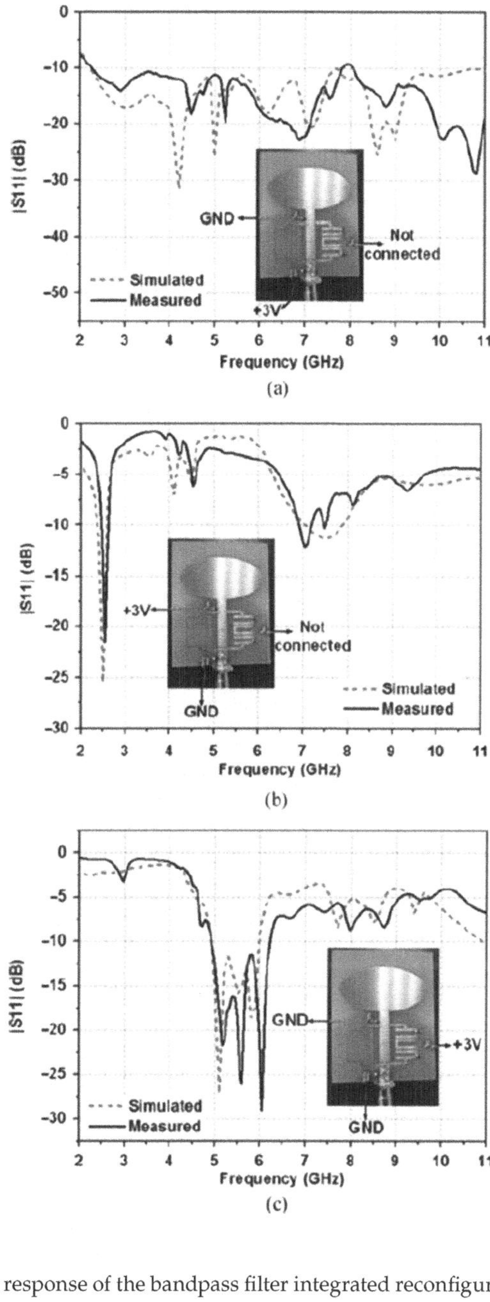

FIGURE 5.20
Simulated and measured S_{11} response of the bandpass filter integrated reconfigurable monopole antenna [48].

spectrum while the narrowband response can be used as the "communicating" antenna. Figure 5.24 shows a flowchart of a dual reconfigurable antenna clearly indicating two different kinds of antenna functionality and reconfigurability [50].

A single coplanar waveguide (CPW)-fed printed monopole antenna provides a reconfigurable ultrawideband (UWB)-notched or narrowband (NB) response, depending on

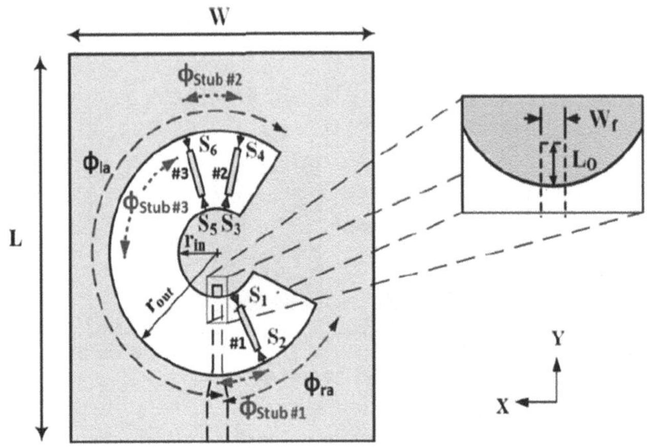

FIGURE 5.21
Schematic diagram of stub and PIN loaded frequency reconfigurable annular sector slot antenna [49].

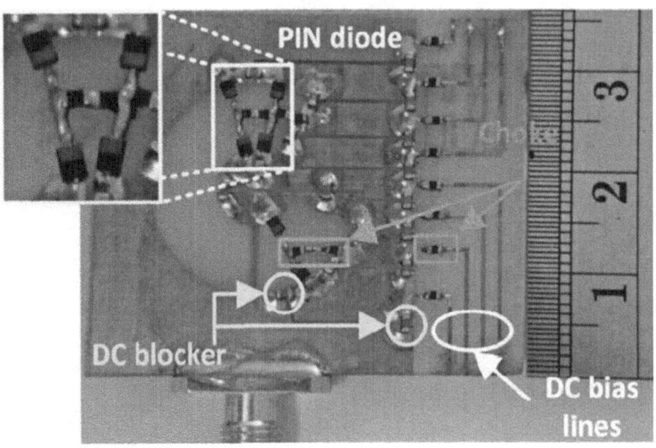

FIGURE 5.22
Fabricated prototype of stub and PIN loaded frequency reconfigurable annular sector slot antenna [49].

the status of the switches and the position of the split-ring resonators (SRRs) in the CPW feed region of the antenna. Mechanical movement of the superstrate, printed with three pairs of SRR of different dimensions in the feed region of the radiator, provides the desired tunability of the antenna. Figure 5.25 shows the geometrical lay out of the antenna along with that of the split-ring resonator. Figure 5.26 shows the fabricated prototypes of the antenna with and without copper switch loading along with the superstrate layer printed with three pairs of split-ring resonators of various size. The basic radiator consists of a popularly known circular disc monopole antenna (CMPA) and is fed by a CPW line. As discussed in Chapters 3 and 4, loading of the SRRs [17] and/or shunt strip [37–39] in the feed region of the antenna provides frequency-notched UWB and narrowband impedance characteristics. The axial magnetic field of the propagating EM wave along the CPW line induces an oscillating current between two rings of SRR and thereby yields a resonance

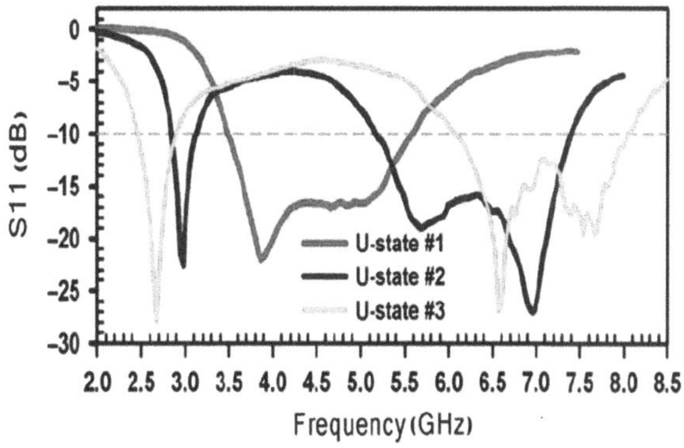

FIGURE 5.23
Measured S_{11} response of the bandpass filter integrated reconfigurable monopole antenna for three different states [49].

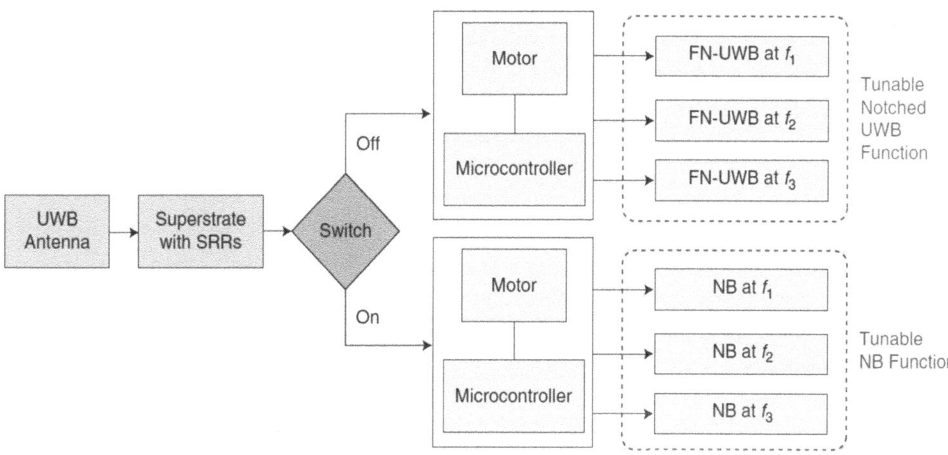

FIGURE 5.24
Flow chart representation of the dual reconfigurable antenna [17].

phenomenon and a strong notch frequency. Transformation of this SRR-loaded CPW from a notch filter to a narrow band pass filter, when loaded with a pair of switches, is the key design concept in achieving the complementary impedance profile. Mechanical movement of the superstrate is realized using an indigenously designed rack and pinion arrangement and actuated through a servo motor controlled by an Arduino Uno microcontroller. Alignment of the particular SRR pairs' center with that of the center of the slot lines of the CPW excites the SRR's resonance which in turn provides the frequency notch in the UWB antenna. Thus, based on various positions of the superstrate, achieved through suitably programming the Arduino Uno microcontroller, the proposed antenna can provide a tunable notch function. Rectilinear movement of the rectangular superstrate is achieved through gear support mechanism, i.e. rack and pinion arrangements where the rotational motion of the pinion gives rise to rectilinear motion of the rack. The rack and pinion are

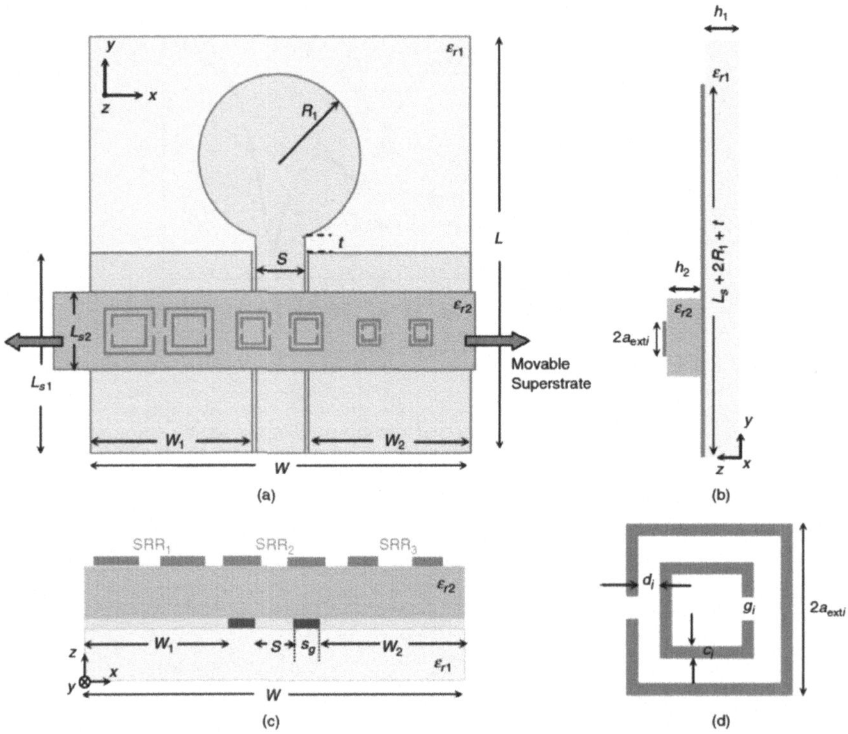

FIGURE 5.25
Geometrical layout of the proposed antenna: (a) top view along with the movable superstrate printed with SRR, (b) side view of the antenna, (c) cross-section view with copper strips (pale purple color) placed in the slots for narrowband operation, (d) unit cell of the SRR.

Reproduced by permission of © 2018 The IEEE [17].

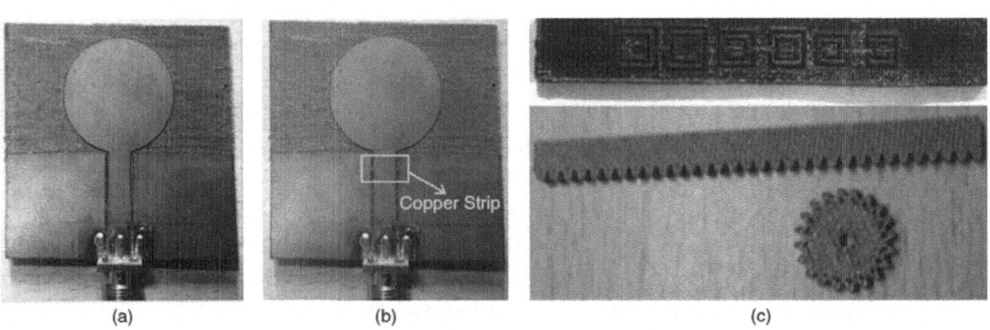

FIGURE 5.26
Fabricated prototype: (a) circular monopole antenna, (b) circular monopole antenna with copper strips in the slots of the feed region, (c) superstrate of the antenna with three different pairs of SRRs printed along with the rack and pinion used for mechanical reconfigurability.

Reproduced by permission of © 2018 The IEEE [17].

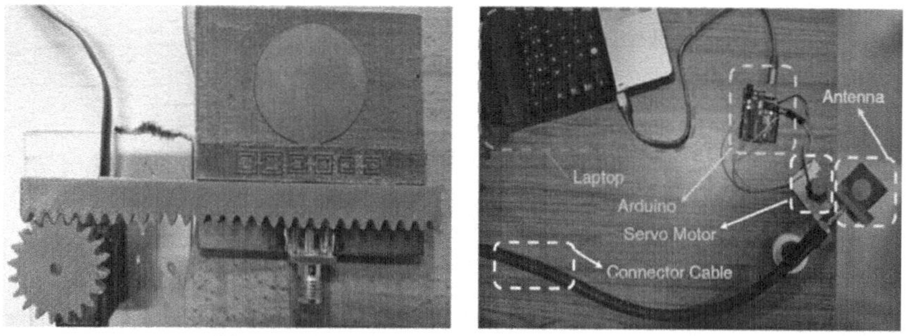

(a) (b)

FIGURE 5.27
(a) Fabricated prototype of the antenna indicating the superstrate placement, along with rack and pinion mounted on servo motor for reconfigurability, (b) complete measurement setup indicating the microcontroller and personal computer to control the reconfigurability.

Reproduced by permission of © 2018 The IEEE [17].

3D-printed using ABS plastic and are as shown in Figure 5.26c. To provide the mechanical movement, the superstrate is attached to the rack and the pinion is placed on the spindle of the servo motor as shown in Figure 5.27a. The servo motor utilized is the Towerpro SG90 commercial servo [51] which can provide 180^0 rotation and has minimum footprint area. The motor is controlled using the Arduino Uno microcontroller which provides the control on the movement of the superstrate. The required ~5 V DC supply of the motor is provided by the microcontroller and the microcontroller is powered by a USB cable attached to a personal computer and which provides the control code programmed in Arduino. The overall setup is shown in Figure 5.27b. The software coding to control the servo motor, thereby the rectilinear motion of the superstrate, is done in Arduino. The superstrate position can be controlled by specifying the angle of rotation of the motor (either discrete movement or continuous) so that one SRR pair is aligned with the slots of the feeding section. For discrete movement, the angle of rotation is specified, depending on which the motor rotates thereby moving the superstrate which in turn contributes to the precise alignment of the SRR pairs. Alternatively, the superstrate can be moved continuously from one position to the other with a small delay between the switching. Initially, SRR2 is aligned to the slots; when the angle is specified, the superstrate moves to align SRR3, then back to SRR2 and then to SRR1 as the servo motor can move 90° in either direction. The software code to control the servo motor, and thereby the rectilinear motion of the superstrate, is shown in Figure 5.28.

The same antenna, when loaded with a pair of switches in the CPW slots and similar mechanical movement of the SRR loaded superstrate is performed, results in a reconfigurable narrowband antenna. Thus, based on mechanical movement of the superstrate layer and status of the switches (ON/OFF), the proposed antenna can provide a dual reconfigurable characteristic. Table 5.2 summarizes the detailed design parameters of the dual reconfigurable antenna. Figure 5.29 shows the measured screenshot of the reflection coefficient (S_{11}) of the CMPA loaded with superstrate comprising SRRs with the entire setup. The FN-UWB characteristics can be observed in the vector network analyzer display for a single case where the SRR2 is aligned to the slots of the feed section of the antenna. With the aid of the Arduino coding and the servo motor, the superstrate is moved laterally so that different SRRs printed on it can be aligned, resulting in three different FN-UWB responses.

```
#include <Servo.h>
Servo myservo;
// create servo object to control a servo
// a maximum of eight servo objects can be created
int pos = 0;
// variable to store the servo position
void setup()
{
  myservo.attach(9);
  // attaches the servo on pin 9 to the servo object
}
void loop()
{
  for(pos = 90; pos >32; pos -= 5)
  // goes from 0 degrees to 180 degrees
  // in steps of 1 degree
  {
    myservo.write(pos);
    // tell servo to go to position in variable 'pos'
    delay(100);
    // waits 15ms for the servo to reach the position
  }
  delay(3000);
  for(pos =32; pos<=90; pos+=5)
  // goes from 180 degrees to 0 degrees
  {
    myservo.write(pos);
    // tell servo to go to position in variable 'pos'
    delay(100);
    // waits 15ms for the servo to reach the position
  }
   delay(3000);
  for(pos = 90; pos<160; pos+=5)
  // goes from 180 degrees to 0 degrees
  {
    myservo.write(pos);
    // tell servo to go to position in variable 'pos'
    delay(100);
    // waits 15ms for the servo to reach the position
  }
  delay(3000);
  for(pos = 160; pos>90; pos-=5)
  // goes from 180 degrees to 0 degrees
  {
    myservo.write(pos);
    // tell servo to go to position in variable 'pos'
    delay(100);
    // waits 15ms for the servo to reach the position
  }
  delay(5000);
}
```

FIGURE 5.28
Arduino code for controlling the superstrate movement in continuous motion (clockwise and anticlockwise) with small stop interval.

Figure 5.30 shows the simulated and measured S_{11} of CMPA for three different pairs of SRRs' alignment with respect to the slots in the feed section. It clearly reveals notched UWB response at 6.66, 8.34, and 9.85 GHz in measurement against 6.5, 8.34, and 9.55 GHz for simulation, respectively. Theoretically, computed notch frequency, calculated from SRR's resonance frequencies are 6.67, 8.69, and 10.16 GHz respectively. To realize this

TABLE 5.2

Design Parameters of the Fabricated Antenna and the Superstrate with SRRs

Antenna Design Parameters (all dimensions are in mm)		L	W	L_{s1}	R_1
		50	50	22	12.5
		W_1	W_2	S	sg
		22	22	5	0.5
		t	L_{s2}	h_1	h_2
		0.2	10	1.575	1.575
SRR Parameters (all dimensions are in mm)	#	a_{ext}	c	d	g
	SRR_1	2.25	0.5	0.3	0.5
	SRR_2	1.9	0.5	0.3	0.5
	SRR_3	1.73	0.5	0.3	0.5

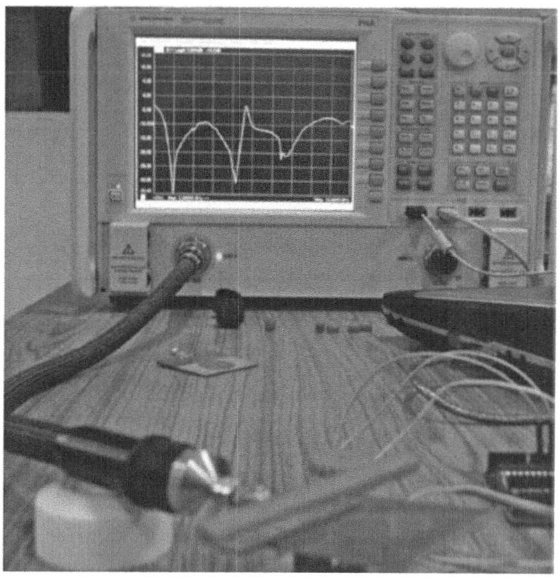

FIGURE 5.29
Measured S_{11} characteristics for SRR_1 pair aligned along the slots of the CPW. Antenna parameters are as detailed in Table 5.1.

configuration, as shown in Figure 5.30b, copper strips, acting as closed switch (the ON state), are added between the ground and signal plane of the CPW. An identical process, as described in the reconfigurable notch case, is used to move the superstrate to align different SRR pairs. Figure 5.31 shows the simulated and measured reflection coefficient (S_{11}) of the antenna in NB configuration. Depending on the alignment of the SRR pairs with the slot lines, tunable NB operations are achieved. For the SRRi (i = 1, 2, and 3), the NB response at 6.69, 8.45, and 9.89 GHz is obtained in measurement against 6.57, 8.34, and 9.58 GHz for simulation, respectively. It can be observed from Figures 5.30 and 5.31 that the obtained tunable notches and tunable NB response are complementary, indicating dual reconfigurable characteristics of the proposed antenna. The measured normalized radiation patterns of the CMPA antenna for the FN-UWB and narrowband response in the x-y plane

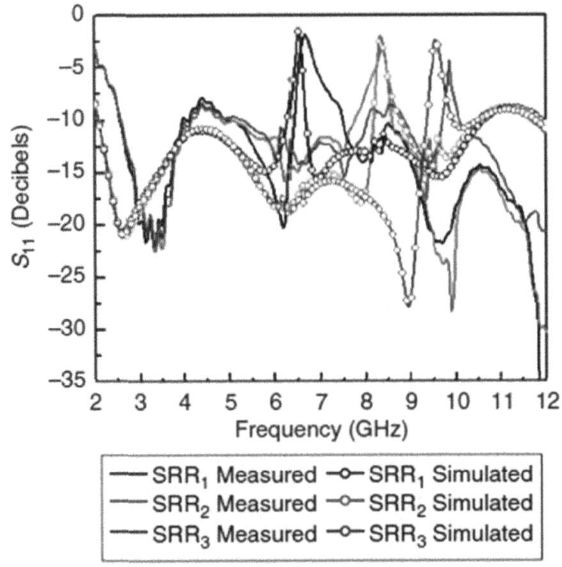

FIGURE 5.30
Simulated and measured S_{11} characteristics of the circular monopole antenna loaded with SRR printed superstrate along the feeding microstrip line for different SRRs aligned along the slots.

Reproduced by permission of © 2018 The IEEE [17].

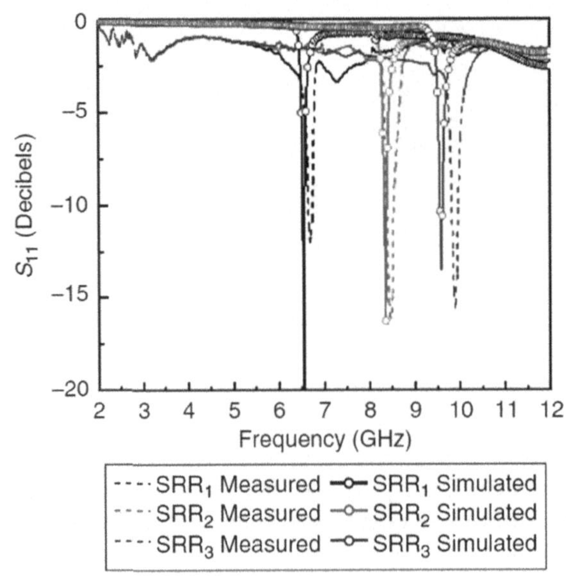

FIGURE 5.31
Simulated and measured S_{11} characteristics of the circular monopole antenna loaded with SRR printed superstrate and copper strip along the feeding microstrip line for different SRRs aligned along the slots.

Reproduced by permission of © 2018 The IEEE [17].

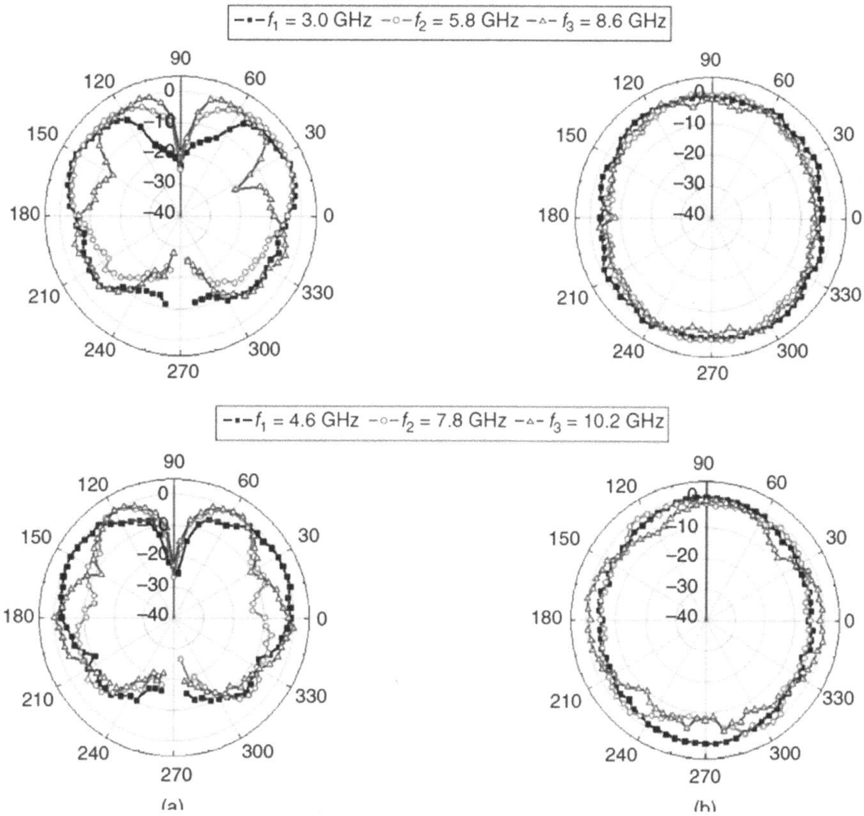

FIGURE 5.32
Measured (a) E-plane (*x-y*) (b) H-plane (*y-z*) radiation patterns in frequency-notched UWB operation.

Reproduced by permission of © 2018 The IEEE [17].

and *y-z* plane are presented in Figure 5.32 and 5.33, respectively. As revealed from the plot, for both the configurations, the E-plane (*x-y* plane) radiation pattern reveals monopole type directional pattern with axial null, whereas the H-plane (*y-z* plane) pattern is omnidirectional in nature. The simulated radiation efficiency computed as a ratio of the gain and directivity is plotted for both the configurations (FN-UWB and NB) in Figure 5.34. The radiation efficiency is at its minimum at the resonance frequency of the SRRs, confirming the presence of the notch and acceptable radiation performance at other frequencies.

To demonstrate the dual reconfigurable antenna functionality change (from a notched UWB to an NB) in a more practical manner, the copper switches are replaced with the PIN diode parameters in a simulation. To demonstrate the switching response in the simulation, the parameters of a silicon PIN diode (SMP1345–040LF) are utilized [52]. The proposed diode loaded CMPA is validated using both the EM solver and circuit solver, employing the EM co-simulation platform of CST Microwave Studio. The antenna is modeled in the 3-D layout and, to incorporate the diodes for the schematic simulation, a discrete port is placed at the diode's location. With this, the 3-D layout looks as shown in Figure 5.35a, where "1" indicates the excitation port for the antenna, and "2" and "3" indicate the discrete ports where the PIN diodes are connected. Once the 3-D layout simulation

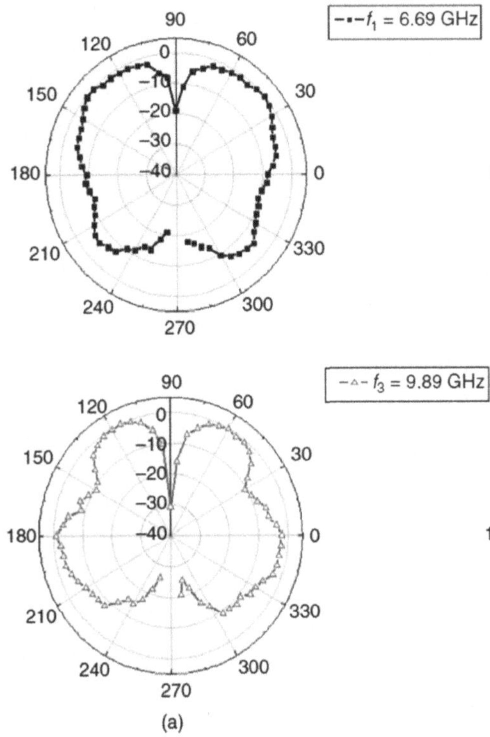

FIGURE 5.33
Measured (a) E-plane (*x-y*) (b) H-plane (*y-z*) radiation patterns for narrowband operation.

Reproduced by permission of © 2018 The IEEE [17].

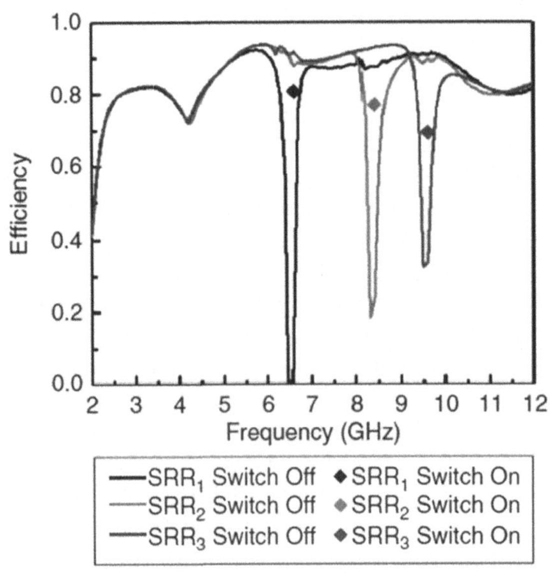

FIGURE 5.34
Simulated efficiency of the proposed antenna for frequency-notched UWB operation and narrowband operation.

Reproduced by permission of © 2018 The IEEE [17].

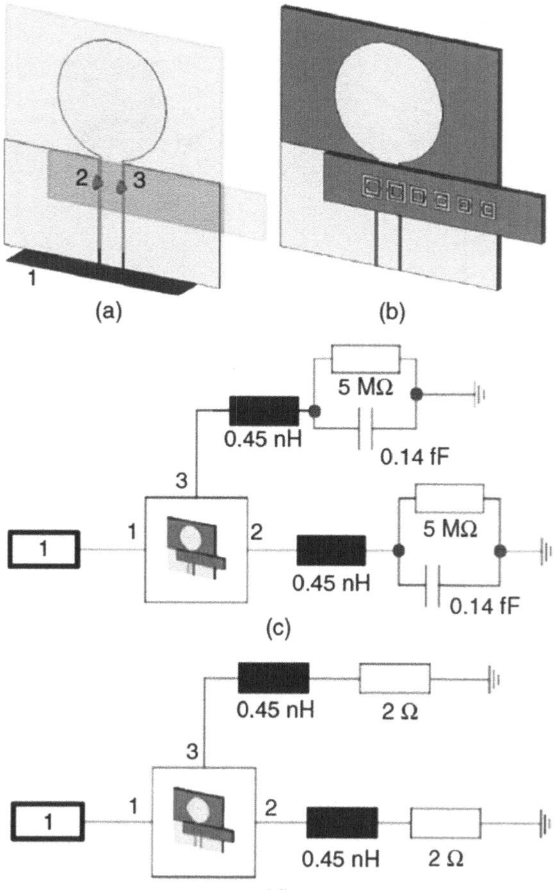

FIGURE 5.35
Simulation setup for circuit simulation in CST: (a) port definition of wave port (1) and discrete ports (2 and 3) in order to facilitate PIN diode connection in schematic, (b) perspective view of the structure (c) and (d) schematic layout showing PIN lumped element equivalent connected to antenna in OFF and ON state, respectively.

Reproduced by permission of © 2018 The IEEE [17].

is executed, the schematic simulation window is employed, where the diodes are replaced by their equivalent circuit between terminals "2" and "3".

The simulated reflection coefficient of the diode loaded CMPA for one set of SRRs is shown in Figure 5.36 for the diode in both OFF and ON conditions. The plot clearly reveals a complementary impedance profile, reconfirming the frequency-notched/NB performance of the proposed antenna evolved from the basic CMPA.

5.5 Filtenna as UWB Reconfigurable Antenna

A filtering antenna or filtenna is basically an antenna in which a filter section is embedded within the antenna. As we have discussed previously, the reconfigurable feature in

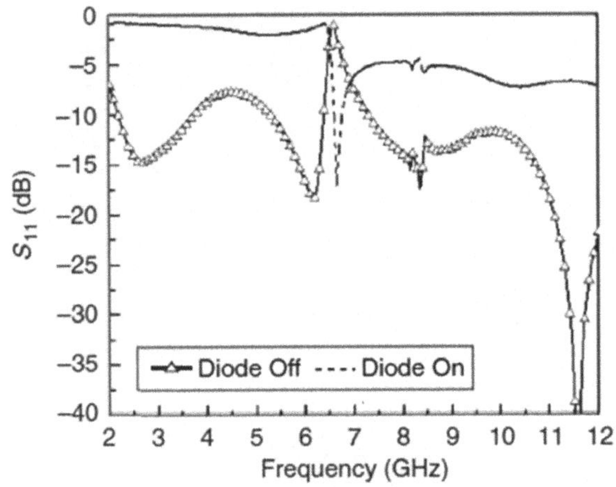

FIGURE 5.36
Simulated S_{11} characteristics of the circular monopole antenna loaded with SRR printed superstrate along the feeding microstrip line for SRR_1 aligned along the slots with the diode OFF and ON.

Reproduced by permission of © 2018 The IEEE [17].

FIGURE 5.37
Conceptual block diagram of a reconfigurable UWB antenna using filtenna.

the filter section, loaded in the feed region of the antenna, can contribute to an excellent reconfiguration in the antenna. Figure 5.37 shows a conceptual block diagram of a filtering reconfigurable antenna.

In filtenna, unlike regular reconfigurable antennas, tuning elements are loaded only in the feed region. This, for many applications, is considered a better choice as it doesn't negatively impact the radiation performance of the antenna. The reconfigurable features in a filtenna are mainly contributed by electronic and optical tuning techniques, discussed in Section 5.3. Based on the type of the filter section, reconfigurable filtenna can be used for various UWB applications.

References

1. C.G. Christodoulou, Y. Tawk, S.A. Lane, and S.R. Erwin. "Reconfigurable antennas for wireless and space applications", *Proceedings of the IEEE*, vol. 100, no. 7, 2012, 2250–2261.
2. D.B. Schaubert, "Frequency-agile polarization diversity microstrip antennas and frequency scanned arrays", U.S. Patent, 4, no. 367, issued 1983, 474.
3. J.K. Smith "Reconfigurable aperture antenna (RECAP)", *DARPA*, 1999 [Online]. Available: www.darpa.mil.
4. FCC, "Report of the spectrum efficiency working group", FCC Spectrum Policy Task Force Technical Report, 2002.

5. J.T. Bernhard, "Reconfigurable antennas", *Synthesis Lectures on Antennas*, vol. 2, no. 1, 2007, 1–66.
6. G. Washington, H.S. Yoon, M. Angelino, and W.H. Theunissen, "Design, modeling, and optimization of mechanically reconfigurable aperture antennas", *IEEE Transaction on Antennas Propagat*, vol. 50, no. 5, 2002, 628–637.
7. L.N. Pringle, P.H. Harms S.P. Blalock, G.N. Kiesel, E.J. Kuster, P.G. Friederich, R.J. Prado, J.M. Morris, and G.S. Smith, "A reconfigurable aperture antenna based on switched links between electrically small metallic patches", *IEEE Transaction on Antennas Propagat*, vol. 52, no. 6, 2004, 1434–1445.
8. K. Chang, M. Li, T.-Y. Yun, and C.T. Rodenbeck, "Novel low-cost beam steering techniques", *IEEE Transaction on Antennas Propagating*, vol. 50, no. 5, 2002, 618–627.
9. C.T. Rodenbeck M. Li, and K. Chang, "Design and analysis of a reconfigurable dual-beam grating antenna for low-cost millimeter-wave beam-steering", *IEEE Transaction on Antennas Propagating*, vol. 52, no. 4, 2004, 999–1006.
10. P. Bhartia and I. Bahl, "A frequency agile microstrip antenna", *Antennas and Propagation Society International Symposium*, vol. 20, 1982, 304–307, IEEE.
11. K.C. Gupta J. Li, R. Ramadoss, and C. Wang, "Design of frequency- reconfigurable slot ring antennas", *Proceeding of IEEE/URSI International Symposium on Antennas and Propagation*, vol. 1, 2000, 326.
12. D. Peroulis K. Sarabandi, and L.P.B. Katehi, "Design of reconfigurable slot antennas", *IEEE Transaction on Antennas Propagating*, vol. 53, 2005, 645–654.
13. D.J. Roscoe, L. Shafai, A. Ittipiboon, M. Cuhaci, and R. Douville, "Tunable dipole antennas", *Proceeding of IEEE International Symposium on Antennas and Propagation*, vol. 2, 1993, 672–675.
14. J. Kiriazi, H. Ghali, H. Radaie, and H. Haddara, "Reconfigurable dual-band dipole antenna on silicon using series MEMS switches", *Proceeding of IEEE International Symposium on Antennas and Propagation*, vol. 1, 2003, 403–406.
15. Y.E. Erdemli, K. Sertel, R.A. Gilbert, D.E. Wright, and J.L. Volakis, "Frequency-selective surfaces to enhance performance of broad-band reconfigurable arrays", *IEEE Transaction on Antennas Propagating*, vol. 50, 2002, 1716–1724.
16. Y.E. Erdemli, R.A. Gilbert, and J.L. Volakis, "A reconfigurable slot aperture design over a broad-band substrate/feed structure", *IEEE Transaction on Antennas Propagating*, vol. 52, no. 11, 2004, 2860–2870.
17. C. Saha, A. Latheef, R. Shaik, Y.M. Muntha, M. Antar, and J. Y. Siddiqui, "A dual reconfigurable printed antenna", *IEEE Antennas and Propagation Magazine*, vol. 60, no. 3, 2018, 66–74.
18. E. Erdil, K. Topalli, M. Unlu, O.A. Civi, and T. Akin, "Frequency tunable microstrip patch antenna using rf mems technology", *IEEE Transactions on Antennas and Propagation*, vol. 55, no. 4, 2007, 1193–1196.
19. G.H. Huff, and J.T. Bernhard, "Integration of packaged rf mems switches with radiation pattern reconfigurable square spiral microstrip antennas", *IEEE Transactions on Antennas and Propagation*, vol. 54, no. 2, 2006 464–469.
20. C. Won Jung, M.-J. Lee, G. Li, and F. De Flaviis, "Recofigurable scan beam single-arm spiral antenna integrated with rf-mems switches", *IEEE Transactions on Antennas and Propagation*, vol. 54, no. 2, 2006, 455–463.
21. D. Peroulis, K. Sarabandi, and L.P. Katehi, "Design of reconfigurable slot antennas", *IEEE Transactions on Antennas and Propagation*, vol. 53, no. 2, 2005, 645–654.
22. T.-Y. Han, and Y.-J. Liao, "A frequency reconfigurable half annular ring slot antenna design", *IEEE Transactions on Antennas and Propagation*, vol. 62, no. 6, 2014, 3428–3431.
23. J.-S. Row, W.-L. Liu, and T.-R. Chen, "Circular polarization and polarization reconfigurable designs for annular slot antennas", *IEEE Transactions on Antennas and Propagation*, vol. 60, no. 12, 2012, 5998–6002.
24. N. Behdad and K. Sarabandi, "A varactor-tuned dual-band slot antenna", *IEEE Transactions on Antennas and Propagation*, vol. 54, no. 2, 2006, 401–408.

25. C.R. White, and G.M. Rebeiz, "Single-and dual-polarized tunable slotring antennas", *IEEE Transactions on Antennas and Propagation*, vol. 57, no. 1, 2009, 19–26.
26. C.J. Panagamuwa, A. Chauraya, and J.C. Vardaxoglou, "Frequency and beam reconfigurable antenna using photoconducting switches", *IEEE Transactions on Antennas and Propagation*, vol. 54, no. 2, 2006, 449–454.
27. Y. Tawk, A.R. Albrecht, S. Hemmady, G. Balakrishnan, and C.G. Christodoulou, "Optically pumped frequency reconfigurable antenna design", *IEEE Antennas and Wireless Propagation Letters*, vol. 9, 2010, 280–283.
28. J. Costantine, Y. Tawk, J. Woodland, N. Flaum, and C.G. Christodoulou, "Reconfigurable antenna system with a movable ground plane for cognitive radio", *IET Microwaves, Antennas and Propagation*, vol. 8, no. 11, 2014, 858–863.
29. Firgelli Technologies Inc. Miniature linear motion Series L12, Jun. 25, 2013 [Online]. Available: http://www.firgelli.com/pdf/L12_datasheet.pdf.
30. S. Mazlouman, M. Jalali, A. Soleimani, C.M. Mahanfar, and R.G. Vaughan, "Pattern reconfigurable square ring patch antenna actuated by hemispherical dielectric elastomer", *Electronics Letters*, vol. 47, no. 3, 2011, 164–165.
31. W.-L. Liu, T.-R. Chen, S.-H. Chen, and J.-S. Row, "Reconfigurable microstrip antenna with pattern and polarisation diversities", *Electronics Letters*, vol. 43, no. 2, 2007, 77–78.
32. J. Perruisseau-Carrier M. Tamagnone, J.S. Gomez-Diaz, M. Esquius-Morote, and J.R. Mosig, "Resonant and leaky-wave reconfigurable antennas based on graphene plasmonics", *Proceeding of IEEE International Symposium on Antennas and Propagation*, Orlando, FL, 2013, 136–137.
33. V. Sathi N. Ehteshami, and J. Nourinia, "Optically tuned frequency reconfigurable microstrip antenna", *IEEE Antennas and Wireless Propagation Letters*, vol. 11, 2012, 1018–1020.
34. W. Hu, M.Y. Ismail, R. Cahill, J.A. Encinar, V.F. Fusco, H.S. Gamble, D. Linton, R. Dickie, N. Grant, and S.P. Rea, "Liquid-crystal-based reflectarray antenna with electronically switchable monopulse patterns", *Electronics Letters*, vol. 43, no. 14, 2007.
35. L. Liu, and R.J. Langley, "Liquid crystal tunable microstrip patch antenna", *Electronics Letters*, vol. 44, no. 20, 2008, 1.
36. D. Rodrigo, L. Jofre, and B. Cetiner, "Circular beam-steering reconfigurable antenna with liquid metal parasitics", *IEEE Transaction on Antennas Propagating*, vol. 60, no. 4, 2012, 1796–1802.
37. J.Y. Sidiqui, C. Saha, and Y.M.M. Antar, "Compact SRR loaded UWB circular monopole antenna with frequency notch characteristics", *IEEE Transaction on Antennas Propagating*, vol. 62, no. 8, 2014, 4015–4020.
38. J.Y. Sidiqui, C. Saha, and Y.M.M. Antar, "Compact dual-SRR loaded UWB monopole antenna with dual frequency and wideband notch characteristics", *IEEE Antennas Wireless Propagating Letter*, vol. 14, 2015, 100–103.
39. L.A. Shaik, C. Saha, J.Y. Siddiqui, and Y.M.M. Antar, "Ultrawideband monopole antenna for multiband and wideband frequency notch and narrowband applications", *IET Microwave Antennas Propagating*, vol. 10, no. 11, 2016, 1204–1211.
40. Symeon Nikolaou, Nickolas D. Kingsley, George E. Ponchak, John Papapolymerou, and Manos M. Tentzeris, "UWB elliptical monopoles with a reconfigurable band notch using MEMS switches actuated without bias lines", *IEEE Transactions on Antennas and Propagation*, vol. 57, no. 8, 2009, 2242–2251.
41. T. Li, H. Zhai, L. Li, C. Liang, and Y. Han, "Compact UWB antenna with tunable band-notched characteristic based on microstrip open-loop resonator", *IEEE Antennas and Wireless Propagation Letters*, vol. 11, 2012, 1584–1587.
42. S.H. Zheng, X. Liu, and M.M. Tentzeris, "Optically controlled reconfigurable band-notched UWB antenna for cognitive radio systems", *Electronics Letters*, vol. 50, no. 21, 2014, 1502–1504.
43. D. Zhao, L. Lan, Y. Han, F. Liang, Q. Zhang, and B.-Z. Wang, "Optically controlled reconfigurable band-notched UWB antenna for cognitive radio applications", *IEEE Photonics Technology Letters*, vol. 26, no. 21, 2014, 2173–2176.

44. A.C.K. Mak, C.R. Rowell, R.D. Murch, and C.-L. Mak, "Reconfigurable multiband antenna designs for wireless communication devices", *IEEE Transactions on Antennas and Propagation*, vol. 55, no. 7, 2007, 1919–1928.
45. A.R. Qin, Y. Pei-Yuan, J.G. Weily, T.S. Bird, and C.-H. Liang, "Frequency reconfigurable quasi-Yagi folded dipole antenna". ,*IEEE Transactions on Antennas and Propagation*, vol. 58, no. 8, 2010, 2742–2747.
46. S. Shelley, Joseph Costantine, Christos G. Christodoulou, Dimitris E. Anagnostou, and James C. Lyke, "FPGA-controlled switch-reconfigured antenna", *IEEE Antennas and Wireless Propagation Letters*, vol. 9, 2010, 355–358.
47. T. Li, H. Zhai, L. Li, and C. Liang, "Frequency-reconfigurable bow-tie antenna with a wide tuning range", *IEEE Antennas and Wireless Propagation Letters*, vol. 13, 2014, 1549–1552.
48. J. Deng, S. Hou, L. Zhao, and L. Guo, "Wideband-to-narrowband tunable monopole antenna with integrated bandpass filters for UWB/WLAN applications", *IEEE Antennas and Wireless Propagation Letters*, vol. 16, 2017, 2734–2737.
49. M. Gholamrezaei, F. Geran, and R.A. Sadeghzadeh, "Completely independent multi-ultrawideband and multi-dual-band frequency reconfigurable annular sector slot antenna (FR-ASSA)", *IEEE Transactions on Antennas and Propagation*, vol. 65, no. 2, 2017, 893–898.
50. J. Deng, S. Hou, L. Zhao, and L. Guo, "A reconfigurable filtering antenna with integrated bandpass filters for UWB/WLAN applications", *IEEE Transactions on Antennas and Propagation*, vol. 66, no. 1, 2018, 401–404.
51. Micro Servo 5-10g: SG90 Digital, 2014, [online] Available: http://www.towerpro.com.tw/product/sg90-7/.
52. Skyworks. SMP1345 series: Very low capacitance, plastic packaged silicon PIN diodes. Skyworks Solutions, Inc. Irvine, MA Technical Report: 200046M, 2012, May 9. [Online]. Available: http://www.skyworksinc.com/uploads/ documents/200046M.pdf.

6

UWB MIMO Antennas

6.1 Introduction

We are already aware that due to its vast available spectrum from 3.1 to 10.6 GHz, UWB technology is highly efficient in supporting high data rate wireless communication. However, constraint on the maximum allowed power spectral density of the UWB signal imposed by the FCC in the US [1] and other regulatory bodies in other parts of the world, limits the data transmission rate and hence curtails the full potential of UWB technology. Moreover, inevitable multi-path fading in a practical wireless channel severely impacts the quality and reliability of wireless communications, which in turn imposes restrictions on the maximum transmission rate.

On the other hand, various upcoming technologies, for example, in home audio/video networks such as high-definition television (HDTV), high-speed internet access, live streaming/downloading in real time, would require data rates of several gigabits per second (GB/s). Achieving such high data rate communication even using UWB technology is quite challenging. Multiple-input-multiple-output (MIMO) technology, which utilizes multiple transmitting antenna at the transmitter and multiple receiving antenna at the receiver, emerged as a breakthrough to mitigate this bottleneck on high data rate communication [2]. The combined capability of UWB and MIMO technology, known as an UWB-MIMO system, can provide significantly improved data rate over standalone UWB and has emerged as the most intriguing area of research for communication and EM engineers over the last few years [3]. The inherent principle of supporting high data rate by combining UWB and MIMO technology lies in the fact that channel capacity of the UWB system is proportional to the number of transmit/receive antennas [4]. Another advantage of UWB-MIMO technology is the trade-off between bandwidth and number of antennas to meet a particular data rate. In another sense, the same data rate can be achieved with lower bandwidth, but with more antennas deployed on both the transmitter and receiver sides.

In this chapter, we focus exclusively on MIMO antennas for UWB applications. As MIMO systems deal with multiple antennas installed in a common platform, along with the conventional antenna figure of merits, a MIMO antenna must qualify against a few additional parameters, such as envelope correlation coefficient (ECC), diversity gain (DG), mean effective gain (MEG), total active reflection coefficient (TARC), etc. Another important point when dealing with the design of MIMO antenna is to ensure sufficient isolation (S_{ij}, $i \neq j$) between the individual antenna elements. This, in turn, contributes to reduced mutual coupling between the antenna elements. After a brief overview of additional MIMO antenna parameters, the design of UWB-based MIMO antennas is thoroughly discussed in this chapter. This is based on various contemporary contributions and can be broadly classified into planar UWB MIMO antennas and dielectric resonator-based UWB MIMO antennas. In addition, the chapter includes two sections dedicated to frequency-notched MIMO antenna design and isolation-enhancement techniques in MIMO antennas.

6.2 Introduction to MIMO Technology and MIMO Techniques

As briefly described in the previous section, MIMO technology has drawn significant attention from wireless engineers over last few years, predominantly due to two benefits and appealing features of this technology, namely, the potential to combat against multi-path fading which impairs the wireless channels and the potential to support high data rate with the same available resources (bandwidth). Multi-path, caused by scattering of the electromagnetic signals in the environment, causes differing angles and/or different time delays and/or different frequency shifts (doppler) of the transmitted signal at the receiver end. Since the EM environment, known as the channel, is dynamic, the received signal power, contributed by various multi-path components, fluctuates in space (due to angle spread) and/or frequency (due to doppler spread) and can severely impact the quality and reliability of the wireless communication. By exploiting diversity, a MIMO system can mitigate this effect of multi-path fading. Before discussing the principle supporting the high data rate communication in UWB-MIMO, let us have a comprehensive look at the maximum data rate supported by UWB and other currently available data communication standards as shown in Table 6.1.

As can be observed from Table 6.1, among various standards, UWB (802.15.3a) supports a maximum data rate over 100 MBPS. As indicated previously, by incorporating MIMO technologies in UWB communications, a further increase in data rate up to several GB/s is possible. The benefit of UWB-MIMO over standalone UWB can be understood by comparing the data rate vs signal-to-noise ratio (SNR) for various cases. Figure 6.1 plots the data rate versus SNR in a 100 KHz channel for an M×M fading link consisting of M-transmit and M-receive antennas, with M = 1, 2, and 4. Here, the case of M = 1 indicates a conventional system, named as single-input-single-output (SISO), and M = 2 and 4, correspond to two different MIMO cases. As can be observed from Figure 6.1, for a fixed SNR, compared to the SISO case (M = 1), the MIMO system provides a higher data rate. For example, for SNR = 25 dB, the SISO system can deliver a data rate of 0.7 MBPS whereas for M = 2 and 4, the system can deliver a data rate of 1.4 MBPS and 2.8 MBPS, respectively.

Another interesting feature revealed from Figure 6.1 is that the difference in data rate between SISO and MIMO systems increases with SNR. Thus, if the system can be designed

TABLE 6.1

Capabilities of UWB and other IEEE Standards

	IEEE Standards						
	Wireless LAN (WLAN)			Bluetooth	WPAN	UWB	Zigbee
	802.11a	**802.11b**	**802.11g**	**802.15.1**	**802.15.3**	**802.15.3a**	**802.15.4**
Operational Frequency	5 GHz	2.4 GHz	2.4 GHz	2.4 GHz	2.4 GHz	3.1–10.6 GHz	2.4 GHz
Maximum Data Rate	54 MBPS	11 MBPS	54 MBPS	1 MBPS	55 MBPS	>100 MBPS	250 KBPS
Maximum Range	100 meters	100 meters	100 meters	10 meters	10 meters	10 meters	50 meters

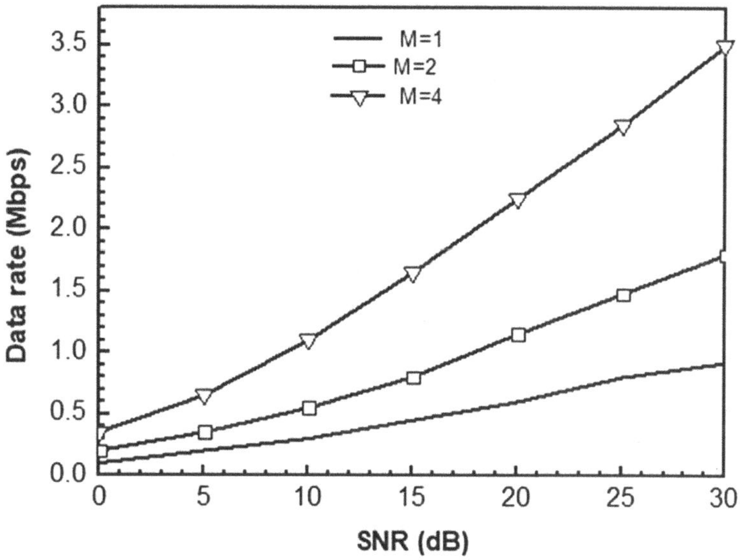

FIGURE 6.1
Data rate vs SNR in an M×M channel.

for higher SNR, more benefits can be derived from a MIMO system. One possible technique to enhance the SNR of the system is designing high gain antennas, i.e. using directional antennas instead of omnidirectional ones. Directional antennas efficiently focus the radiated energy in a desired direction by narrowing its main beam pattern and therefore contribute to a higher SNR as EM signals from such antennas pick up less spatial noise. The benefits of using directional antenna, even for SISO systems, can be understood as follows: for additive white Gaussian noise (AWGN) interference, the capacity or spectral efficiency of an SISO system, comprising omnidirectional antennas is given by,

$$C = \log_2(1 + \rho_0) \quad \text{bits / s / Hz} \tag{6.1}$$

where ρ_0 is the average signal-to-noise ratio (SNR) at the receiver. For the same SISO system using directional antennas at both transmitter and receiver, capacity is expressed as,

$$C = \log_2(1 + G^T \rho_0 G^R) \quad \text{bits / s / Hz} \tag{6.2}$$

where G^T and G^R are gain of the transmitting and receiving antennas in linear scale and their maximum radiation points toward each other. Comparing Equations 6.2 and 6.1, we understand that for the same SNR, by designing high gain transmit and receive antennas, spectral efficiency can be significantly increased. Moreover, as evident from the plot of Figure 6.1, spectral efficiency and hence the data rate can be further increased by using various cases of multiple antennas (M=2, 3, …) at the transmitter and receiver.

6.2.1 Multi-Path and Diversity

In a practical wireless communication channel dealing with non-line-of-slight (NLOS) propagation, signal radiated by the transmitting antenna can reach the receiving antenna through various paths. Various propagation mechanisms, such as reflections, diffraction

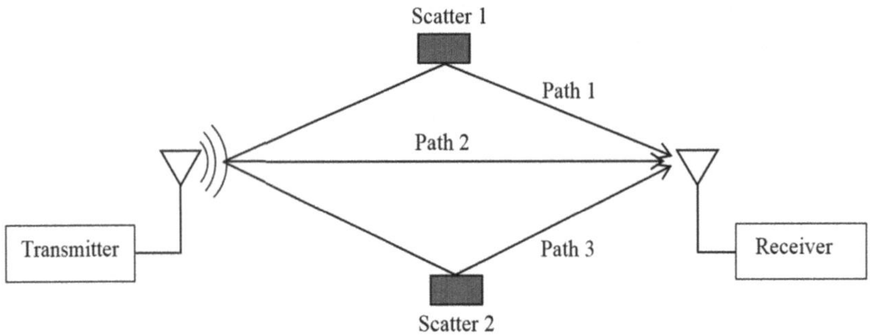

FIGURE 6.2
Simple illustration of a multi-path channel.

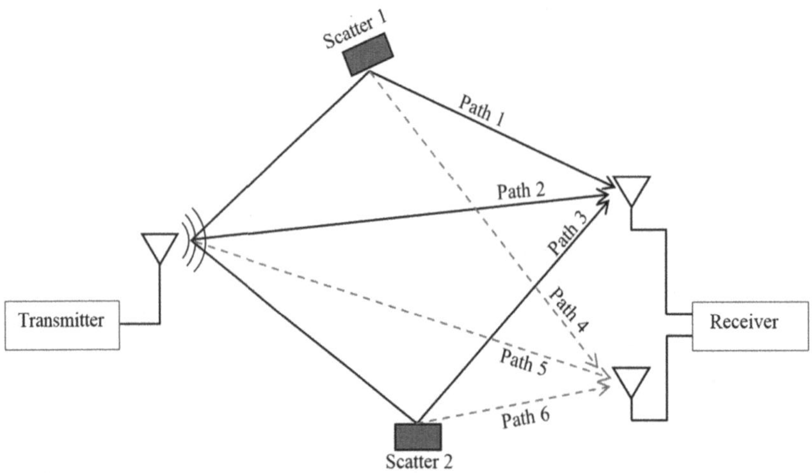

FIGURE 6.3
Example of 2×1 receive-diversity system (SIMO).

and scattering in various objects, contribute to such multiple paths through which the signal reaches the receiving antenna. A simple illustration of a multi-path reception is shown in Figure 6.2. Since various multi-paths reaching the receiving antennas are of different path lengths and have encountered different scatterers, in general they are of different amplitudes and phases. In some cases, the multi-path signals can add constructively which contributes to good received signal strength. On the other hand, if the signals add destructively, the resultant received signal is very weak or highly attenuated. Moreover, as the intermediate medium between the transmit antenna (Tx) and receiving antenna (Rx) is dynamic (movement of various possible scatterers and change of refractive index of the medium due to various atmospheric effects), the strength of the received signal fluctuates with time. Thus, multi-path interference and the "dynamic" nature of the intermediate channel are the biggest threats to a reliable wireless communication. To mitigate this issue of multi-path interference, we can employ multiple antennas at the receiver, with each antenna separated by an optimum distance. Figure 6.3 shows a simplified block diagram of such a receiving stage consisting of a two-element antenna array. The advantage of deploying multiple antennas at the receiver is that, based on the amplitude and phases of the incoming multiple signals, at least one of the antennas will have the minimum effect of multi-path fading. Thus, by increasing the number of antennas, we can minimize the chances of poor reception caused by fading. Moreover, by combining the signals from various antenna elements intelligently through proper signal processing, we can increase the overall signal strength at the receiver. This concept of employing multiple antennas at the receiving end for a reliable reception is known as "receive diversity". A similar concept of employing multiple antennas at the transmitter side to mitigate the effect of multi-path fading is known as "transmit diversity". Deploying multiple antennas both at the transmitter and receiver, reliability on diversity reception, and hence the chance of getting a stronger received signal is further increased. Such a system, in addition to the diversity reception, can contribute to a higher spectral efficiency too.

6.2.2 Evolution of MIMO Systems

Based on the configuration and number of antennas used in the transmitter and receiver in a wireless communication system, it can be classified into following four categories:

1. Single-input-single-output (SISO) system
2. Single-input-multiple-output (SIMO) system
3. Multiple-input-single-output (MISO) system
4. Multiple-input-multiple-output (MIMO) system

Figure 6.4 shows a basic schematic of these systems. While SISO is considered as the conventional communication system, SIMO and MISO architectures employ the diversity either in the receiver or in the transmitter side. On the other hand, MIMO architecture employs combined transmit and receive diversity and allows parallel transmission of data or spatial multiplexing [4]. Due to this property, MIMO technology can support very high data rate communication even with a very limited bandwidth. Figure 6.5 shows a basic block diagram of a MIMO system consisting of M-transmit and N-receive antennas. In a MIMO system, different signals are transmitted from each transmitting element and as a result, the receiving antenna system always receives a superposition of all the transmitted signals. The capacity of a MIMO system depends on the antenna configurations and/or types and typically grows with $m = \min(M, N)$ for uncorrelated channels [5]. To ensure independent fading across the different antennas, the antenna spacing should be larger than the coherence distance [6] or orthogonal polarizations should be used [7].

6.2.3 Benefits of MIMO Technology

Before going into the more technical details of MIMO antennas, let us summarize the key benefits of MIMO technology which contribute to a significant performance enhancement of a wireless communication system employing SISO configurations. These are as follows:

- *Array gain*

  By ensuring proper spatial processing of the receive and/or transmit antenna array, overall gain of a MIMO system can be increased which results in an increased receive SNR. Array gain improves the noise immunity of the system which, in turn, can improve the coverage and range of the wireless network.

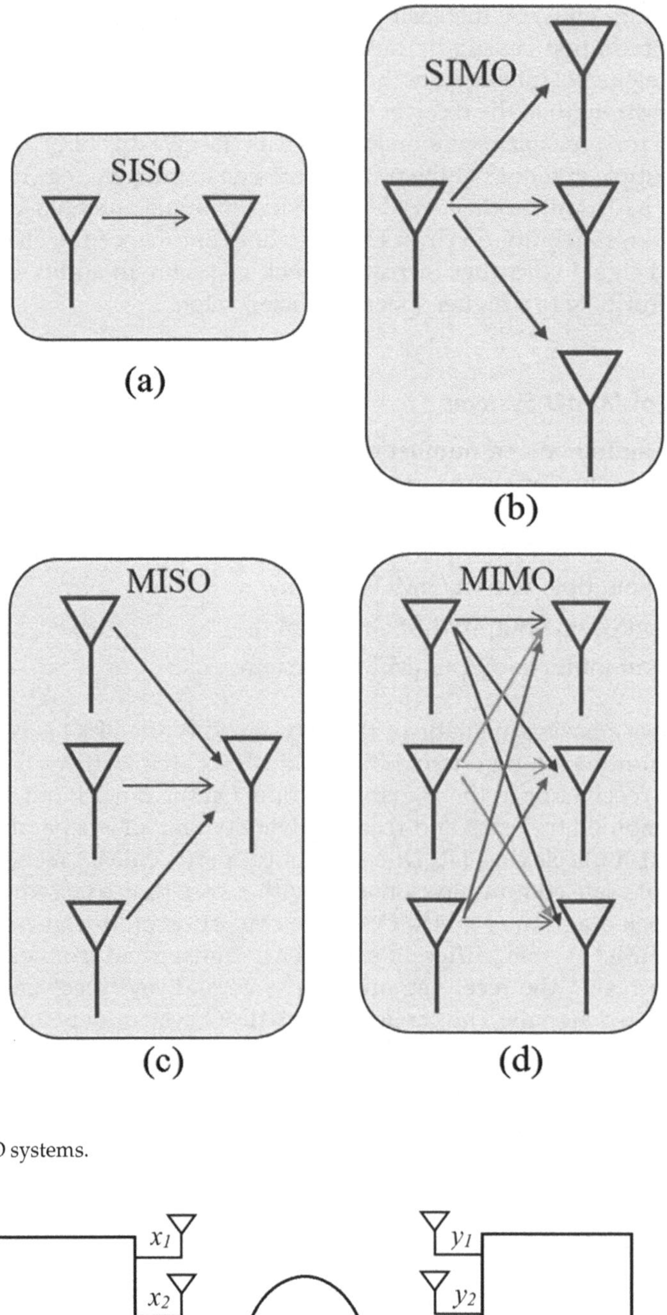

FIGURE 6.4
Evolution of MIMO systems.

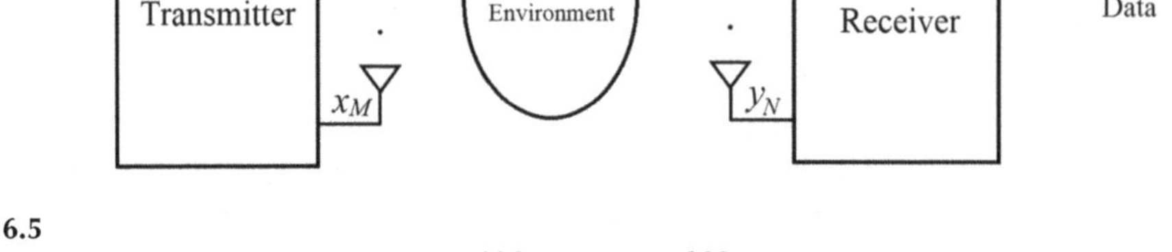

FIGURE 6.5
Block diagram of a MIMO system consisting of M-transmit and N-receive antenna.

- *Spatial diversity gain*

  As discussed earlier, due to multi-path interference and randomness of the channel, received signal level in a wireless communication system fluctuates and fades. Due to deployment of multiple antennas at the transmitter and receiver, a MIMO system offers multiple copies of the transmitted signal in space, frequency, and time to the receiver. A MIMO channel with M-transmit antennas and N-receive antennas can potentially offer M×N independent fading links. With this increased number of diversity order, the probability of at least one of the copies not undergoing a deep fade increases. This contributes to improved quality and reliable reception in a MIMO system.
- *Spatial multiplexing gain*

  With multiple transmit antennas, a MIMO system can transmit multiple independent data streams concurrently within the bandwidth of operation. Due to this spatial multiplexing phenomenon, a MIMO system offers a linear increase in data rate over a conventional SISO system [5]. The number of data streams that can be reliably supported by a MIMO channel is equal to the minimum of the number of antennas used as the transmitter and receiver. Thus, spatial multiplexing capability of a MIMO system contributes to a higher channel capacity or bit rate in a wireless communication network.
- *Interference reduction and avoidance*

  A MIMO system is capable of mitigating unwanted interference from undesired users by exploiting the spatial dimension to adjust the separation between users. With enhanced array gain, it is capable of increasing the signal-to-interference-plus-noise (SINR) ratio.

6.3 Characterization of MIMO Antenna

In the previous section, we discussed the basic concepts, evolution, and benefits of MIMO technology in a brief manner. Due to the limited scope of this book and the specific focus on antennas, we will confine our discussion to "antenna aspects" of MIMO technology. More specifically, our goal is to discuss the design of various families of antennas for UWB-MIMO applications.

Each antenna in a MIMO, whether deployed in the transmitting side or at the receiving side, must satisfy the fundamental requirements of input matching, i.e. have input impedance close to 50 Ω, a good directional pattern as required for a particular application, and a good efficiency. In addition, an antenna/array in a MIMO system, customarily known as MIMO antenna which is basically an array/cluster of identical antennas, must conform to a few additional figures of merit.

6.3.1 Envelope Correlation Coefficient

The correlation coefficient is a measure that describes how much the communication channels are isolated or correlated with each other. The square of the correlation coefficient is known as the envelope correlation coefficient. The envelope correlation for a two-antenna system is computed as,

$$\rho_e = \frac{\left| \iint_{4\pi} [F_1(\theta,\phi) * F_2(\theta,\phi)] d\Omega \right|^2}{\iint_{4\pi} \left| [F_1(\theta,\phi)] \right|^2 d\Omega \iint_{4\pi} \left| [F_2(\theta,\phi)] \right|^2 d\Omega} \tag{6.3}$$

where $F_i(\theta,\phi)$ is the field radiation pattern of the antenna system when port i is excited, and * denotes the Hermitian product. However, computation of ECC using Equation 6.3 requires the radiation pattern of the antenna system and we need to perform numerical integration which is quite a cumbersome process. To reduce the effort in evaluating ECC, an alternative S-parameter based expression proposed by Blanch et al. [8] can be used for a two-port MIMO system. This is given by,

$$\rho_e = \frac{\left| S_{11}{}^* S_{12} + S_{21}{}^* S_{22} \right|^2}{\left(1 - |S_{11}|^2 - |S_{21}|^2\right)\left(1 - |S_{21}|^2 - |S_{22}|^2\right)} \tag{6.4}$$

This is a simplified method of obtaining ECC compared to evaluating using the field radiation pattern and it holds good when a lossless single mode antenna is used. Ideally the value of ECC should be zero.

A modified expression for correlation considering the antenna efficiency [9] yields a more accurate result. This can be expressed as,

$$|\rho_{ij}|^2 = \rho_{eij} = \left| \frac{\left| S_{ii}{}^* S_{ij} + S_{ji}{}^* S_{jj} \right|}{\sqrt{\left(1 - |S_{ii}|^2 - |S_{ij}|^2\right)\left(1 - |S_{jj}|^2 - |S_{ij}|^2\right) \eta_{\mathrm{radi}} \eta_{\mathrm{radj}}}} \right|^2 \tag{6.5}$$

where:

ρ_{ij} is the correlation coefficient between elements i and j

ρ_{eij} is the envelop correlation coefficient

S_{ij} is the S-parameter between i^{th} and j^{th} ports

η_{radi} and η_{radj} indicates the radiation efficiencies of ith and j^{th} antenna, respectively.

For an ideal MIMO system, ECC is zero. However, due to mutual coupling between the antenna elements, in a practical MIMO antenna system it is greater than zero. According to the rule of thumb for good diversity/MIMO performance, ECC should be less than 0.5 ($\rho_e < 0.5$).

6.3.2 Diversity Gain

As discussed previously, in MIMO systems diversity corresponds to receiving multiple versions/copies of the transmitted data through different channel paths contributed by multiple antennas. If the signals are uncorrelated, the combined signals at the receiver provide higher signal-to-noise ratio levels, and thus better signal reception.

Diversity gain can be defined as the increase in time-averaged signal-to-noise ratio due to the diversity scheme over that of a single antenna system (i.e. one diversity channel), provided the SNR is above the reference channel. Alternatively, it can be defined as the amount of reduction in transmitted power when a diversity scheme is introduced over that of a single antenna system, without any loss of performance. Diversity gain of a diversity system can be computed as,

$$\text{Diversity Gain} = \left[\frac{\gamma_c}{SNR_c} - \frac{\gamma_1}{SNR_1}\right]_{P(\gamma_c<\gamma_s/SNR)} \tag{6.6}$$

where:

γ_c and SNR_c are the instantaneous and mean SNR for the diversity system
γ_1 and SNR_1 are the instantaneous and mean SNR for the single antenna system.

Here, γ_s/SNR is the reference level and $P(\gamma_c < \gamma_s/SNR)$ indicates the probability that instantaneous SNR of the diversity system is less than the reference level and can be expressed as,

$$P(\gamma_c < \gamma_s/SNR) = \left[1 - e^{-\frac{\gamma_s}{SNR}}\right]^M, \tag{6.7}$$

where M indicates the number of antennas. Diversity gain of a MIMO system is directly related to the number of antennas deployed in the system as an increase in the number of antennas leads to an increased combined received power and hence the diversity gain of the system.

6.3.2.1 Mean Effective Gain (MEG)

Conventional gain of standalone antenna, obtained from an electromagnetic field analysis or from an EM simulator or obtained by measurement in an anechoic chamber, doesn't reflect its actual directional performance, as in real time the antenna is never operated in an ideal environment considered in an EM simulator nor inside an anechoic chamber. The antenna, rather, is used in a real environment which contributes to multi-paths due to reflection, refraction, diffraction, and scattering. Thus, to evaluate the real performance of an antenna, studying its radiation performance considering the impact of the surrounding is crucial. A useful and widely used measure of an antenna to quantify this is the mean effective gain (MEG). MEG is a statistical measure of an antenna gain considering a mobile environment and is defined as the ratio of mean received power by an antenna to the total mean incident power.

Real time radiation performance and hence calculation of MEG of popular handset antennas (helix, whip and patch) considering environmental effects was reported in Nielsen et al. [10]. However, such processing of evaluating an antenna's radiation pattern is time-consuming, costly, and moreover, it might lead to a multi-iterative cycle of design-fabrication testing to ensure a desired performance. MEG, proposed in Taga [11], uses a probabilistic model to mimic the environments. Using the 3-D radiation pattern and this probabilistic model, MEG can be numerically obtained by solving a mathematical expression that involves these two quantities [12]. The mathematical expressions for calculation of MEG are as follows:

$$MEG = \int_0^{2\pi}\int_0^{\pi}\left[\frac{XPR}{XPR+1}G_\theta(\theta,\phi)P_\theta(\theta,\phi) + \frac{1}{XPR+1}G_\theta(\theta,\phi)P_\theta(\theta,\phi)\right]\sin\theta d\theta d\phi \tag{6.8}$$

$$\int_0^{2\pi}\int_0^{\pi}\left[G_\theta(\theta,\phi)+G_\phi(\theta,\phi)\right]\sin\theta d\theta d\phi = 4\pi \tag{6.9a}$$

$$\int_0^{2\pi}\int_0^{\pi} P_\theta(\theta,\phi)\sin\theta d\theta d\phi = \int_0^{2\pi}\int_0^{\pi} P_\phi(\theta,\phi)\sin\theta d\theta d\phi = 1 \tag{6.9b}$$

$$XPR = \frac{P_V}{P_H} \tag{6.10}$$

In the above expressions, *XPR* is the cross-polarization ratio, also known as cross-polarization discrimination, that represents the ratio of vertical and horizontal component of the mean incoming power.

$G_\theta(\theta,\phi)$ and $G_\phi(\theta,\phi)$ are the antenna gain components and $P_\theta(\theta,\phi)$ and $P_\phi(\theta,\phi)$ represent the θ - and ϕ - components of angular density functions of the incoming plane waves in the environment, obtained from a statistical model in which the angular density functions are assumed to be Gaussian in elevation plane and uniform in azimuth plane (i.e. assuming zero correlation between the two).

Thus, using Equations 6.8 to 6.10, we can calculate the MEG using gain pattern of an antenna in an ideal environment obtained from an anechoic chamber and a suitable statistical model to mimic the environments/channel. However, considering a suitable channel pertinent to a particular environment is critical in calculating the MEG of an antenna. More information on calculation MEG cab be found in Jamaly et al., Kalliola et al. and Ando et al. [13–15].

6.3.2.2 Total Active Reflection Coefficient

Information on scattering matrix or *S*-parameters are not sufficient to fully characterize a MIMO antenna. For proper characterization of a MIMO antenna system, we introduce a new parameter, known as total active reflection coefficient (TARC). TARC of the MIMO antenna is defined as the ratio of square root of the total reflected power to the square of root of total incident power [16].

For a MIMO antenna consisting of *N*-elements, TARC can be expressed as,

$$\Gamma_a^t = \frac{\sqrt{\sum_{i=1}^{N}\left|b_i\right|^2}}{\sqrt{\sum_{i=1}^{N}\left|a_i\right|^2}}, \tag{6.11}$$

where a_i and b_i represent incident and reflected signals at i^{th} port of the antenna. As a_i and b_i are related by *S*-parameters, TARC of antenna can be conventionally expressed in terms of *S*-parameters. With simple calculation using the relation,

$$b = Sa \tag{6.12}$$

TARC for a two-port MIMO antenna system can be expressed as [17],

$$\left(\Gamma_a^t\right)_{2-\text{port}} = \frac{\sqrt{\left|S_{11}+S_{12}e^{j\theta}\right|^2+\left|S_{21}+S_{22}e^{j\theta}\right|^2}}{\sqrt{2}} \tag{6.13}$$

In a similar manner, TARC for a four-port MIMO antenna system is expressed as,

$$\left(\Gamma_a^t\right)_{4-\text{port}} = \frac{\sqrt{\sum_{i=1}^{4}\left|S_{i1} + S_{i2}e^{j\theta_1} + S_{i3}e^{j\theta_2} + S_{i4}e^{j\theta_3}\right|^2}}{2} \tag{6.14}$$

Here,

θ is the phase of S-parameters

S_{ii} is the reflection coefficient at i^{th} port

S_{ij} is the coupling between i^{th} and j^{th} port.

6.3.3 Branch Power Ratio

Another important parameter to evaluate the performance of a MIMO antenna system can be obtained by comparing the power levels associated with various antenna branches. To ensure a good diversity, the power level of various branches should be as close as possible. The power level difference of various branches is estimated by power level ratio k, which is defined as the ratio of minimum and maximum power levels of two antennas.

Thus, the branch power level ratio is expressed as,

$$k = \frac{p_{\min}}{p_{\max}}, \tag{6.15}$$

Where $p_{\min}$ and $p_{\max}$ represent lowest and highest power associated with two antennas of a MIMO system.

Power level ratio k directly affects the diversity gain of an antenna as the effective diversity gain is the product of diversity gain and inverse of k. Considering k in (6.7), it modified to,

$$P(\gamma_c < \gamma_s/SNR) \approx \frac{1}{k}\left[1 - e^{-\frac{\gamma_s}{SNR}}\right]^M, \tag{6.16}$$

Alternatively, branch power ratio k can be calculated from the MEG of the antenna system [18]. For a two-element MIMO system,

$$k = \min\left(\frac{MEG_2}{MEG_1}, \frac{MEG_1}{MEG_2}\right), \tag{6.17}$$

Where MEG_2 and MEG_1 are mean effective gain of antennas 1 and 2, respectively. For a good MIMO system with high diversity, gain k should be in the range of $0 \le k \le 0.5$, i.e. between 0 to −3 dB.

6.3.4 System Capacity

One of the main goals of introducing a multiple antenna system (SIMO, MISO, and MIMO) in modern wireless applications is to achieve a higher channel capacity in a particular fading environment over a conventional single-input-single-output (SISO) system. This allows faster communication with enhanced data rate with same available bandwidth. Therefore, channel capacity of a MIMO system is another significant figure of merit. Accurate estimation of channel capacity of a MIMO system is not straightforward as it depends highly on the channel environment (which again is not constant, rather dynamic) along with the number and radiation characteristics of the antennas deployed in the MIMO system. A simplified expression for channel capacity of an N-element MIMO antenna, considering equal power feeding to each antenna is expressed as,

$$C = \log_2\left[\det\left(I_N + \frac{\rho}{N} HH^T\right)\right] \tag{6.18}$$

where C is in bits/s/Hz.

In Equation 6.18, ρ is average SNR, H is the normalized channel covariance matrix, I_N is an $N \times N$ identity matrix and N is the number of antennas both at the transmitter and receiver side. Under further assumption of zero correlation coefficient of antennas at the transmitter and receiver side, completely uncorrelated transmitting and receiving wave, and similar power and MEG values of the antenna, Equation 6.18 reduces to,

$$C = N\log_2\left(1 + \frac{\rho}{N}\right) \tag{6.19}$$

It can be noted that, Equation 6.19 indicates a linear increase in channel capacity over a SISO system. However, it must be remembered that this is an ideal value and practically can never be reached due to unavoidable finite correlation coefficient between the channels and non-zero correlation coefficient between the antenna elements. A more practical way to estimate the channel capacity in the multi-path environment is to consider a proper channel matrix [19] and non-zero correlation between the antenna elements.

6.4 Printed UWB-MIMO Antennas

In Section 6.3, we discussed various antenna parameters to be considered while characterizing the performance of MIMO antennas. Now, we will focus on the design of UWB-MIMO antennas. While discussing this, we will consider all classes of UWB antennas, i.e. complying with the FCC definition (BW: 3.1–10.6 GHz), general UWB definition (BW $\geq$ 500 MHz), and alternative definition (ratio bandwidth $\geq$ 0.25). To focus the discussion specifically on design aspects and/or functional requirements, the discussion on UWB-MIMO antenna design is divided into three sections: printed UWB-MIMO antennas, discussed in this section; dielectric resonator-based UWB-MIMO antennas, discussed in Section 6.5; and frequency-notched UWB-MIMO antennas, discussed in Section 6.6.

Printed UWB-MIMO antenna, as the name suggests, consists of multiple printed antenna elements accommodated on a common dielectric substrate and preferably with a common connected ground plane [20]. The antenna elements should maintain an optimum spatial separation to ensure sufficiently high-isolation and low-correlation coefficient, and at the same time ensuring its compactness. The basic radiating elements such as dipoles [21], monopoles [22–24], loops [25], open ended and short ended slots [26], PIFA [27], etc. are utilized in designing most of the printed MIMO antennas. Figure 6.6 shows various popular candidates used in realizing MIMO antennas for various applications.

One of the early demonstrations and realizations of diversity/MIMO antenna dates back to 2001 when Ko and Murch [28] proposed a new concept of diversity reception by

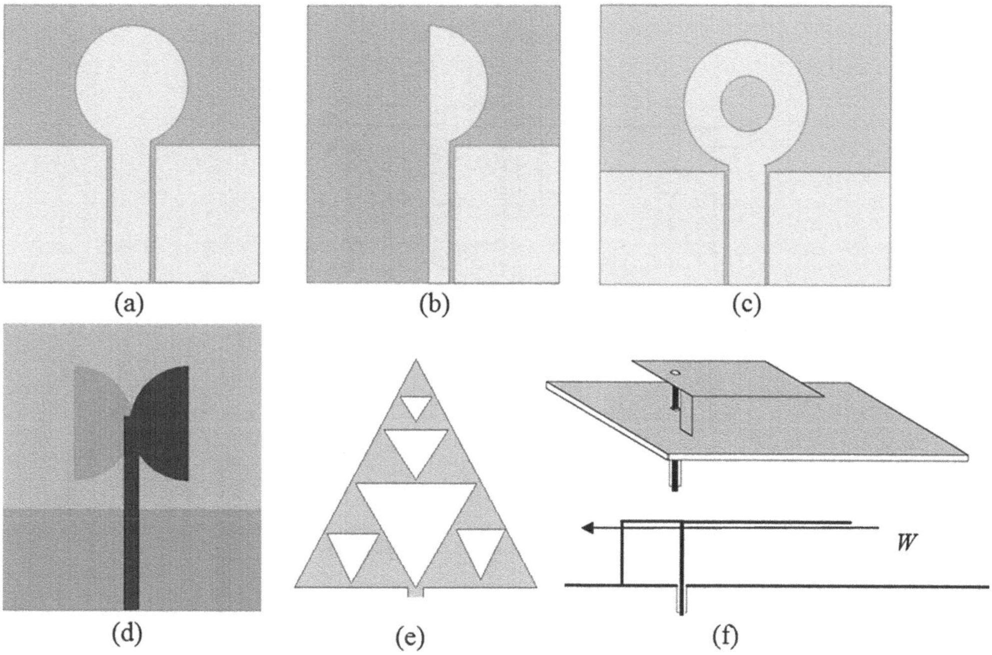

FIGURE 6.6
Popular antenna elements for UWB-MIMO applications: (a) circular monopole, (b) half-circular monopole, (c) ring monopole, (d) antipodal bow tie (e) fractal monopole, and (f) PIFA.

merging two patch antennas in combination with capacitive loading. Along with desired impedance matching, the antenna reported by Ko and Murch provides good isolation and low ECC (<0.1) at operating frequency of 2.2 GHz. A printed UWB diversity monopole antenna reported in Wong et al. [29] is another early addition to the UWB-MIMO antenna family. This antenna, consisting of truncated square monopoles printed on the two sides of a T-shaped protruded ground plane, provides a very wide impedance bandwidth (2.3–7.7 GHz), good port isolation ($S_{21} < -20$dB), excellent spatial diversity and stable gain over the entire impedance bandwidth. A two-element diversity/MIMO antenna system using inverted-F antenna as the basic element proposed in Karaboikis et al. [30] is another popular candidate that emerged during the early development phase of the UWB-MIMO antenna. In this antenna, low mutual coupling was achieved by using corrugated ground planes. The antenna, successfully characterized for various MIMO parameters, was proposed for 5.2 GHz ISM band.

Two orthogonal cone-shaped radiators placed inside the T-shaped protruded region of the ground plane is designed for a two-element UWB-MIMO antenna operating in the frequency range of 3.1–5.8 GHz. The designed antenna [31] shown in Figure 6.7 is printed on Rogers 4003 substrate ($\varepsilon_r = 3.38$) having an overall dimension of 60×62×1.52 mm^3. The position of the radiating elements in this antenna is optimized to ensure a good spatial diversity. In addition, the optimized design exhibits good isolation between the two radiating elements. Figure 6.8 shows the simulated and measured return loss and isolation of this antenna.

Figure 6.9a,b shows the measured radiation pattern of the antenna at 3.1 GHz in three principal planes (x-y plane, $\phi = 45°/225°$ plane, and $\phi = 135°/315°$) for port 1 and port 2

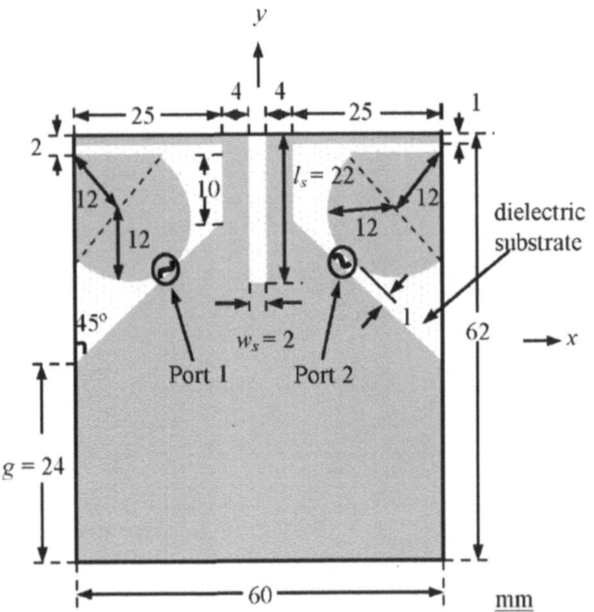

FIGURE 6.7
A two-element diversity/MIMO antenna consisting of two cone-shaped radiators placed in the T-shaped protruded ground plane.

Reproduced by permission of © 2006 The IEEE [31].

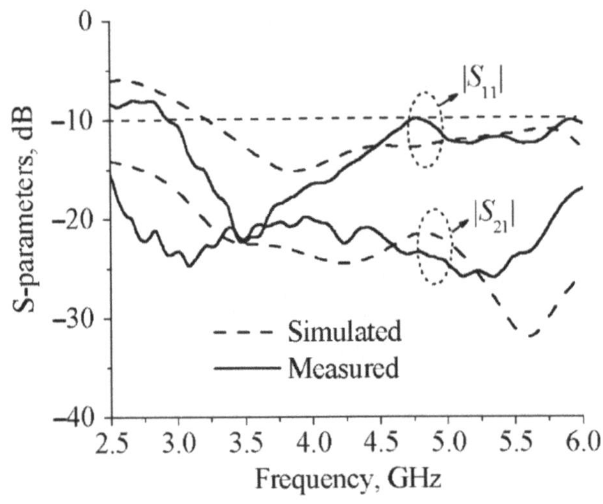

FIGURE 6.8
Simulated and measured S_{11} and S_{21} of the two-element diversity/MIMO antenna consisting of two cone-shaped radiators placed in the T-shaped protruded ground plane.

Reproduced by permission of © 2006 The IEEE [31].

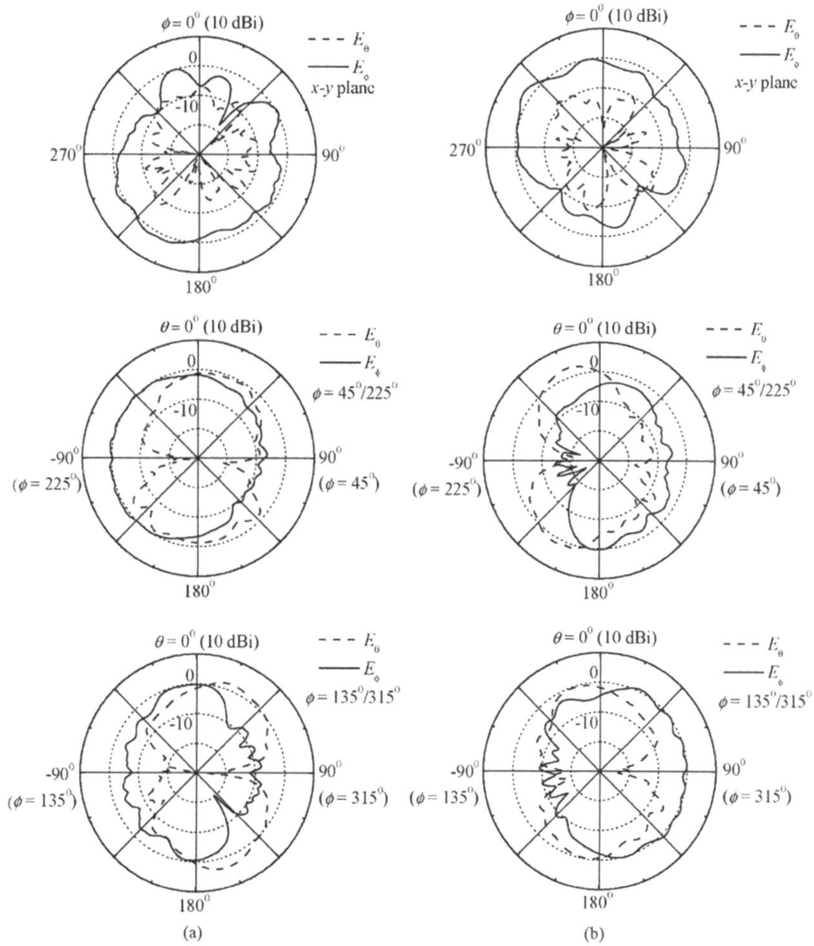

FIGURE 6.9
Measured radiation pattern of the two-element diversity/MIMO antenna consisting of two cone-shaped radiators placed in the T-shaped protruded ground plane: (a) for port 1 excitation, (b) for port 2 excitation.

excitation, respectively. The measured results reveal stable radiation pattern with acceptable gain and cross-polar discrimination.

Two notched triangular radiating structures fed by a 50 Ω microstrip line [32] is another excellent candidate for two-element UWB-MIMO applications. The antenna is compact in size and exhibits good MIMO performance metrics (Figure 6.10).

As revealed in the *S*-parameters plot of Figure 6.11, the antenna provides an impedance bandwidth from 3.1–5 GHz along with good isolation ($S_{21} < -20$ dB) for the entire impedance bandwidth. Stable gain, omnidirectional radiation pattern along with acceptable efficiency makes this antenna a good choice for wireless portable devices such as USB dongles.

Two-element UWB-MIMO using the classical Yagi-Uda structure reported in Jehangir and Sharawi [33] is presented in Figure 6.12. A loop element, with the half portion in the top side of the substrate and other half portion at the bottom of the substrate, acts as the driven element while an arc-shaped element on the top side plays the role of the director.

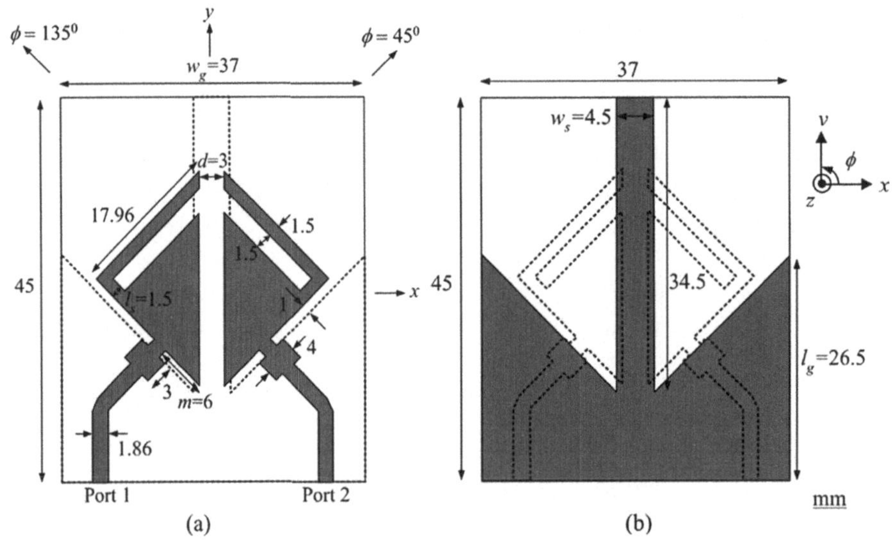

FIGURE 6.10
A two-element diversity/MIMO antenna consisting of two notched triangular radiators for UWB-MIMO applications: (a) top side, (b) bottom side of the MIMO antenna system.

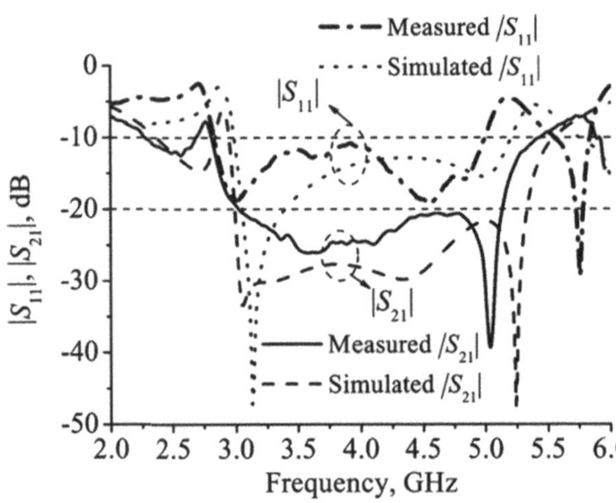

FIGURE 6.11
Measured and simulated *S*-parameters of the two-element diversity/MIMO antenna consisting of two notched triangular radiators for UWB-MIMO applications.

Two such identical units constitute the two-element MIMO antenna. The antenna elements are excited by a 50 Ω microstrip line. As shown in Figure 6.12b, the overall size of the antenna is 80×50×0.76 mm^3. As revealed in Figure 6.14a, the designed MIMO antenna yields good impedance matching and isolation (better than 17 dB) over a wide frequency range of 4.184 GHz to 6.584 GHz (Figure 6.13).

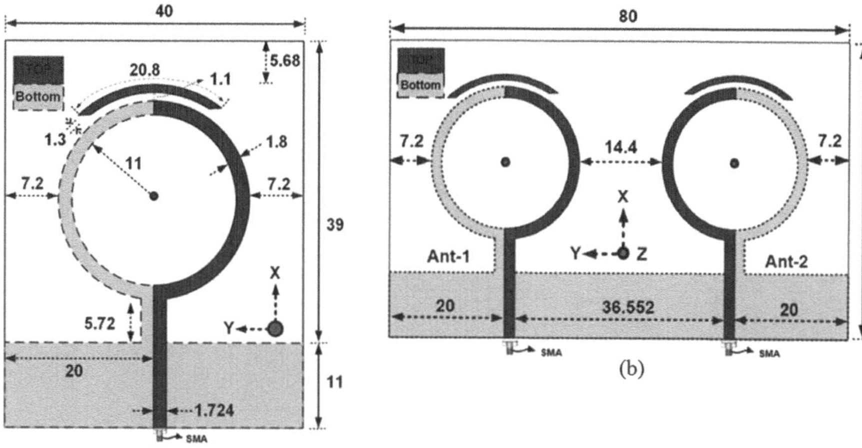

FIGURE 6.12
Two-element UWB-MIMO antenna consisting of two printed Yagi-like radiators.

Reproduced by permission of © 2017 The IEEE [33].

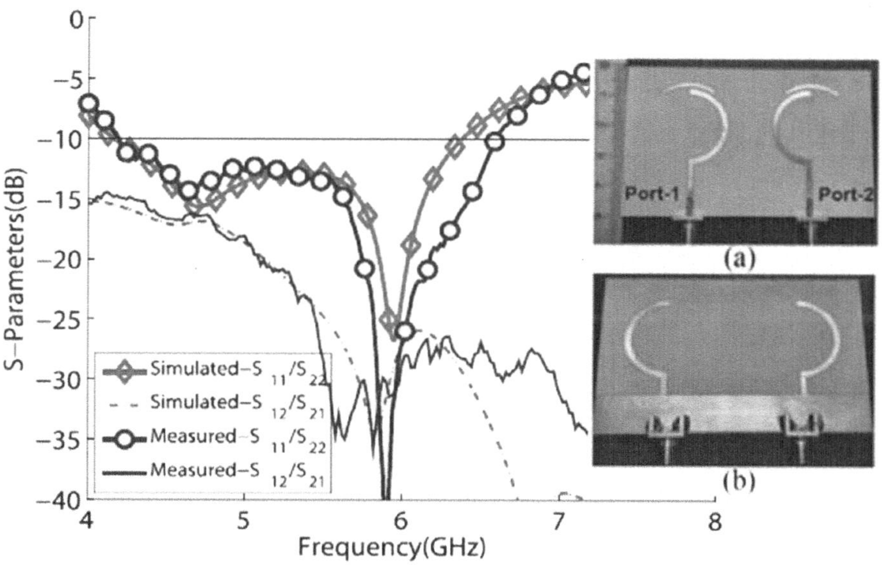

FIGURE 6.13
Simulated and measured *S*-parameters of the two-element UWB-MIMO antenna consisting of two printed Yagi-like radiators.

Reproduced by permission of © 2017 The IEEE [33].

Figure 6.14b shows the simulated and measured gain and efficiency of the UWB-MIMO antenna for both the radiators. It can be noted that both the antenna elements provide a maximum realized gain better than 5 dBi and measured radiation efficiency above 80%.

Very recently, a four-element ultrawide frequency reconfigurable MIMO antenna was reported in Hussain et al. [34]. As shown in Figure 6.15, the antenna consists of four-element

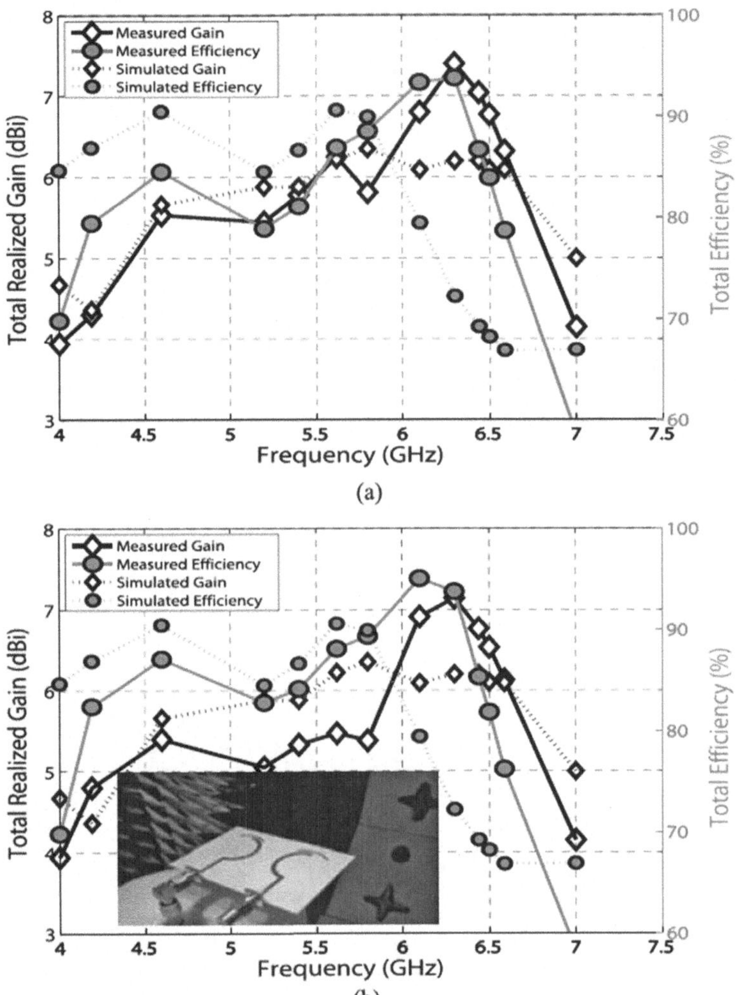

FIGURE 6.14
Simulated and measured gain and efficiency of the two-element UWB-MIMO antenna consisting of two printed Yagi-like radiators: (a) for radiator 1, (b) for radiator 2.

Reproduced by permission of © 2017 The IEEE [33].

pentagonal slots etched from the bottom side of a double-sided copper-clad substrate. The slots are excited by a 50 Ω microstrip line printed on the other side of the substrate. The overall dimensions of the antenna are 60×120×1.56 mm^3. By loading a pair of varactor diodes on each antenna with suitable biasing scheme, the MIMO antenna yields a wide frequency tuning from 1.32–1.49 GHz and 1.75–5.2 GHz. Figure 6.16 shows the simulated and measured S_{11} and S_{12} response of the antenna. The antenna also exhibits good MIMO performance metrics such as ECC and TARC.

UWB-MIMO antenna, along with dynamic spectrum allocation policy of the cognitive radio (CR) environment in software-defined radio (SDR) for efficient utilization of the spectrum, can further enhance the potential of modern wireless radio. Such technology demands UWB-MIMO antenna to be frequency reconfigurable over a wide range.

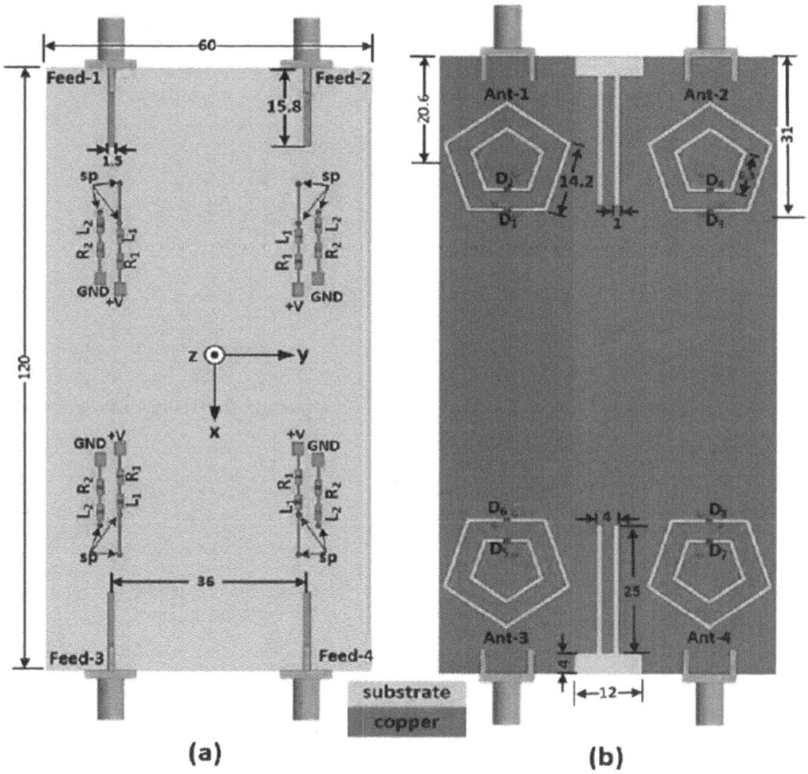

FIGURE 6.15
Schematic diagram of a four-element reconfigurable UWB-MIMO antenna consisting of pentagonal slot-shaped radiators: (a) top surface, (b) bottom surface of the antenna.

Reproduced by permission of © 2018 The IEEE [34].

Such antenna design utilizes the concept of various frequency reconfigurable mechanisms [35–40] and techniques while designing UWB-MIMO antennas.

6.5 Dielectric Resonator-Based UWB-MIMO Antennas

Dielectric resonator antennas (DRA), originating from simple dielectric resonators (DR) [41], has evolved as an important class of antenna due to its various features and advantages. Various geometries of DRAs, such as rectangular, hemispherical, cylindrical, cylindrical-ring, conical, and hybrid shapes clustering multiple basic geometries, have been studied extensively over the last two decades. Proper excitation of the dielectric material of a suitable geometry and shape can contribute to multiple modes which in turn can make the DRA multiband and/or extremely wide band. Apart from these multiband/wideband features, DRAs offer very high efficacy due to almost negligible conduction and surface wave loss, unlike the microstrip-based printed antennas. Because of these credentials and attractive features, DRAs appeared as an alternative for wireless engineers for diversity/MIMO applications. Exciting various modes in a single dielectric resonator is one of the

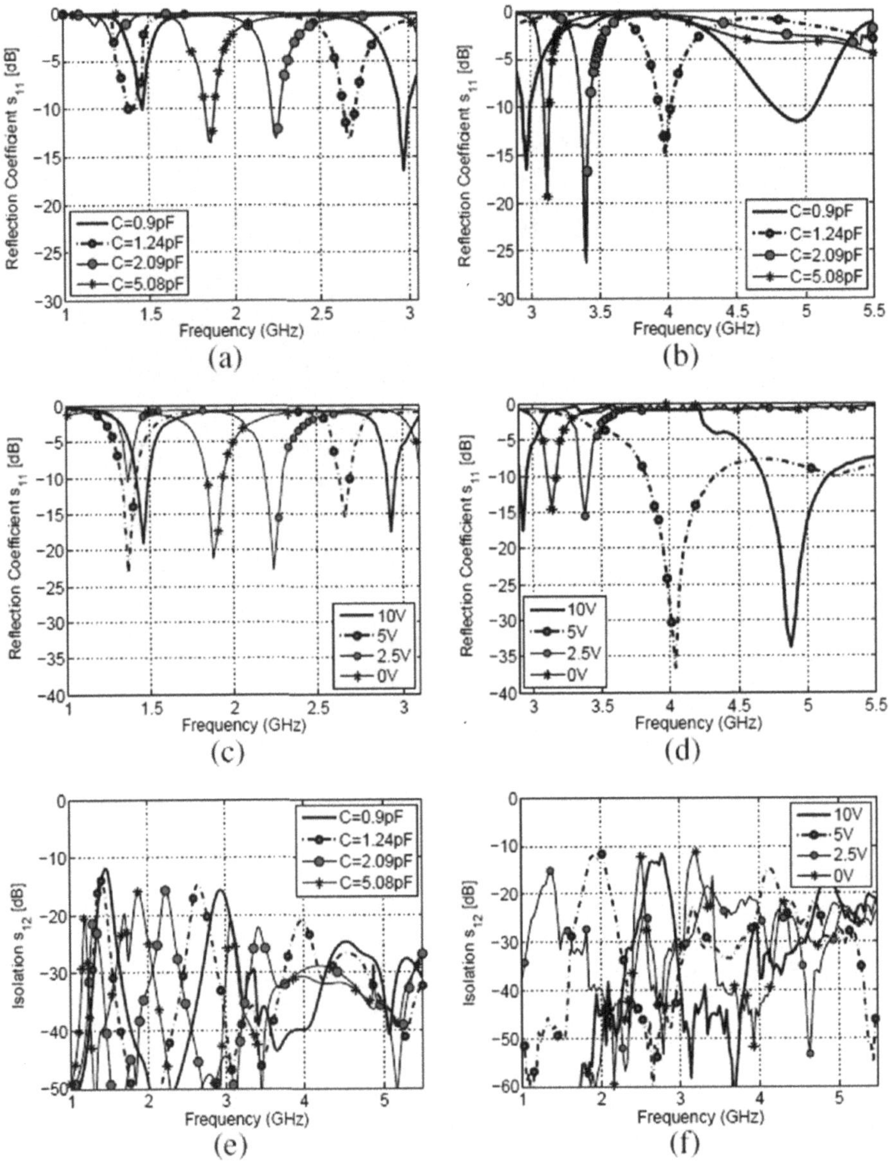

FIGURE 6.16
Simulated and measured S_{11} and S_{12} of the four-element reconfigurable UWB-MIMO antenna consisting of pentagonal slot-shaped radiators: (a) and (b) S_{11} simulated, (c) and (d) S_{11} measured, (e) S_{12} simulated, (f) S_{12} measured.

most popular techniques for diversity applications, and as such, a dielectric resonator can effectively work as a cluster of multiple antennas. Such orthogonal modes excited in single DR can contribute to pattern and polarization diversity along with low correlation-coefficient. However, a major challenge in designing MIMO antennas based on this concept lies in bringing their resonance frequency to the same value. Such modes, excited in a single DR element with the same resonance frequency, are known as degenerated modes.

In addition to utilizing the degenerated modes, which can be excited in a structurally symmetric DR, based on various other techniques and concept involving multiple modes, multiple DR elements, utilization of hybrid shapes, and array-based design, DRA-MIMO antenna research has evolved over the last decade.

The dielectric resonator element was added to the MIMO antenna family in 2008 by Ishimiya et al. [42, 43]. The antenna proposed in Ishimiya et al. [42] and Ishimiya, Ying and Takada [43] comprises a cubic-shaped resonator, having $\varepsilon_r = 21$ and $tan\delta$ = 0.0001352, two silver painted patch elements, and a rod monopole. The patch elements, pasted on two adjacent sides of the DR, excite the patch modes at two ports while the monopole excites the monopolar mode of the DRA from the other port. The entire assembly is mounted on a ground plane of 80×80 mm^2. The designed antenna, aimed at the universal mobile telecommunication system (UTMS) [42] and WLAN band [43], provides good impedance matching at all three ports at targeted 2.6 GHz, acceptable inter-port isolation, and good MEG and DG. When the antenna is excited from port 1, it provides dipole-like radiation, while for excitation from port-2 and port-3, a more directive patch-type radiation pattern is observed. Superiority of a six-port dielectric resonator-based array as a receiving antenna system in indoor NLOS propagation environments in term of rich spatial, polarization, and angular diversity over conventional sparsely separated monopoles reported in Tian et al. [44], greatly escalated DRA-based MIMO antenna research. The six-port DRA array in Tian et al. [44] is realized using two rectangular DRAs each having three different ports excited in various modes. As demonstrated in Ishimiya et al. [42] and Ishimiya, Ying and Takada [43], this excitation is achieved by using two metallic patches on the sides of the DR which results in fundamental TE excitation and a vertically erected monopole inside the DR which excites the additional monopolar modes. The TE mode results in a radiation pattern like a magnetic dipole while the monopole excites the TM mode. Theoretical analysis on channel characteristics employing dual-polarized cylindrical DRA for a 2×2 MIMO system [45] reveals a great capacity of such a system. In this design, two orthogonal $HE_{11\delta}$ modes are excited in both the elements of a two-element MIMO system using monopole excitation from two orthogonal feeds. Application of DRA-based MIMO antenna for long-term evolution (LTE) technology is explored in [46]. The MIMO-DRA antenna is capable of transmitting/receiving independently using TE and HE modes in the LTE bands 12 and 17. The DRA-MIMO antennas, in this case, are designed using perturbation theory to achieve degenerate modes by exciting $TE_{011+\delta}$ and $HE_{11\delta}$ modes at the same frequency. The dimension of the perturbed cylindrical DRA, resulting in an elliptical cylinder, is 80×84×31 mm^3. Such a degenerate mode excited MIMO antenna exhibits very low inter-mode coupling (−40 dB) and a great channel capacity. A channel capacity as high as 11.1 b/s/Hz against the theoretical limit of 13. 14 b/s/Hz for a 2×2 MIMO system was achieved in this technique.

An excellent frequency-agile MIMO dielectric resonator antenna based on perturbation theory proposed in Yan and Bernhard [47] and tuning and switching elements such as varactor and PIN diodes are shown in Figure 6.17. Figure 6.18 shows the fabricated prototype of the antenna which is connected with two 50 Ω orthogonal commercial connectors to excite the degenerate $TE_{011+\delta}$ and $HE_{11\delta}$ modes. Figures 6.19 and 6.20 reveal the frequency reconfigurable functionalities of the dual-mode DRA antenna for HE-mode excitation and TE-mode excitation, respectively. Adjusting DC bias voltages to systematically turn on the varactors and/or the PIN diodes, independent control the TE and HE modes of the antenna, and good isolation, are achieved.

Dielectric resonator-based MIMO antenna can be used for 4G [48] and WiMAX/WLAN [49] applications. This antenna, shown in Thamae and Wu [45], excites two orthogonal

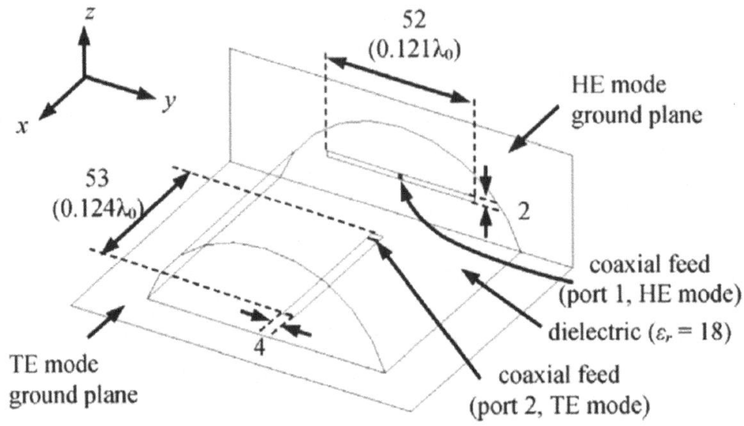

FIGURE 6.17
3-D perspective view of MIMO-DRA excited by two slots.

Reproduced by permission of © 2013 The IEEE [47].

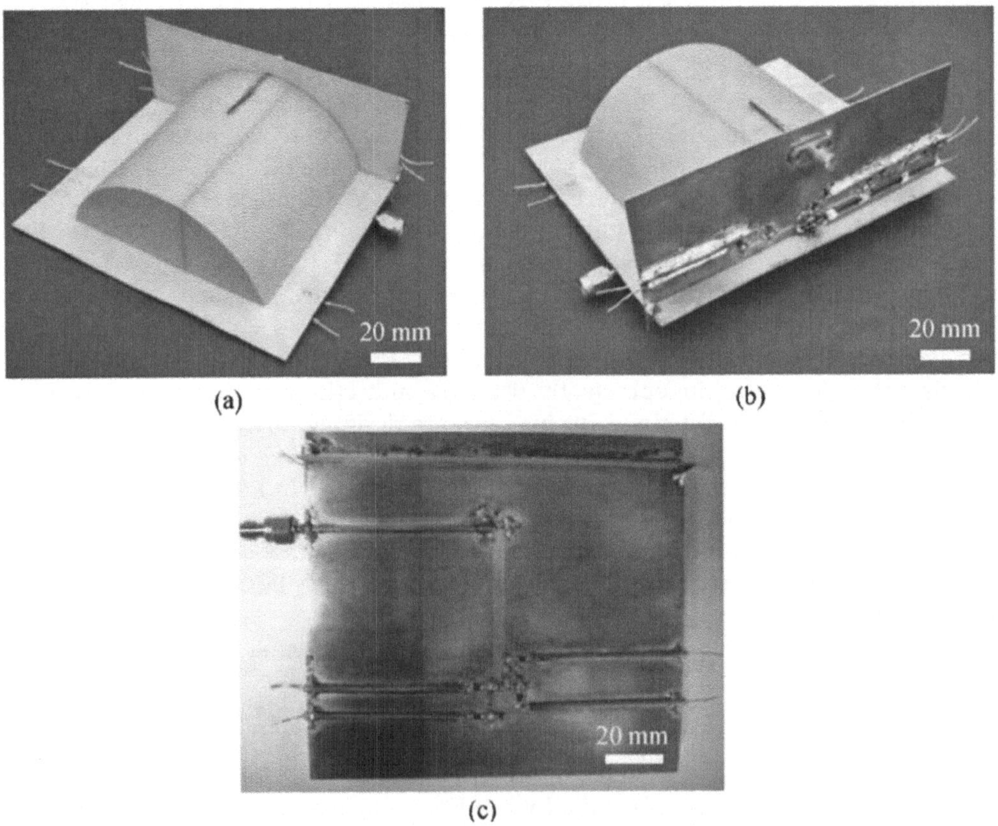

FIGURE 6.18
Fabricated prototype of the frequency-agile MIMO-DRA: (a) front view, (b) back view, (c) bottom view.

Reproduced by permission of © 2013 The IEEE [47].

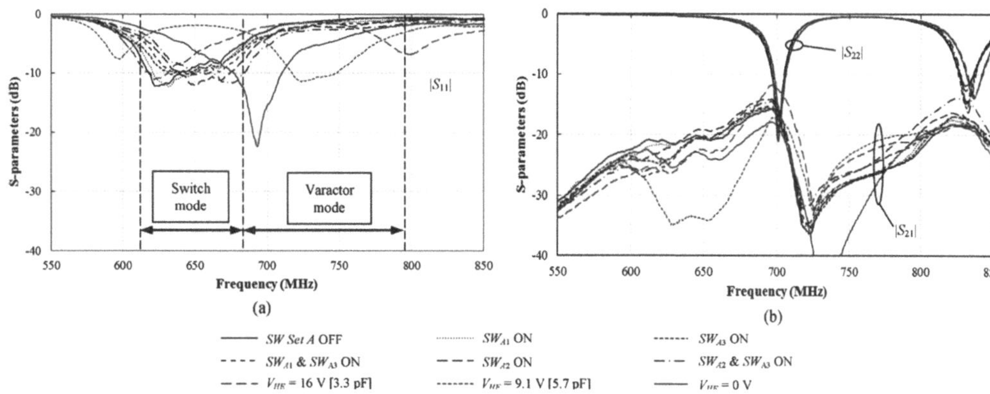

FIGURE 6.19
Measured *S*-parameters of the frequency-reconfigurable MIMO-DRA for HE-mode frequency reconfiguration: (a) S_{11} for HE mode, (b) S_{22} for TE mode.

Reproduced by permission of © 2013 The IEEE [47].

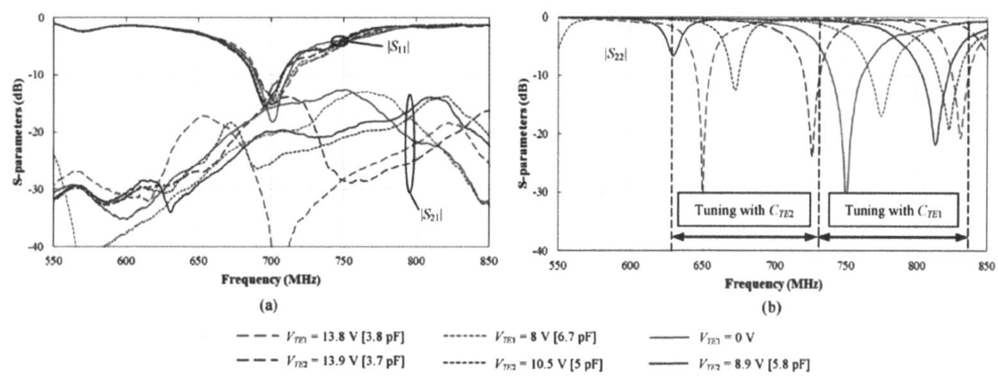

FIGURE 6.20
Measured *S*-parameters of the frequency-reconfigurable MIMO-DRA for TE-mode frequency reconfiguration: (a) S_{11} for HE mode and S_{21}, (b) S_{22} for TE mode.

Reproduced by permission of © 2013 The IEEE [47].

modes with two different feed mechanisms, e.g. CPW and co-axial probes. This MIMO antenna provides a broadband impedance matching at both the ports, gain of 4.97 dBi and 4.51 dBi for port 1 and port 2, respectively, as well as good ECC (0.03) and diversity gain (~10 dB) over the operating bandwidth. Improvised shapes of DRs such as mushroom, pawn, etc., excited by the monopole, are quite popular for achieving very wide band impedance response. Employing the mushroom-shaped DR, fed by a conformal trapezoidal patch, a two-element wideband MIMO-DR antenna is reported in Sharma and Biswas [50]. The reported antenna provides an impedance bandwidth of 4.89–9.61 GHz at port 1 and 5.08–9.50 GHz at port 2 along with good isolation and acceptable MIMO performance metrics.

DRA-based MIMO antenna with multiple frequency bands are of great interest to WLAN/WiMAX and fourth-generation wireless access points. Employing eight-cylindrical dielectric resonators (CDR) in which four CDRs are excited in $HEM_{111+\delta}$ mode and four

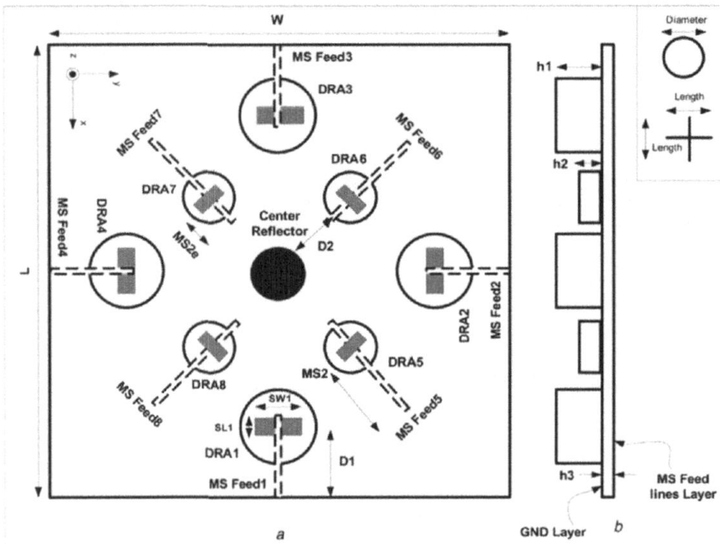

FIGURE 6.21
Schematic diagram of an eight-element dual-frequency DRA-MIMO antenna: (a) top view, (b) cross-sectional view.

Reproduced by permission of © 2017 The IEEE [51].

other CDRs in $TE_{011+\delta}$ mode, a dual-frequency MIMO antenna [51] with very good channel capacity can be achieved. Figure 6.21 shows the schematic diagram of the antenna. Here, all the DR elements are excited by a slot coupling mechanism. Slots are accommodated on the ground plane and are excited by a 50 Ω microstrip line placed at the bottom of the dielectric board which in turn excites CDRs placed above the slots.

As shown in Figure 6.21, DRA1, DRA2, DRA3, and DRA4 have a higher diameter and height, thereby contributing to a lower frequency band, while DRA5, DRA6, DRA7, and DRA8 contribute to a higher frequency band. The CDRA element size for lower and higher bands are obtained using [51],

$$f = \frac{30k_0a}{2\pi h\left(a/h\right)} \tag{6.20}$$

Where

f is the resonance frequency of the modes in GHz
k_0 is the wavenumber in free space and
a and h are the radius and height of the CDRA in centimeters.

Here,

$$\left.\begin{aligned} k_0a &= \frac{6.324}{\sqrt{\varepsilon_r+2}}\left(0.27+0.36\left(\frac{a}{2h}\right)-0.02\left(\frac{a}{2h}\right)^2\right) \\ Q &= 0.01007\varepsilon_r^{1.3}\frac{a}{h}\left(1+100e^{-2.05\left(\left(a/2h\right)-1/80\left(a/2h\right)^2\right)}\right) \end{aligned}\right\} \begin{aligned} &\text{For HEM11+d mode} \\ &\text{and } 0.4 \le \frac{a}{h} \le 6.0 \end{aligned} \qquad (6.21)\ (6.22)$$

and

$$\left.\begin{aligned} k_0 a &= \frac{2.327}{\sqrt{\varepsilon_r+1}}\left(1+0.2123\left(\frac{a}{h}\right)-0.00898\left(\frac{a}{h}\right)^2\right) \\ Q &= 0.078912\varepsilon_r^{1.27}\frac{a}{h}\left(1+17.31\left(\frac{h}{a}\right)-21.57\left(\frac{h}{a}\right)^2\right. \\ &\quad \left.+10.86\left(\frac{h}{a}\right)^3-1.98\left(\frac{h}{a}\right)^4\right) \end{aligned}\right\} \begin{array}{c}\text{For } HEM_{11+\delta} \text{ mode} \\ \text{and } 0.4 \le \frac{a}{h} \le 6.0\end{array} \qquad (6.23)\ (6.24)$$

Where Q is the quality factor of the mode excited.

A metallic reflector, placed at the center of this MIMO antenna reduces the field correlation and enhances isolation and directivity. Figure 6.22a,b shows the top and bottom view

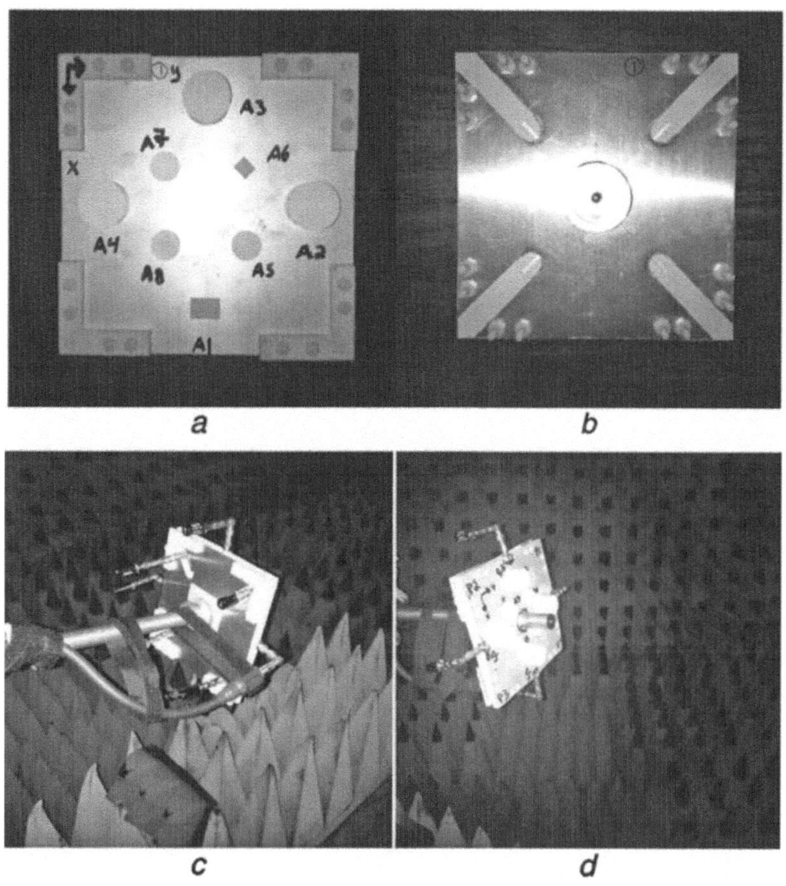

FIGURE 6.22
Fabricated eight-element dual-frequency DRA-MIMO antenna: (a) top view, (b) bottom view, (c) measurement setup with terminations, (d) with large reflector.

Reproduced by permission of © 2017 The IEEE [51].

of the DRA array. Measurement setup of the antenna, along with connectors and mounting fixtures, placed inside the anechoic chamber is shown in Figure 6.22c,d, respectively. Figure 6.23 shows the measured and simulated reflection and isolation at various ports for lower and higher bands. As can be seen from the plots, the eight-element DRA offers minimum impedance bandwidth of 90 MHz and 200 MHz at the lower and higher bands along with a minimum isolation of 16 dB. ECC of the antenna determined using both far-field computation and *S*-parameter values in both frequency bands is plotted in Figure 6.24. ECC values obtained in both the bands are below 0.17 indicating a good MIMO operation and diversity behavior. Wideband/multiband DRA-MIMO antenna employing similar cylindrical DRA and ring DRA [52–54] exhibits good efficiency in addition to good MIMO performance.

Dielectric resonator-based MIMO antenna, due to various inherent advantages and potential in forthcoming 5G communication standards [55], is being considered as one of the strongest candidates in millimeter-wave band. It was reported recently that mm-wave bands with at least 500 MHz of BW in the 28 and 38 GHz frequency bands can be

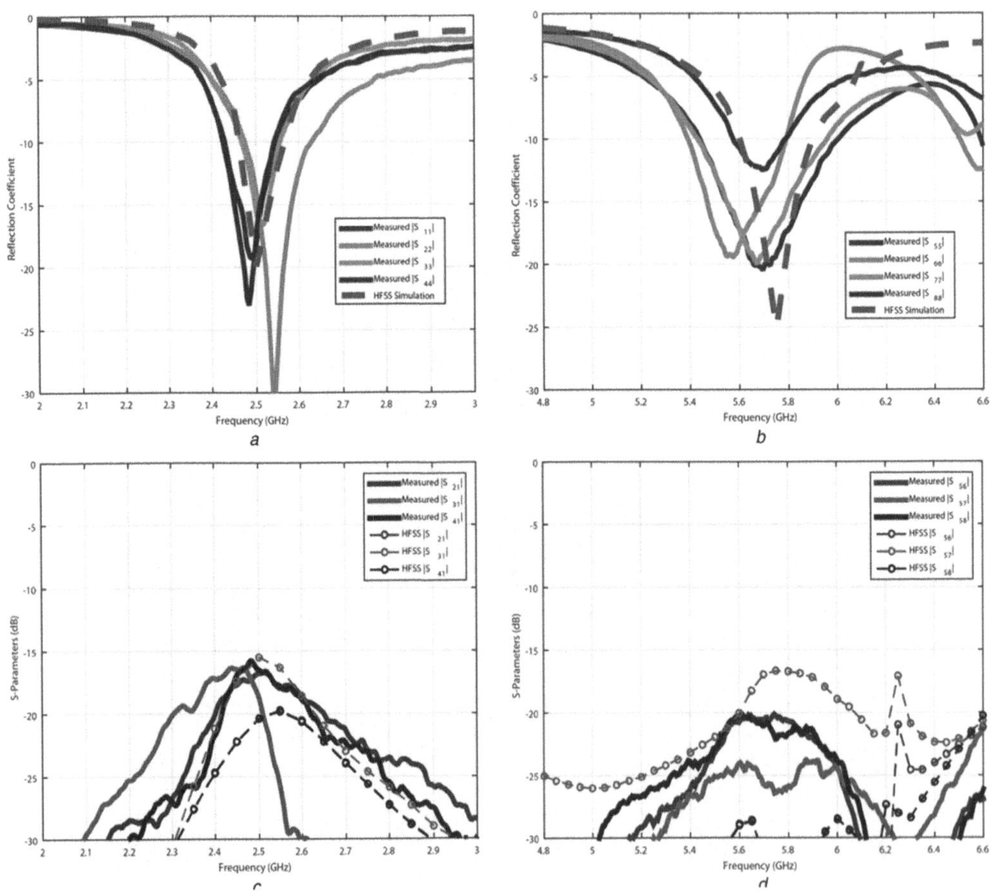

FIGURE 6.23
Measured and simulated *S*-parameters of the eight-element dual-frequency DRA-MIMO antenna: (a) and (b) reflection parameters at lower and higher band, (c) and (d) coupling at lower and higher band.

Reproduced by permission of © 2017 The IEEE [51].

used in wireless cellular communication systems [54]. An interesting two-element linear array-based DRA-MIMO antenna was recently proposed by Sharawi et al. [56]. Each array, consisting of four cylindrical DR antennas, represents a single element of the proposed MIMO antenna. Figure 6.25 shows the schematic diagram of the antenna along with feed network. Similar to Sharawi et al. [51], all the DR elements are excited by a microstrip-line-based 1:4 power divider. Figure 6.26 shows the fabricated prototype of the proposed mm-wave MIMO-DRA antenna. Figure 6.27 shows the measured and simulated *S*-parameters of the antenna. As revealed from Figure 6.27a,b, millimeter-wave DRA offers an impedance bandwidth greater than 1.5 GHz with an isolation better than 25 dB. The measured maximum realized gain values of better than 7 dBi, considering the losses in the feed network along with acceptable MIMO parameter metrics, make this antenna a very good choice in commercial handheld devices for short-range and high-speed data link.

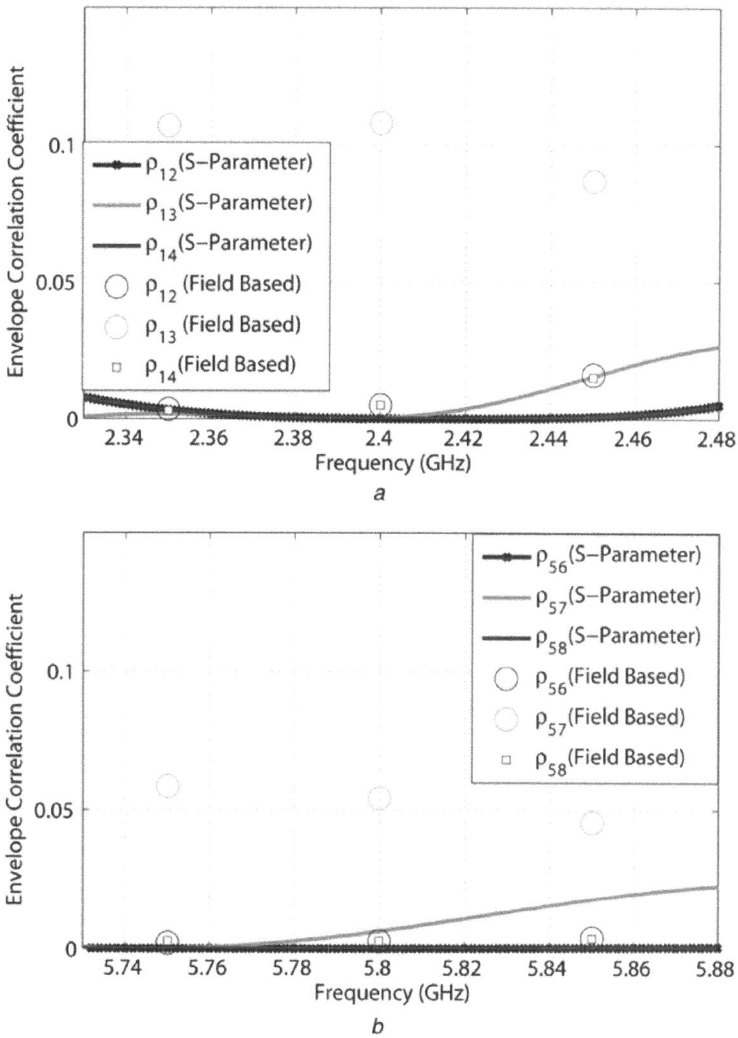

FIGURE 6.24
ECC of eight-element dual-frequency DRA-MIMO antenna at (a) lower and (b) higher frequency band.

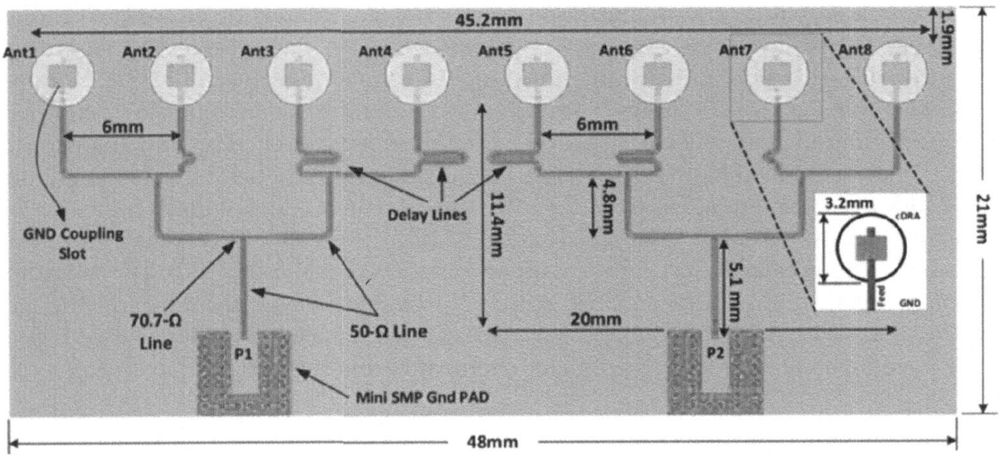

FIGURE 6.25
Cylindrical DRA-based mm-wave MIMO antenna.

Reproduced by permission of © 2017 The IEEE [56].

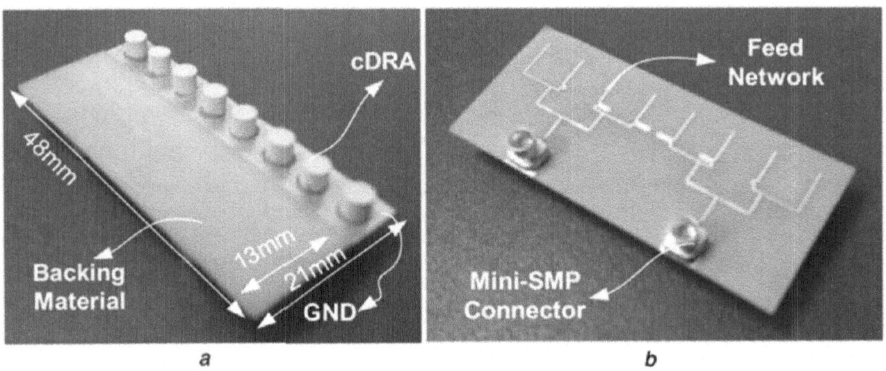

FIGURE 6.26
Fabricated prototype of the cylindrical DRA-based mm-wave MIMO antenna: (a) top view, (b) bottom view.

Reproduced by permission of © 2017 The IEEE [56].

6.6 Frequency-Notched UWB-MIMO Antenna

As discussed in Chapter 3, overlapping of various narrowband services, such as WLAN, WiMAX, X-band satellite downlink, etc., with the FCC-licensed UWB spectrum [1] of 3.1–10.6 GHz, causes possible interference between these narrow bands and UWB spectrum. This interference problem is mitigated when designing the UWB antenna by introducing an intrinsic filtering section inside the UWB antenna. The resultant antenna with inbuilt frequency rejection stage is called a frequency-notched UWB antenna. As discussed in detail in Chapter 3, various techniques are available in open literature to address this frequency-notching issue in UWB antenna design. The same issue of incorporating

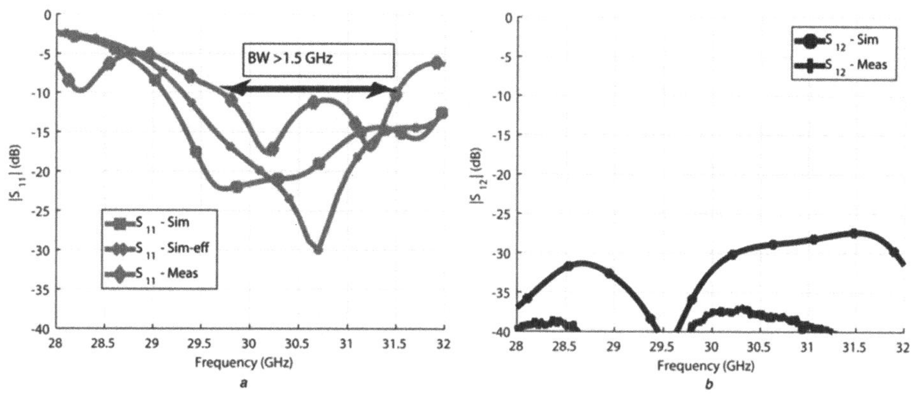

FIGURE 6.27
Measured and simulated, (a) reflection coefficient and (b) coupling, of cylindrical DRA- based mm-wave MIMO antenna.

Reproduced by permission of © 2017 The IEEE [56].

the frequency-notching stage arises when designing UWB MIMO antenna. Each antenna element of the UWB-MIMO array must be integrated with an efficient frequency-notching element to solve this issue. While any of the frequency-nothing techniques discussed in Chapter 3, can, in principle, be used in a MIMO array for this purpose, the antenna designer needs to select the optimum technique based on the applications, type of the antenna element in the MIMO array, size of the array, etc. Table 6.1 summarizes various techniques [57–70] of designing single- and multiple-notched UWB-MIMO antennas. Traditional techniques, such as adding parasitic resonators around the radiating elements or ground plane, slots of various geometries on the radiator and/or ground planes, are mostly used for this purpose. Yoon et al. [57] reported a WLAN band-notched two-element MIMO array by introducing quarter-wavelength slots on the heptagonal monopoles printed on a flexible PET substrate. Similar WLAN-rejected MIMO array design has been reported in various works [58–70] using various printed and DRA-based MIMO antenna. Such frequency notch can also be obtained by loading CSRR on the ground plane [61]. Figure 6.28a shows a dual frequency-notched very compact two-element MIMO antenna with an overall size of 18×34 mm^2 exhibiting a very large impedance bandwidth from 2.9 GHz to 20 GHz [70]. Dual frequency notches at 5.5 GHz and 6.9 GHz are contributed by two L-shaped resonant slits introduced on the radiator. As revealed in the *S*-parameter plots of Figure 6.28b, along with sharp notches, contributed by the L-shaped slits, the MIMO antenna exhibits excellent isolation between the two ports. ECC, DG, and TARC of the antenna shown in Figure 6.29a,b confirm an excellent MIMO performance. Utilizing a similar concept, four-element single/dual frequency-notched UWB-MIMO antennas are reported in Khan et al., Tripathi et al., Zhu et al., and Liu and Tu [62, 64, 66, 68]. In Tripathi et al., [64] the notch is contributed by C-shaped slots loaded inside the radiators. The choice of fractal geometry results in a miniaturized design in addition to good isolation achieved by loading a vertical stub on one side of the separated ground plane. However, as discussed in Sharawi [20], disconnected ground planes in MIMO antenna design, considered in Tripathi et al. [64], is not encouraged due to the requirement of a common system ground (Table 6.2).

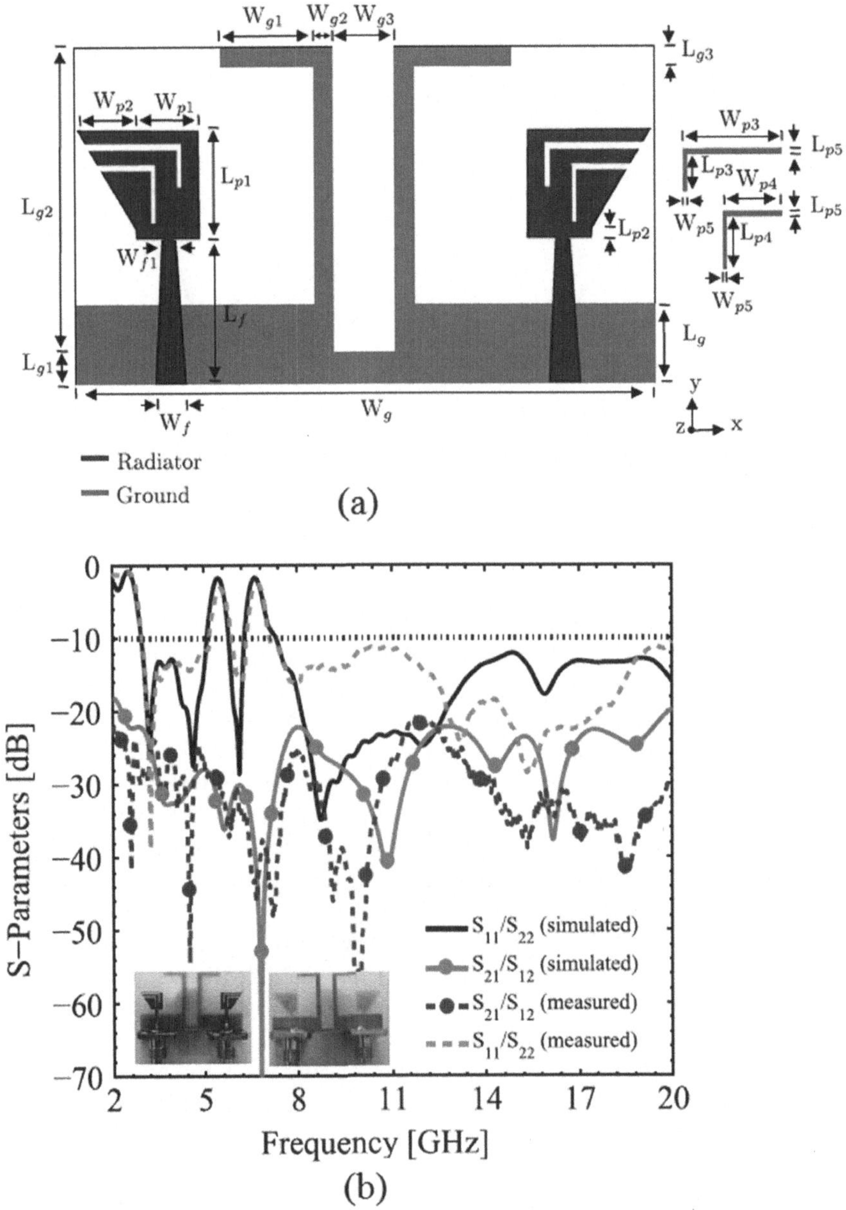

FIGURE 6.28
Dual frequency-notched compact two-element MIMO antenna: (a) schematic diagram, (b) measured and simulated *S*-parameters.

Figure 6.30 shows a WLAN-rejected DR-based two-element MIMO antenna [69]. Two inserted rectangular DRs of dimension 11×7 mm^2 and height 5 mm, machined from ECCOSTOCK HiK material ($\varepsilon_r = 10$ and $tan\delta$ = 0.002), excited by two separates 50 Ω microstrip lines printed on a Taconic substrate ($\varepsilon_r = 3.5$) provide the desired impedance bandwidth. A simple rectangular stub extended from the ground plane provides a good

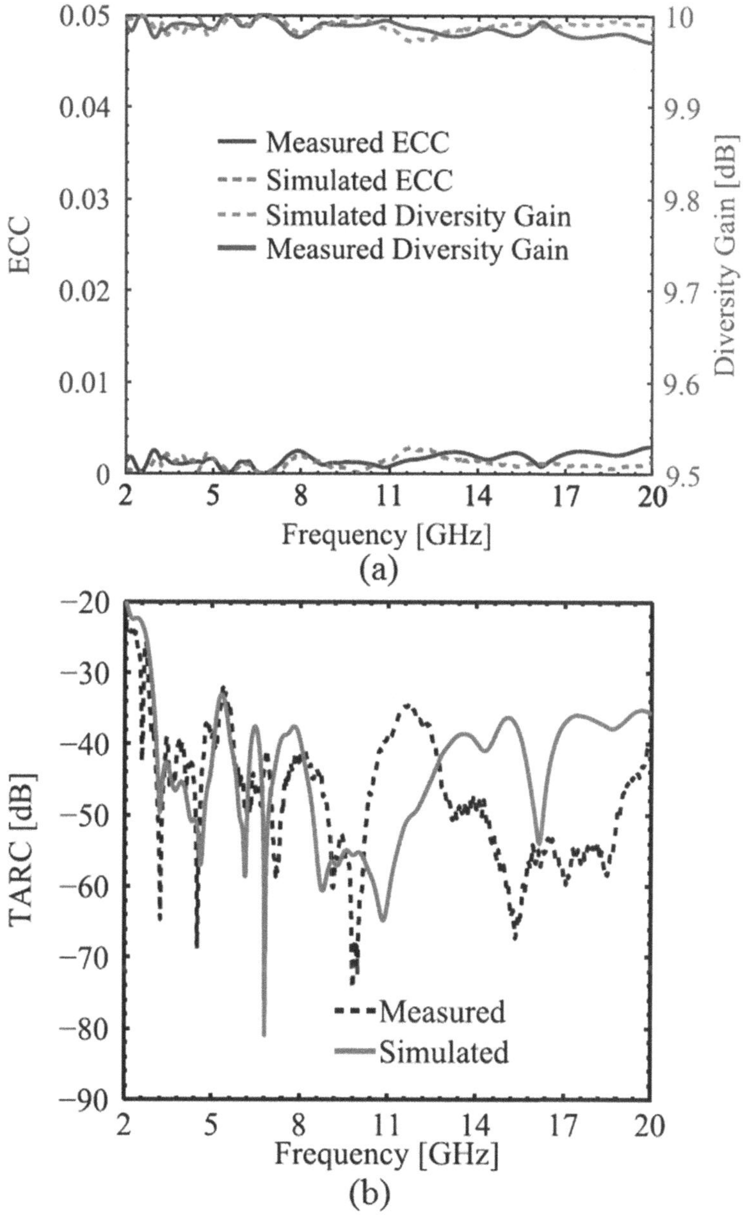

FIGURE 6.29
Simulated and measured (a) ECC, DG, (b) TARC of the two-element frequency-notched MIMO antenna.

Reproduced by permission of © 2017 The IEEE [70].

TABLE 6.2
Summary of Various Techniques of Frequency-notched MIMO Antenna Design

Ref.	MIMO Type Printed/ DRA-Based	Number of Elements in MIMO Array	Notching Technique	Number of Notches	Notch Frequency	Application
Yoon et al. [57]	Printed	2	Quarter-wavelength slot on the radiator	1	5.5 GHz	WLAN rejection
Lee et al. [58]	Printed	2	Open stub on the radiator	1	5.5 GHz	WLAN rejection
Li et al. [59]	Printed	2	Slot on the radiator and metal strip beneath the radiator	2	3.5 GHz and 5.5 GHz	WiMAX and WLAN rejection
Tang et al. [60]	Printed	2	Metallic strips beneath the radiator	2	3.5 GHz and 5.5 GHz	WiMAX and WLAN rejection
Gao et al. [61]	Printed	2	CSRR on the radiator	1	5.5 GHz	WLAN rejection
Khan et al. [62]	Printed	4	Stubs on the ground plane	1	5.5 GHz	WLAN rejection
Kang et al. [63]	Printed	2[a]	L-shaped slits on the ground plane	1	5.5 GHz	WLAN rejection
Tripathi et al. [64]	Printed	4	C-shaped slots on the radiator	1	5.5 GHz	WLAN rejection
Chandel et al. [65]	Printed	2	L-shaped strip beneath the radiator	1	4.17 GHz	C-band rejection
Zhu et al. [66]	Printed	2 and 4	Bent slit on the radiator	1	3.5 GHz and 5.5 GHz	WLAN rejection
Toktas et al. [67]	Printed	2	G-shaped slot form radiator	1	5.5 GHz	WLAN rejection
Liu et al. [68]	Printed	4	U-shaped stubs on the slot-radiator	1	5.5 GHz	WLAN rejection
Abedian et al. [69]	DRA	2	L-shaped parasitic strips near the DRs	1	5.5 GHz	WLAN rejection
Chandel et al. [70]	Printed	2	L-shaped slits on the radiator	2	5.5 GHz and 6.9 GHz	WLAN and IEEE INSAT

[a] A single slot is excited with two microstrip lines.

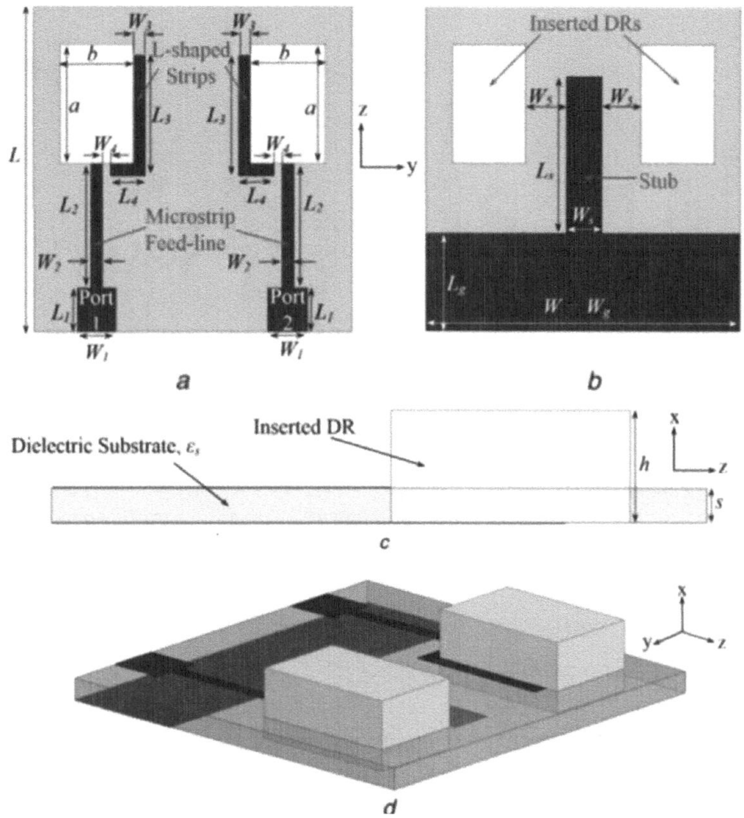

FIGURE 6.30
Schematic diagram of two-element frequency-notched DRA-based MIMO antenna: (a) front view, (b) back view, (c) cross-sectional view, (d) 3-D perspective view.

Reproduced by permission of © 2017 The IET [69].

isolation. The notch frequency is contributed by two L-shaped half-wavelength strips connected to the DRs on the top side of the substrate. The *S*-parameters of the antenna, plotted in Figure 6.31, indicate frequency-notched UWB response of the antenna and also reveal good port isolation.

6.7 Isolation-Enhancement Techniques in UWB-MIMO Antenna

In a MIMO antenna system, close proximity of multiple antenna elements housed on a common substrate can cause a significant amount of undesired field coupling from one radiating element to another. This is contributed by the coupling of surface currents between two antenna elements and near field spurious coupling between the radiating elements which in turn can also generate spurious currents in a radiator. While keeping the radiating elements widely apart can naturally reduce this coupling, this not a feasible solution as the overall antenna system must be compact and meet an extremely stringent

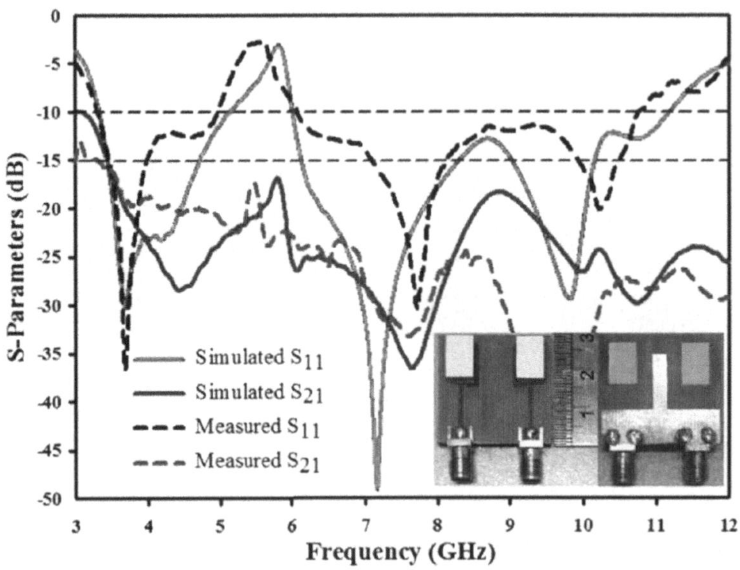

FIGURE 6.31
Simulated and measured *S*-parameters of two-element frequency-notched DRA-based MIMO antenna.

size criterion. Thus, reducing mutual coupling or enhancing the isolation between various radiating elements or ports in the MIMO antenna without sacrificing its compactness is one of the crucial criterion in designing a MIMO array [71]. Isolation enhancement using a metamaterial-based unit cell is demonstrated by Hsu et al. [72] in a narrowband antenna operating at 2.6 GHz for WiMAX application. The most fundamental concept of achieving good isolation is orthogonal placement of the radiating elements [73] which minimizes the near field coupling and therefore contributes to a reasonable isolation. Another advantage of adopting this technique is low correlation and achieving polarization diversity, too. Such a technique, which doesn't introduce additional isolation-enhancing elements, to some extent has the potential to also be adopted into wideband antennas.

However, in many cases, especially for MIMO antennas having very wideband impedance response, e.g. for UWB applications, this alone is not sufficient to achieve the desired level of isolation between the radiating elements for the entire band. Thus, in addition to optimal placement of the antenna elements, various kinds of decoupling structures are used for this purpose. Various techniques, such as introducing various kinds of defects/slots on the ground plane, incorporating stubs on the ground plane, and introducing metamaterial elements between the radiators, etc., are widely used decoupling methods. Using isolated ground planes [64] for individual antenna elements nullifies the mutual current coupling and contributes to a very good isolation. However, such designs are not very useful [20] due to the requirement of a common ground for any practical electronic system. Table 6.3 summarizes various isolation-enhancement techniques [59], [74–82] for antennas with very wideband impedance response. As indicated in the table, incorporation of these techniques contributes to good isolation for a wide bandwidth. Figure 6.32 shows the fabricated prototype of a two-element printed MIMO antenna [81] in which a wideband neutralization line, added between the antennas elements, yields a strong isolation.

TABLE 6.3

Summary of Various Isolation-enhancement Techniques in UWB-MIMO Antenna Design

Ref.	Number of Elements in MIMO Array	Isolation-enhancement Techniques	IBW of the MIMO Antenna	Isolation Achieved
Lee et al. [59]	2	Slot + metal strip on the ground plane	3–11 GHz	Better than 20 dB
Zhang et al. [74]	2	Adding tree-like stub on the ground plane	3.1 10.6 GHz	16–20 dB
Luo et al. [75]	2	Defected ground structures	3.1–10.6 GHz	Better than 18 dB
Salem et al. [76]	8	Adding closed loop FSS, circular arc and strips	3.1–10.6 GHz	~20 dB more, most of the IBW
Deng et al. [77]	2	Etching slots on the ground	2.4–10.6 GHz	~15–20 dB
Lin et al. [78]	2	Loading carbon black film	2.5–11 GHz	Better than 15 dB
Reddy et al. [79]	2	Frequency selective surfaces	2.2–4 GHz and 5–8 GHz	Better than 14 dB
Roshna et al. [80]	2	Metal strip acting as reflector	3.1–10.6 GHz	Better than 14 dB
Zhang et al. [81]	2	Wideband neutralization line	3.1–5 GHz	Better than 22 dB
Liu et al. [82]	2	Using quasi self complementary concept	2.19–11.07 GHz	Better than 20 dB

FIGURE 6.32
Fabricated prototype of the two-element printed MIMO antenna with wideband neutralization lines.

Reproduced by permission of © 2016 The IEEE [81].

As indicated in Figure 6.33a, measured S_{21} better than 25 dB is achieved from 3.1 GHz to 5 GHz. Surface current distributions of the antenna, with and without the neutralization lines, shown in Figure 6.33b, clearly reveal excellent decoupling of the antenna when loaded with the neutralization line.

Isolation enhancement in a two-element dielectric resonator-based mm-wave MIMO antenna employing meta-surface is shown in Figure 6.34 [83]. The meta-surface comprising an array of split-ring resonators is placed vertically in between two cylindrical DRAs excited by microstrip lines. As indicated in Figure 6.35, the meta-surface shield provides an isolation of 30 to 46.5 dB for the operating frequency range of the antenna.

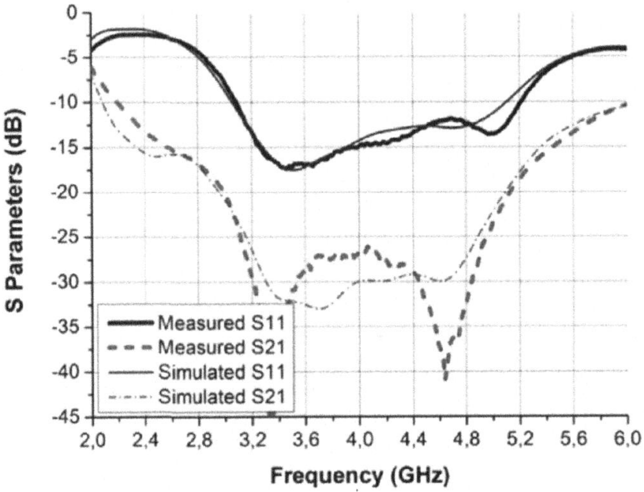

(a)

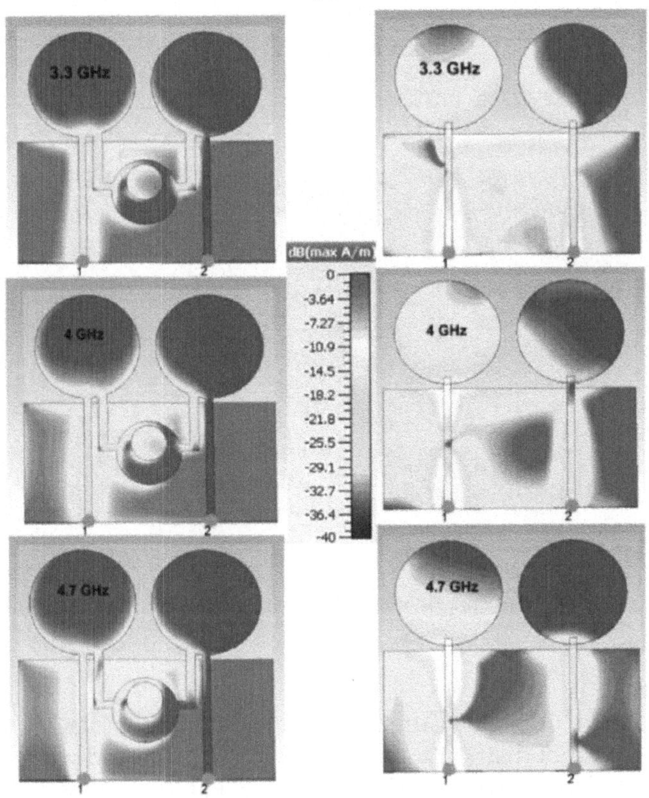

(b)

FIGURE 6.33
(a) *S*-parameters and (b) surface current distribution of the two-element printed MIMO antenna with wideband neutralization lines in Roshna et al. [80].

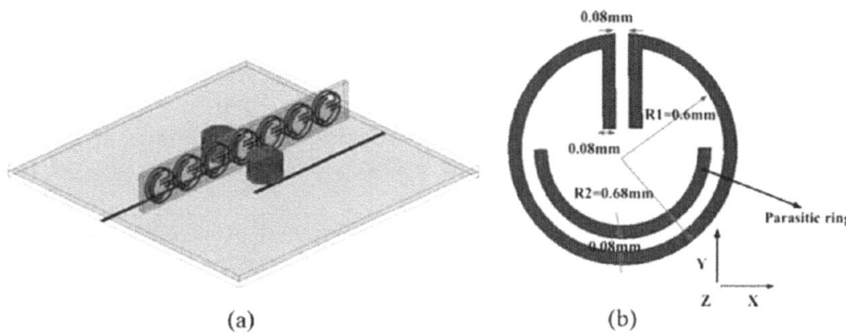

FIGURE 6.34
(a) Schematic diagram of two-element mm-wave DRA-based MIMO antenna with meta-surface shield, (b) enlarged view of the SRR element of the meta-surface shield.

Reproduced by permission of © 2017 The IEEE [83].

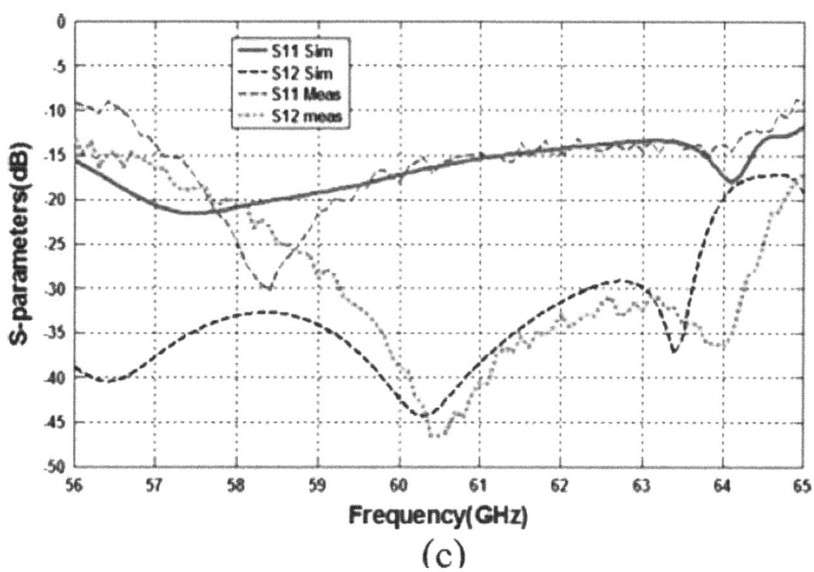

FIGURE 6.35
Simulated and measured S-parameters of the two-element DRA-based MIMO antenna.

Reproduced by permission of © 2017 The IEEE [83].

References

1. US Federal Communications Commission (FCC), Part 15, October 2003 [Online]. Available: http ://www.fcc.gov/oet/info/rules.
2. E. Biglieri, R. Calderbank, A. Constantinides, A. Goldsmith, A. Paulraj, and H. Vincent Poor, *MIMO Wireless Communications*, Cambridge University Press, 2007.
3. M.S. Sharawi, *Printed MIMO Antenna Engineering*, Artech House, 2014.
4. N. Costa and S. Haykin, *Multiple-Input Multiple-Output Channel Models: Theory and Practice*, vol. 65, John Wiley & Sons, 2010.

5. G.J. Foschini, "Layered space-time architecture for wireless communication in a fading environment when using multi-element antennas", *Bell Labs Technical Journal*, vol. 1, no. 2, 1996, 41–59.
6. E. Telatar, "Capacity of multi-antenna Gaussian channels", *European Transactions on Telecommunications*, vol. 10, no. 6, 1999, 585–595.
7. G.J. Foschini and Michael J. Gans, "On limits of wireless communications in a fading environment when using multiple antennas", *Wireless Personal Communications*, vol. 6, no. 3, 1998, 311–335.
8. S. Blanch, J. Romeu, and I. Corbella, "Exact representation of antenna system diversity performance from input parameter description", *Electronics Letters*, vol. 39, no. 9, 2003, 705–707.
9. P. Hallbjorner, "The Significance of Efficiencies when Using S-Parameters to Calculate the Received Signal Correlation from Two Antennas", *IEEE Antennas and Wireless Propagation Letters*, vol. 4, 2005, 97–99.
10. J.O. Nielsen, G.F. Pederson, K. Olesen, and I.Z. Kovacs, "Computation of mean effective gain for 3D measurements", *IEEE 49th Vehicular Technology Conference*, vol. 1, Texas, USA, May 1999, 787–791.
11. T. Taga, "Analysis of mean effective gain of mobile antennas in land mobile radio environments", *IEEE Transactions on Vehicular Technology*, vol. 39, no. 2, May 1990, 117–131.
12. M.S. Sharawi, "Printed multi-band MIMO antenna systems and their performance metrics", *IEEE Antennas and Propagation Magazine*, vol. 55, no. 5, 2013, 218–232.
13. N. Jamaly, A. Derneryd, and T. Svensson, "Analysis of antenna pattern overlap matrix in correlated non-uniform multipath environments", *IEEE 7th European Conference on Antennas and Propagation (EuCAP)*, Gothenburg, Sweden, April 2013, 2113–2117.
14. K. Kalliola, K. Sulonen, H. Laitinen, O. Kivekäs, J. Krogerus, and P. Vainikainen, "Angular power distribution and mean effective gain of mobile antenna in different propagation environments", *IEEE Transactions on Vehicular Technology*, vol. 51, no. 5, 2002, 823–838.
15. A. Ando, T. Taga, A. Kondo, K. Kagoshima, and S. Kubota, "Mean effective gain of antennas in line-of-sight street microcells with low base station antennas", *IEEE Transactions on Antennas and Propagation*, AP-56, vol. 11, 2008, 3552–3565.
16. M. Manteghi and Y. Rahmat-Samii, "Multiport characteristics of a wideband cavity backed annular patch antenna for multipolarization operations", *IEEE Transactions on Antennas and Propagation*, AP-53, vol. 1, 2005, 466–474.
17. S. Chae, W. Kawk, S.-O. Park, and K. Lee, "Analysis of mutual coupling in mimo antenna array by tarc calculation", *Microwave Conference*, APMC 2006, Asia-Pacific, 2090–2093, IEEE, 2006.
18. R. Vaughan and J.B. Andersen, "Antenna diversity in mobile communications", *IEEE Transactions on Vehicular Technology*, vol. 36, no. 4, 1987, 149–172.
19. Z. Li, Z. Du, M. Takahashi, K. Saito, and K. Ito, "Reducing mutual coupling of MIMO antennas with parasitic elements for mobile terminals", *IEEE Transactions on Antennas and Propagation*, AP-60, vol. 2, 2012, 473–481.
20. M.S. Sharawi, "Current misuses and future prospects for printed multiple-input, multiple-output antenna systems [wireless corner]", *IEEE Antennas and Propagation Magazine*, vol. 59, no. 2, 2017, 162–170.
21. W.Y. Li and W.J. Chen, "Concurrent 2-port/3-port MIMO antenna system for UMTS/LTE 2500 operation in the mobile phone", *IEEE International Symposium on Antennas and Propagation (APS/URSI)*, Washington, USA, July 2011, 1918–1921.
22. C.H. See, R.A. Abd-Alhameed, Z.Z. Abidin, N.J. McEwan, and P.S. Excell, "Wideband printed MIMO/diversity monopole antenna for WiFi/WiMAX applications", *IEEE Transactions on Antennas and Propagation*, AP-60, vol. 4, 2012, 2028–2035.
23. N. Ojaroudi and M. Ojaroudi, "A novel design of triple-band monopole antenna for multi-input multi-output communication", *Microwave and Optical Technology Letters*, vol. 55, no. 6, 2013, 1258–1262.
24. X.M. Ling and R.L. Li, "A novel dual-band MIMO antenna array with low mutual coupling for portable wireless devices", *IEEE Antennas and Wireless Propagation Letters*, vol. 10, 2011, 1039–1042.

25. X. Zhou, X. Quan, and R.L. Li, "A dual-broadband MIMO antenna system for GSM/UMTS/LTE and WLAN Handsets", *IEEE Antennas and Wireless Propagation Letters*, vol. 11, 2012, 551–554.
26. R. Karimian, H. Oraizi, S. Fakhte, and M. Farahani, "Novel F-shaped quad-band printed slot antenna for WLAN and WiMAX MIMO systems", *IEEE Antennas and Wireless Propagation Letters*, vol. 12, 2013, 405–408.
27. M. Han and J. Choi "Dual-band MIMO antenna using a symmetric slotted structure for 4G USB dongle application", *IEEE International Symposium on Antennas and Propagation (APS/URSI)*, Washington, USA, July 2011, 2223–2226.
28. S.C. Ko and R.D. Murch, "Compact integrated diversity antenna for wireless communications", *IEEE Transactions on Antennas and Propagation*, vol. 47, 2001, 954–960.
29. K.L. Wong, S.W. Su, and Y.L. Kuo, "A printed ultra-wideband diversity monopole antenna", *Microwave and Optical Technology Letters*, vol. 38, no. 4, 2003, 257–259.
30. M. Karaboikis, C. Soras, G. Tsachtsiris, and V. Makios, "Compact dual-printed inverted-F antenna diversity systems for portable wireless devices", *IEEE Antennas and Wireless Propagation Letters*, vol. 3, no. 1, 2004, 9–14.
31. H. Zhao, L. Liu, T.S.P See, and Z.N. Chen, "A printed ultra-wideband diversity antenna", *The 2006 IEEE 2006 International Conference on Ultra-Wideband*, 351–356, IEEE, 2006.
32. T.S.P. See and Z.N. Chen, "An ultrawideband diversity antenna", *IEEE Transactions on Antennas and Propagation*, vol. 57, no. 6, 2009, 1597–1605.
33. S.S. Jehangir and M.S. Sharawi "A miniaturized UWB biplanar Yagi-Like MIMO antenna system", *IEEE Antennas and Wireless Propagation Letters*, vol. 16, 2017, 2320–2323.
34. R. Hussain, M.S. Sharawi, and A. Shamim, "4-Element concentric pentagonal slot-line-based ultra-wide tuning frequency reconfigurable MIMO antenna system", *IEEE Transactions on Antennas and Propagation*, 2018.
35. N. Behdad and K. Sarabandi, "Dual-band reconfigurable antenna with a very wide tunability range", *IEEE Transactions on Antennas and Propagation*, vol. 54, no. 2, 2006, 409–416.
36. Z. Hu, C. Song, J. Kelly, P. Hall, and P. Gardner, "Wide tunable dual- band reconfigurable antenna", *Electronics Letters*, vol. 45, no. 22, 2009, 1109–1110.
37. Y. Yu, J. Xiong, H. Li, and S. He, "An electrically small frequency reconfigurable antenna with a wide tuning range", *IEEE Antennas and Wireless Propagation Letters*, vol. 10, 2011, 103–106.
38. L. Xing, Y. Huang, Q. Xu, and S. Aljaafreh, "A transparent dielectric- loaded reconfigurable antenna with a wide tuning range", *IEEE Antennas and Wireless Propagation Letters*, vol. 15, 2016, 1630–1633.
39. M. Gholamrezaei, F. Geran, and R.A. Sadeghzadeh, "Completely independent multi-ultra wideband and multi-dual-band frequency reconfigurable annular sector slot antenna (frassa)", *IEEE Antennas and Wireless Propagation Letters*, 2016.
40. T. Li, H. Zhai, L. Li, and C. Liang, "Frequency-reconfigurable bow-tie antenna with a wide tuning range", *IEEE Antennas and Wireless Propagation Letters*, vol. 13, 2014, 1549–1552.
41. S.A. Long, M.W. McAllister, and L.C. Shen, "The resonant cylindrical dielectric cavity antenna", *IEEE Transactions on Antennas and Propagation*, vol. 31, 1983, 406–412.
42. K. Ishimiya, J. Langbacka, Z. Ying, and J.-I. Takada, "A compact MIMO DRA antenna", *Proceeding of IEEE International Workshop on Antenna Technology: Small Antennas and Novel Metamaterials (IWAT '08)*, Chiba, Japan, 2008, 286–289.
43. K. Ishimiya, Z. Ying, and J.-I. Takada, "A compact MIMO DRA for 802.11n application", *Presented at the IEEE Antennas and Propagation Society International Symposium*, San Diego, CA, Jul. 2008.
44. R. Tian, V. Plicanic, B.K. Lau, and Z. Ying, "A compact six-port dielectric resonator antenna array: MIMO channel measurements and performance analysis", *IEEE Transactions on Antennas and Propagation*, 58 (4), 2010, 1369–1379.
45. L.Z. Thamae and Z. Wu, "Dielectric resonator-based multiple-input multiple-output antennas and channel characteristic analysis", *IET Microwaves, Antennas and Propagation*, vol. 6, no. 9, 2012, 1084–1089.

46. J.B. Yan, and J.T Bernhard, "Design of a MIMO dielectric resonator antenna for LTE femtocell base stations", *IEEE Transactions on Antennas and Propagation*, 60(2), 2012, 438–444.
47. J.B. Yan, J.T. Bernhard, "Implementation of a frequency-agile MIMO dielectric resonator antenna", *IEEE Transactions on Antennas and Propagation*, vol. 61, no. 7, 2013, 3434–3441.
48. S.F. Roslan, M.R. Kamarudin, M. Khalily, and M.H. Jamaluddin, "An MIMO rectangular dielectric resonator antenna for 4G applications", *IEEE Antennas and Wireless Propagation Letters*, vol. 13, 2014, 321–324.
49. A.A Khan, M.H. Jamaluddin, S. Aqeel, J. Nasir, and O. Owais, "Dual-band MIMO dielectric resonator antenna for WiMAX/WLAN applications", *IET Microwaves, Antennas and Propagation*, vol. 11, no. 1, 2017, 113–120.
50. A. Sharma and A. Biswas, "Wideband multiple-input–multiple-output dielectric resonator antenna", *IET Microwaves, Antennas and Propagation*, vol. 11, no. 4, 2016, 496–502.
51. M.S. Sharawi, S.K. Podilchak, M.U. Khan, and Y.M. Antar, "Dual-frequency DRA-based MIMO antenna system for wireless access points", *IET Microwaves, Antennas and Propagation*, vol. 11, no. 8, 2017, 1174–1182.
52. R.K. Mongia and P. Bhartia, "Dielectric resonator antennas – a review and general design relations for resonant frequency and bandwidth", *International Journal of RF and Microwave Computer-Aided Engineering*, vol. 4, no. 3, 1994, 230–247.
53. A. Sharma, G. Das, and R.K. Gangwar, "Design and analysis of tri-band dual-port dielectric resonator based hybrid antenna for WLAN/WiMAX applications", *IET Microwaves, Antennas and Propagation*, vol. 12, no. 6, 2017, 986–992.
54. G. Das, A. Sharma, and R.K. Gangwar, "Wideband self-complementary hybrid ring dielectric resonator antenna for MIMO applications", *IET Microwaves, Antennas and Propagation*, vol. 12, no. 1, 2017, 108–114.
55. T.S. Rappaport, S. Sun, R. Mayzus, R. Mayzus, H. Zhao, Y. Azar, K. Wang, G.N. Wong, J.K. Schulz, M. Samimi, and F. Gutierrez Jr., "Millimeter wave mobile communications for 5G cellular: It will work!", *IEEE Access*, vol. 1, 2013, 335–349.
56. M.S. Sharawi, S.K. Podilchak, M.T. Hussain, and Y.M.M. Antar, "Dielectric resonator based MIMO antenna system enabling millimetre-wave mobile devices", *IET Microwaves, Antennas and Propagation*, vol. 11, no. 2, 2017, 287–293.
57. H.K. Yoon, Y.J. Yoon, H. Kim, and C.-H. Lee, "Flexible ultra-wideband polarisation diversity antenna with band-notch function", *IET Microwaves, Antennas and Propagation*, vol. 5, no. 12, 2011, 1463–1470.
58. J.-M. Lee, K.-B. Kim, H.-K. Ryu, and J.-M.Woo, "A compact ultrawideband MIMO antenna with WLAN band-rejected operation for mobile devices," *IEEE Antennas and Wireless Propagation Letters*, vol. 11, 2012, 990–993.
59. J.-F. Li, Q.-X. Chu, Z.-H. Li, and X.-X. Xia, "Compact dual band notched UWB MIMO antenna with high isolation", *IEEE Transactions on Antennas and Propagation*, vol. 61, no. 9, 2013, 4759–4766.
60. T.-C. Tang and K.-H. Lin, "An ultrawideband MIMO antenna with dual bandnotched function", *IEEE Antennas and Wireless Propagation Letters*, vol. 13, 2014, 1076–1079.
61. P. Gao, S. He, X. Wei, Z. Xu, N. Wang, and Y. Zheng, "Compact printed UWB diversity slot antenna with 5.5-GHz band-notched characteristics", *IEEE Antennas and Wireless Propagation Letters*, vol. 13, 2014, 376–379.
62. M.S. Khan, A.D. Capobianco, S. Asif, A. Iftikhar, B. Ijaz, and B.D. Braaten, "Compact 4 × 4 UWB-MIMO antenna with WLAN band rejected operation", *Electronics Letters*, vol. 51, 2015, 1048–1050.
63. L. Kang, H. Li, X.H. Wang, and X.W. Shi, "Compact offset microstrip fed MIMO antenna for band-notched UWB applications", *IEEE Antennas and Wireless Propagation Letters*, vol. 14, 2015, 1754–1757.
64. S. Tripathi, A. Mohan, and S. Yadav, "A compact Koch fractal UWB MIMO antenna with WLAN band-rejection", *IEEE Antennas and Wireless Propagation Letters*, vol. 14, 2015, 1565–1568.
65. R. Chandel and A.K. Gautam, "Compact MIMO/diversity slot antenna for UWB applications with band-notched characteristic", *Electronics Letters*, vol. 52, no. 5, 2016, 336–338.

66. J. Zhu, S. Li, B. Feng, L. Deng, and S. Yin, "Compact dual polarized UWB quasi-self-complementary MIMO/diversity antenna with band-rejection capability", *IEEE Antennas and Wireless Propagation Letters*, vol. 15, 2016, 905–908.
67. A. Toktas, "G-shaped band-notched ultra-wideband MIMO antenna system for mobile terminals", *IET Microwaves, Antennas and Propagation*, vol. 11, no. 5, 2017, 718–725.
68. Y. Liu and Z. Tu, "Compact differential band-notched stepped-slot UWBMIMO antenna with common-mode suppression", *IEEE Antennas and Wireless Propagation Letters*, vol. 16, 2017, 593–596.
69. M. Abedian, S.K.A. Rahim, C. Fumeaux, S. Danesh, Y.C. Lo, and M.H. Jamaluddin, "Compact ultrawideband MIMO dielectric resonator antennas with WLAN band rejection", *IET Microwaves, Antennas and Propagation*, vol. 11, no. 11, 2017, 1524–1529.
70. R. Chandel, A.K. Gautam and K. Rambabu, "Tapered fed compact UWB MIMO-diversity antenna with dual band-notched characteristics", *IEEE Antennas and Wireless Propagation Letters*, vol. 66, 2018, 1677–1684.
71. A.C.K. Mak, C.R. Rowell, and R.D. Murch, "Isolation enhancement between two closely packed antennas", *IEEE Transactions on Antennas and Propagation*, vol. 56, no. 11, 2008, 3411–3419.
72. C.-C. Hsu, K.-H. Lin, H.-L. Su, H.-H. Lin, and C.-Y. Wu, "Design of MIMO antennas with strong isolation for portable applications", *Proceedings of IEEE Antennas and Propagation Society International Symposium*, Charleston, SC, 2009, 1–4.
73. S. Deepika, Mahesh P. Abegaonkar, and Shiban Kishen Koul, "Easily extendable compact planar UWB MIMO antenna array", *IEEE Antennas and Wireless Propagation Letters*, vol. 16, 2017, 2328–2331.
74. S. Zhang, Z. Ying, J. Xiong, and S. He, "Ultrawideband MIMO/diversity antennas with a tree-like structure to enhance wideband isolation", *IEEE Antennas and Wireless Propagation Letters*, vol. 8, 2009, 1279–1282.
75. C. Luo, J. Hong, and L. Zhong, "Isolation enhancement of a very compact UWB-MIMO slot antenna with two defected ground structures", *IEEE Antennas and Wireless Propagation Letters*, vol. 14, 2015, 1766–1769.
76. R. Saleem, M. Bilal, K.B. Bajwa, and M.F. Shafique, "Eight-element UWB-MIMO array with three distinct isolation mechanisms", *Electronics Letters*, vol. 51, no. 4, 2015, 311–313.
77. J.Y. Deng, L.X. Guo, and X.L. Liu, "An ultrawideband MIMO antenna with a high isolation", *IEEE Antennas and Wireless Propagation Letters*, vol. 15, 2016, 182–185.
78. G.-S. Lin, C.-H. Sung, J.-L. Chen, L.-S. Chen, and M.-P. Houng, "Isolation improvement in UWB MIMO antenna system using carbon black film", *IEEE Antennas and Wireless Propagation Letters*, vol. 16, 2017, 222–225.
79. T.S. Reddy, Rajkishor Kumar, and Raghvendra Kumar Chaudhary, "Isolation enhancement and radar cross section reduction of MIMO antenna with frequency selective surface", *IEEE Transactions on Antennas and Propagation*, vol. 66, no. 3, 2018, 1595–1600.
80. T.K. Roshna, U. Deepak, V.R. Sajitha, K. Vasudevan, and P. Mohanan, "A compact UWB MIMO antenna with reflector to enhance isolation", *IEEE Transactions on Antennas and Propagation*, vol. 63, no. 4, 2015, 1873–1877.
81. S. Zhang and G.F. Pedersen, "Mutual coupling reduction for UWB MIMO antennas with a wideband neutralization line", *IEEE Antennas and Wireless Propagation Letters*, vol. 15, 2016, 166–169.
82. X–L. Liu, Z–D. Wang, Y–Z. Yin, J. Ren, and J–J. Wu, "A compact ultrawideband MIMO antenna using QSCA for high isolation", *IEEE Antennas and Wireless Propagation Letters*, vol. 13, 2014, 1497–1500.
83. A. Dadgarpour, B. Zarghooni, B.S. Virdee, T.A. Denidni, and A.A. Kishk "Mutual coupling reduction in dielectric resonator antennas using metasurface shield for 60-GHz MIMO systems", *IEEE Antennas and Wireless Propagation Letters*, vol. 16, 2017, 477–480.

Index